T0315025

FUNDAMENTALS OF SPACECRAFT CHARGING

FUNDAMENTALS OF SPACECRAFT CHARGING

Spacecraft Interactions with Space Plasmas

SHU T. LAI

PRINCETON UNIVERSITY PRESS
Princeton and Oxford

Published by Princeton University Press, 41 William Street,
Princeton, New Jersey 08540
In the United Kingdom: Princeton University Press, 6 Oxford
Street, Woodstock, Oxfordshire OX20 1TW
press.princeton.edu

Library of Congress Cataloging-in-Publication Data

Lai, Shu T.
Fundamentals of spacecraft charging : spacecraft interactions with space plasmas / Shu T. Lai.
p. cm.
Includes bibliographical references and index.
ISBN 978-0-691-12947-1 (hardback)
1. Space vehicles—Electrostatic charging. 2. Space plasmas. I. Title.
TL1492.L35 2011
629.4'16—dc22
2011003777

British Library Cataloging-in-Publication Data is available

This book has been composed in ITC Stone Serif
Printed on acid-free paper. ∞
Printed in the United States of America
10 9 8 7 6 5 4 3 2 1

Contents

Preface

The field of spacecraft charging addresses the fundamental interaction of spacecraft surfaces and materials with the ambient plasma and artificial charged particle environments. It is basic space science as well as space engineering. At the time of this writing, there is no textbook devoted entirely to this field, although there are books with a chapter or two on topics related to spacecraft charging. Review papers, research papers, and technical reports on various aspects of spacecraft charging appear from time to time. This book is intended to assimilate the widespread knowledge of spacecraft charging into a textbook for the education of graduate students and for consultation by researchers.

Spacecraft charging was in its infancy in the mid-twentieth century. It has progressed much since then, as reflected by the successive conferences on the subject. The first Spacecraft Technology Conference was sponsored by the Air Force Cambridge Research Laboratory (AFCRL), Hanscom AFB, Massachusetts, and the National Aeronautics and Space Administration (NASA) and was held at the U.S. Air Force Academy in Colorado Springs, Colorado, in 1978. In fact, the first four conferences were all held at the U.S. Air Force Academy. The fifth was held at the Naval Postgraduate School in Monterey, California. The sixth conference was at Hanscom AFB, Massachusetts, and the seventh was at NASA Marshall Spaceflight Center in Huntsville, Alabama. The eighth, ninth, and tenth conferences were held at Noordwijk, the Netherlands, 2001; Tsukuba, Japan, 2005; and Biarritz, France, 2007, respectively. The eleventh conference was at Kirtland AFB, Albuquerque, New Mexico, 2010. Attendance, especially international attendance, has gradually increased, as has the number of journal papers on this subject. Today, spacecraft charging has become a field of its own.

PROLOGUE

The Earth's Space Plasma Environment

Space is not empty—plasma is nearly everywhere in space.[1,2] Spacecrafts interact with the space environment and may become charged. (A spacecraft is *charged* if it has a net amount of positive or negative electrical charge.) The Sun controls the weather of the Earth's space plasma environment.[3,4] Some fundamental properties[5] of the solar wind, magnetosphere, geosynchronous environment, ionosphere, auroral region, and radiation belts are introduced concisely in the following sections.

P.1 The Solar Wind

The Sun continuously emits the solar wind. The solar wind consists of plasma electrons and ions. Its intensity varies over time. Normally, the solar wind electrons and ions take about four days to arrive at the Earth's magnetosphere. Around the Earth [at 1 astronomical unit (AU)], the solar wind speed is normally about 400 km s^{-1}, but when the Sun is highly active, the wind may reach 900 km s^{-1} or more. The average solar wind plasma density is about 5 per cm^3. The average solar wind plasma temperature is about 10 eV. The solar wind also carries a weak magnetic field of about 5 nT on average.

From time to time, the Sun sends out energetic plasma shocks called *corotating interaction regions* (CIRs). Occasionally, the Sun also ejects large magnetized clouds called *coronal mass ejections* (CMEs) (figure P.1). When CIRs and CMEs arrive, they greatly disturb the space weather in the Earth's space environment and the space systems in the environment. Usually, the CME-induced disturbances are very strong, but those induced by CIRs last for a longer time period.

P.2 The Magnetosphere

The Earth has a dipole magnetic field. If unperturbed, a dipole field is symmetrical about the axis connecting the poles. The Sun greatly affects the Earth's dipole magnetic field configuration. The configuration is called the *magnetosphere*.[1-5] It is asymmetrical. The solar wind pushes the Earth's dayside magnetic field lines inward toward the Earth and pulls the nightside magnetic field lines outward (figure P.2). The elongated magnetic field configuration on the nightside is called the *magnetotail*. It extends often to 200 Earth radii, which is beyond the Moon's orbit, or more.

P.3 Geomagnetic Substorms

The elongation of the magnetotail is analogous to stretching a rubber band. The situation is unstable. Eventually, the magnetotail snaps back toward the Earth, following a process called *magnetic field reconnection* (figure P.3), which occurs somewhere in the magnetotail. The

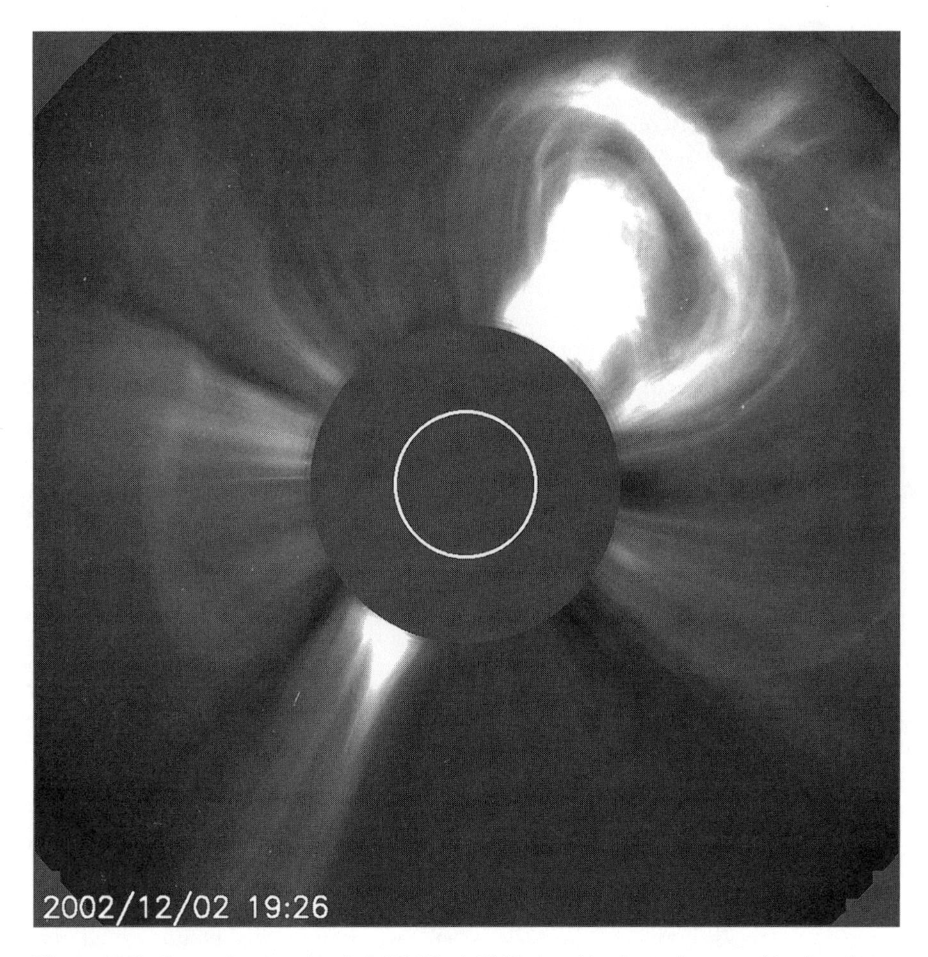

Figure P.1. Coronal mass ejection (CME). A CME cloud is ejected upward in the photo. Solar wind streamers are visible on the two sides. They are much less energetic than a CME. The bright Sun is blocked for better imaging. (Courtesy of SOHO/LASCO consortium. SOHO is a project of international cooperation between ESA and NASA.)

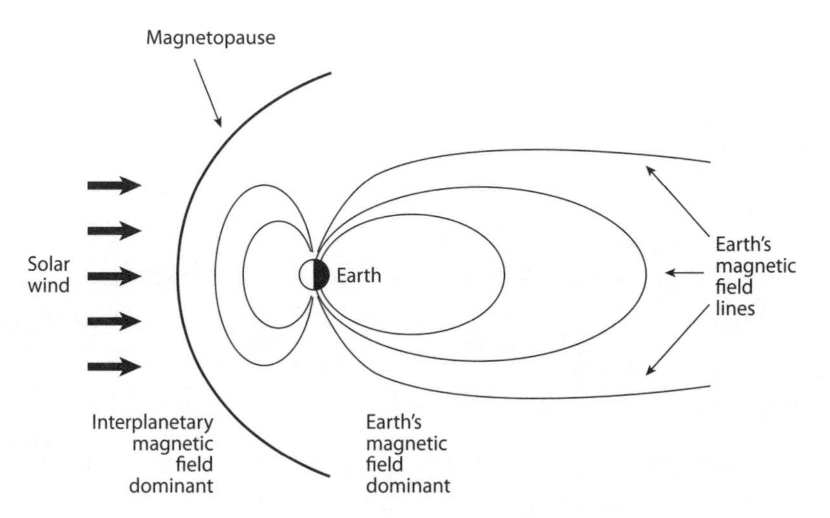

Figure P.2. The Earth's magnetosphere. The solar wind pushes in the dayside magnetic field lines but pulls out the nightside field lines of the magnetosphere.

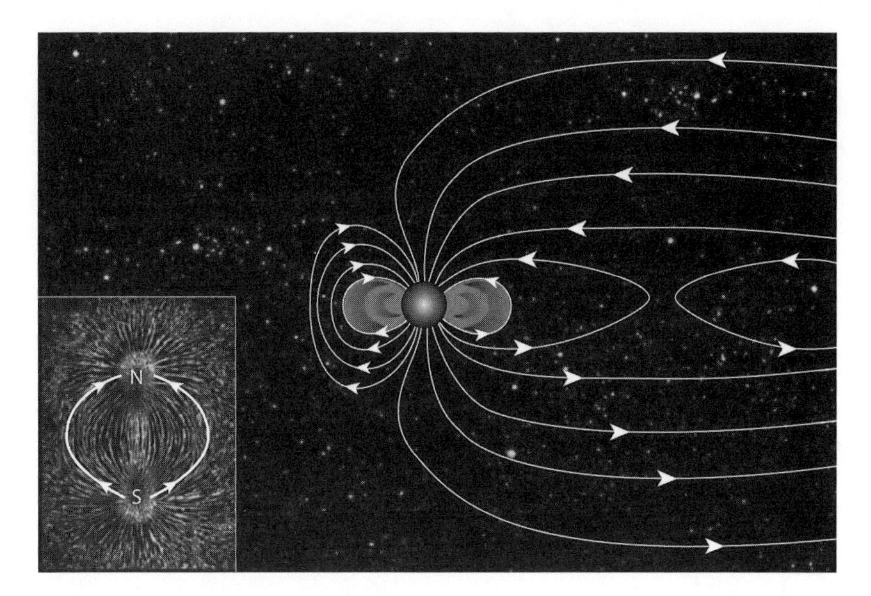

Figure P.3. Reconnection in the magnetotail. The magnetotail is elongated by the solar wind. Magnetic reconnection occurs somewhere in the magnetotail near the Earth. (Courtesy of NASA.)

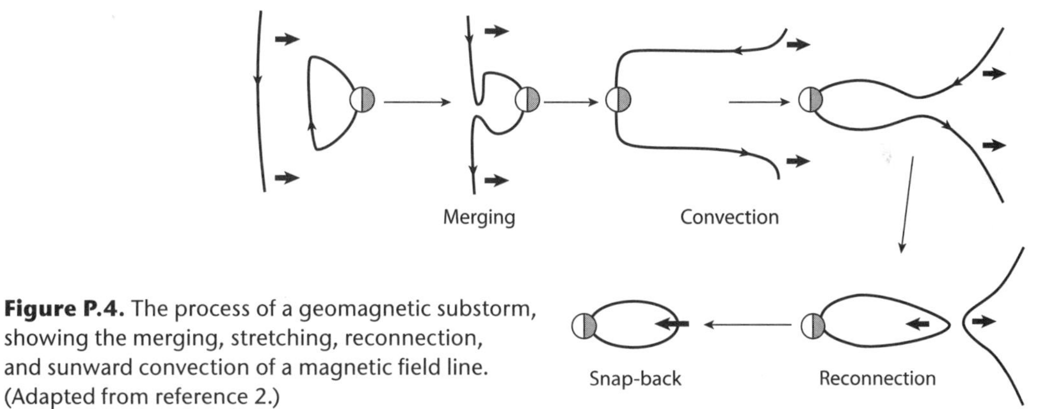

Figure P.4. The process of a geomagnetic substorm, showing the merging, stretching, reconnection, and sunward convection of a magnetic field line. (Adapted from reference 2.)

snap-back compresses the plasma and, as a result, accelerates the electrons and ions to keV (and above) in energy (figure P.4).

The hot (keV or above) electrons and ions arrive at the Earth's geosynchronous orbit (about 6.6 Earth radii) around midnight. There, the electrons and ions are increasingly affected by the Earth's magnetic field gradient and curvature, the corotation with the Earth, and the cross-tail electric fields. As a result, the high-energy electrons tend to drift toward the dawn side. The entire process of pulling and snapping back may last for a few hours only. It may occur more than once in a night. This is called a *geomagnetic substorm*, or simply a substorm.

P.4 Plasma Density

The plasma density is highest in the ionosphere (section P.5). As the radial distance increases, the density in the plasmasphere remains at about 1000 to 100 cm^3 until the rapid drop at the plasmapause (figure P.5). The density (about 0.1 cm^{-3}) is much lower at the geosynchronous

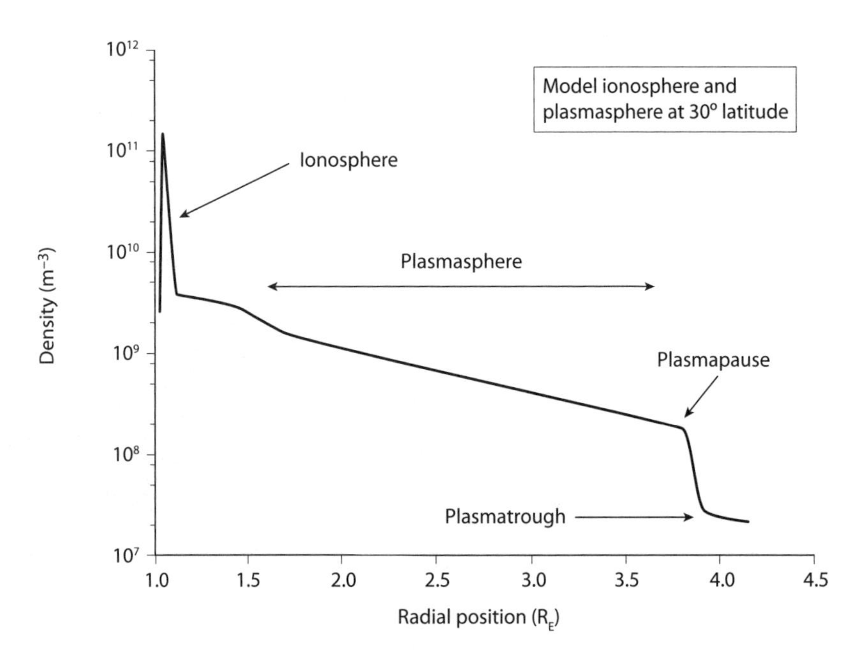

Figure P.5. Electron density as a function of distance from the Earth's surface. The figure shows a model ionosphere and plasmashpere at 30° latitude. (Courtesy of Richard Quinn, 2006.)

orbit (about 6.6 R_E). During quiet days, the plasmasphere expands beyond the geosynchronous orbit, especially on the afternoon side. Normally, the plasma density at or near geosynchronous altitudes is low (about 1 cm^{-3} or less) but the energy is high (up to many keV, during substorms). The hot electrons are responsible for spacecraft charging at geosynchronous altitudes around midnight and in the early morning hours (appendix 1). During quiet days (with negligible solar activity), the plasma density at or near geosynchronous altitudes is about 100 cm^{-3} or higher, but the average energy is a few tens of eV or lower.

P.5 The Ionosphere

In the ionosphere[6,7] (about 100 to 1500 km in altitude), the plasma density is high but the energy is very low. The plasma density is about 10^4 to 10^6 cm^{-3}, the maximum being at about 300 km altitude, where the electron density reaches about 10^5 at night and 10^6 cm^{-3} in the daytime (figures P.6 and P.7). The electron and ion energies in the ionosphere are well below 1 eV.

P.6 The Auroral Region

Aurorae occur mostly at 65° to 75° latitude and 80 to 1000 km altitude. The brightest aurorae are normally seen at about 100 to 130 km. They occur because high-energy electrons impact on the atmospheric molecules, generating emission lines of various colors (wavelengths). The electrons spiral down magnetic field lines in a beam-like manner with energies typically of a few keV (figure P.8). A brief tour of the electron acceleration process is as follows.

Following a reconnection somewhere in the magnetotail, a snap-back of the magnetotail occurs (figure P.4). The snap-back compresses the electrons, which are bottled in the closed magnetic field lines of the nightside magnetosphere. The compression accelerates the

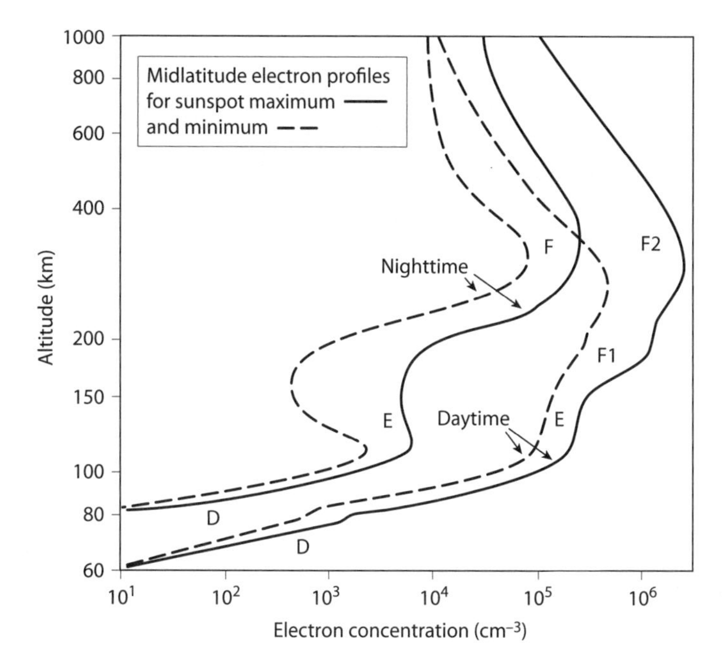

Figure P.6. Electron densities in the ionosphere. (Source of technical data: reference 1.)

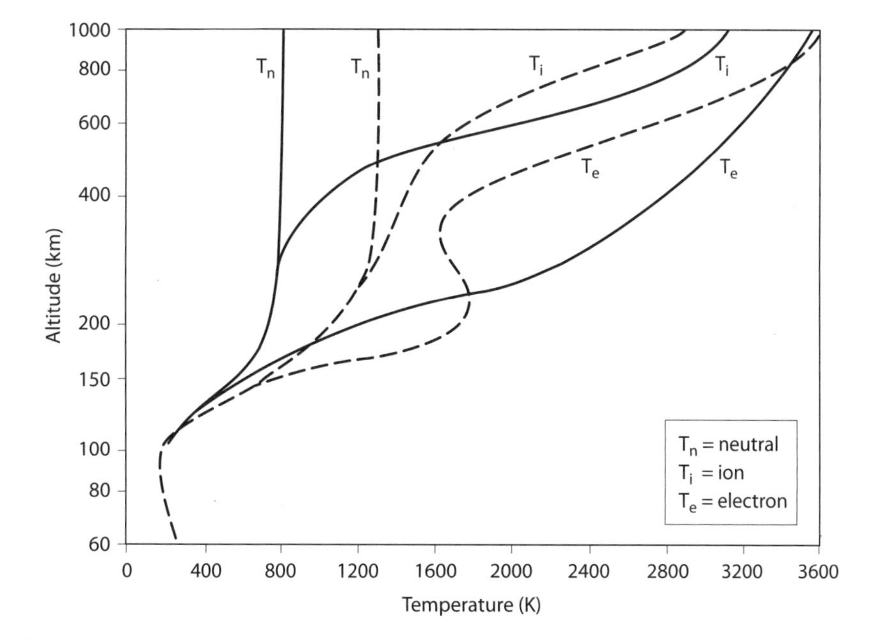

Figure P.7. Daytime electron temperature in the ionosphere. The solid lines are profiles at solar minimum, and the dashed lines are profiles at solar maximum. (Source of technical data: reference 1.)

electrons to high energies. Some electrons come down toward the Earth along the magnetic field lines at the auroral latitudes. When they approach the auroral altitudes, the magnetic field lines gradually bunch together. Some of the electrons and ions mirror back upward at different altitudes depending on their energies, pitch angles, and masses. There, strong electric fields (tens of mV m^{-1}) parallel to the magnetic field lines exist and therefore further

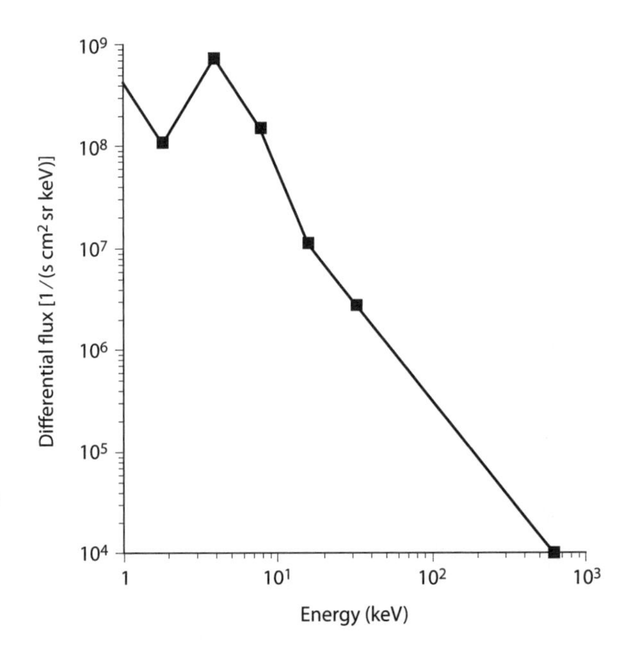

Figure P.8. Measured auroral electron energy spectrum. The peak energy is typically at a few keV. (Available at http://csrsrv1.fynu.ucl.ac.be/csr_web/cpd/tna/node26.html#aurospec.)

accelerate the electrons downward. As they spiral downward, the atmospheric density increases. Electron impacts on atmospheric molecules can excite or ionize the molecules. As a result, aurorae are formed.

P.7 The Radiation Belts

The radiation belts are also called the Van Allen belts or the Van Allen radiation belts.[8–10] Here, *radiation* refers to high-energy (MeV or above) electrons and ions. There are two electron belts and one proton belt (figure P.9). The closed magnetic field lines trap the electrons and ions in the radiation belts. The electrons and ions, in the presence of electromagnetic waves, bounce back and forth. Sometimes, if the solar ejecta compress the front side of the magnetosphere, the electrons and ions in the magnetic bottle may be energized to higher energies.

The inner belt begins at about 1000 km and ends at about 2.5 R_E. The outer belt begins at about 2.8 R_E and ends at about 6.6 R_E. The boundaries, especially those of the outer belt, expand during the Sun's active days. During unusually active events, a third belt may occur in the "slot" region between the two belts.

In the Atlantic region near Southern Brazil, the magnetic field is slightly weaker, and the radiation region dips down to about 250 km altitude. This region is called South Atlantic Anomaly Region (SAAR), where even low-orbit satellites may experience MeV electrons.

The locations of the radiation belts and the plasmasphere sometimes coincide partially. It may sound contradictory that the average electron energy is a few eV while some of the electrons reach MeV and above. The electron energy distribution spans a wide range. While most of the electrons are at low energies, some can have very high energies. If one considers two electron densities, one finds that the density of the high-energy electrons is many orders of magnitude lower than that of the low-energy electrons.

P.8 Relevance of the Space Plasma Environment to Spacecraft Charging

The electrons and ions of energies up to about a few tens of keVs are important players in spacecraft charging, which usually means spacecraft surface charging. The high-energy

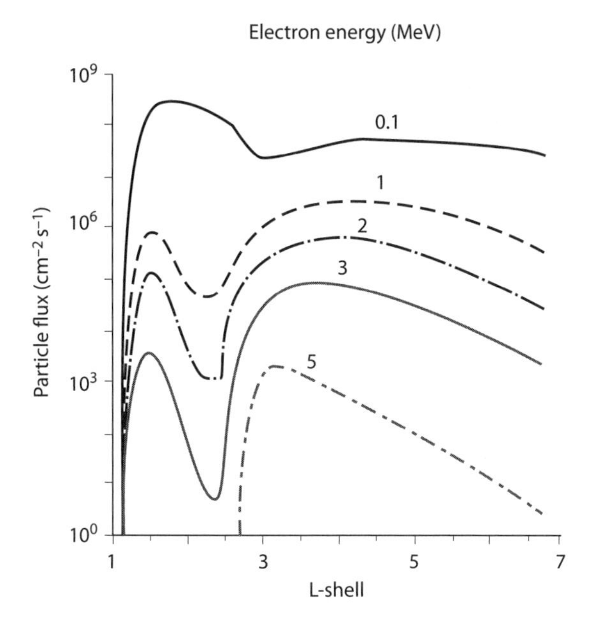

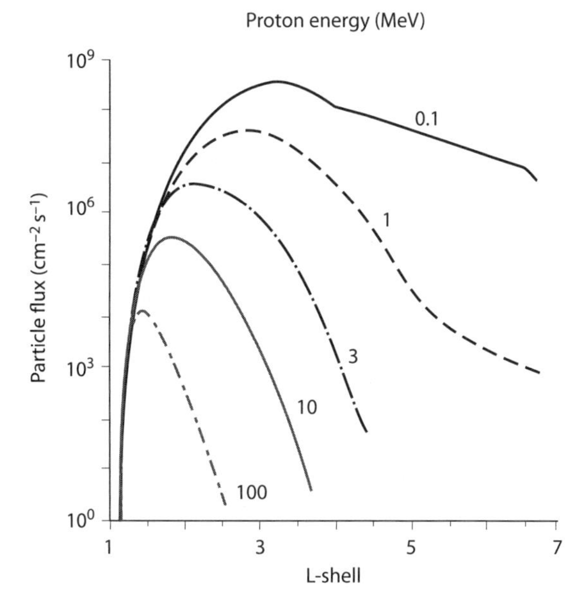

Figure P.9. Typical electron and proton fluxes as functions of L-values in the radiation belts. The electrons form two humps, whereas the ions form only one. The belts vary according to the solar activity (and to the space weather, which is controlled by the Sun). [The L-value is a measure of distance from the center of the Earth. It is defined as $L = 1/\cos^2\Lambda$, where Λ is the magnetic latitude (also called invariant latitude). In the equatorial region, L is a rough approximation of the earth radius R_E, especially near the earth.] The number on each curve labels the energy of the electrons (upper figure) and the ions (lower figure). (Adapted from reference 11.)

(MeV or higher) electrons and ions are of much lower densities and are therefore negligible as far as spacecraft surface charging is concerned. The high-energy fluxes are orders of magnitude lower than the low-energy ones. However, high-energy electrons and ions can penetrate deeply into materials. If the material is nonconducting, the electrons and ions accumulating inside can build up high internal electric fields. This phenomenon is called *deep dielectric charging*, which is also known as *bulk charging*.

Most satellites are in near geosynchronous orbits (GEO) at about 6 to 7 Earth radii and low Earth orbits (LEO) at a few hundred km altitude. As will be discussed in later chapters, spacecraft charging occurs most often at GEO, especially during geomagnetic storms or substorms. Deep dielectric charging may occur if the spacecraft is in a high-energy (MeVs) region in the radiation belts and the electron and ion fluxes have been at high values for many days. More details will be discussed in later chapters.

P.9 References

1. Russell, C. T., "The solar wind interactions with the Earth's magnetosphere: a tutorial," *IEEE Trans. Plasma Sci* 28, no. 6, 1818–1830 (2000).
2. Jursa, A. S., Ed., *Handbook of Geophysics and the Space Environment*, ADA 167000, Air Force Geophysics Laboratory, Hanscom AFB, MA (1985).
3. Akasofu, S.-I., and S. Chapman, *Solar-Terrestrial Physics*, Oxford University Press, Oxford, UK (1972).
4. Kivelson, M. G., and C. T. Russell, *Introduction to Space Physics*, Cambridge University Press, Cambridge, UK (1995).
5. Tribble, A. C., *The Space Environment*, Princeton University Press, Princeton, NJ (2003).
6. Kelley, M. C., *The Earth's Ionosphere*, Academic Press, New York (1989).
7. Shunk, R. W., and A. F. Nagy, *Ionospheres*, Cambridge University Press, Cambridge, UK (2000).
8. Burch, J. L., R. L. Carovillano, and S. K. Antiochos, Eds., *Sun-Earth Plasma Connections*, Geophysical Monograph Series 109, American Geophysical Union, Washington, DC (1999).
9. Song, P., H. J. Singer, and G. L. Siscoe, Eds., *Space Weather*, Geophysical Monograph Series 125, American Geophysical Union, Washington, DC (2001).
10. Daglis, I. A., *Space Storms and Space Weather Hazards*, NATO Science Series, Kluwer Academic Publishers, The Netherlands (2001).
11. Panasyuk, M. I., "Cosmic ray and radiation belt hazards for space missions," p. 253 in reference 10.

FUNDAMENTALS OF SPACECRAFT CHARGING

1

Introduction to Spacecraft Charging

We begin by asking four fundamental questions: What is spacecraft charging? What are the effects of spacecraft charging? How does spacecraft charging occur? Where and when does spacecraft charging occur?

1.1 What Is Spacecraft Charging?

When a spacecraft has a net charge, positive or negative, the net charge generates an electric field according to Gauss's law. Space plasmas are assumed neutral. Although the plasma densities may fluctuate, their time scales (inverse of plasma frequencies) are much faster than spacecraft potential variations. Spacecraft charging takes a longer time than the ambient plasma fluctuations because it takes time to fill up capacitances. In the spacecraft charging community, the potential, ϕ_p, of the ambient space plasma is traditionally defined as zero:

$$\phi_p = 0 \tag{1.1}$$

Since potential is not absolute but relative, it is not surprising that in some other plasma sciences, the plasma potential is sometimes defined relative to that of a surface. In the field of spacecraft charging, the spacecraft potential is relative to the space plasma potential, which is defined as zero. The spacecraft potential is *floating* relative to the ambient plasma potential (figure 1.1). When a spacecraft potential, ϕ_s, is nonzero relative to that of the ambient plasma, the spacecraft is charged:

$$\phi_s \neq 0 \tag{1.2}$$

The basic terminology of spacecraft charging is introduced here:

- For a conducting spacecraft, the charges are on the surfaces. This charging situation is called *surface charging.*
- A uniformly charged spacecraft has only one potential, ϕ_s. This situation is called *uniform charging* or *absolute charging.*
- For a spacecraft composed of electrically separated surfaces, the potentials may be different on different surfaces. The potentials depend on the surface properties and on the environment, which may be nonisotropic. In this case, we have *differential charging.*
- If a spacecraft is covered with connected conducting surfaces (i.e., spacecraft ground or frame) and some unconnected or nonconducting surfaces, the charging of the frame is called *frame charging.*
- When the ambient electrons and ions are very energetic (MeV or higher), they can penetrate deep into dielectrics, which are nonconductors. This situation is called *deep dielectric charging,* or *bulk charging.**

*For a conducting material, an electron penetrating into it moves to the surface because of Coulomb repulsion. Therefore, for conductors, surface charging can occur, but deep conductor charging does not

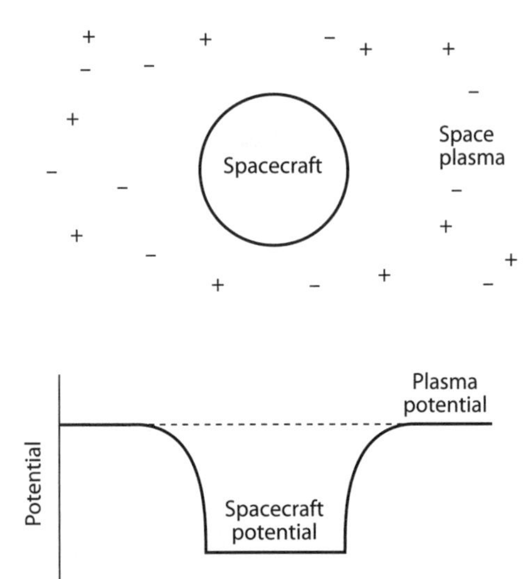

Figure 1.1 Floating potential of a spacecraft. The space plasma potential is defined as zero. A potential sheath is formed around the spacecraft.

1.2 What Are Some Effects of Spacecraft Charging?

Spacecraft charging manifests itself in two types of effects: (1) damage to onboard electronic instruments and (2) interference with scientific measurements. The first type is very rare but may be harmful. The second type is very common. These effects are discussed in the following sections,

1.2.1 Damage to Onboard Electronic Instruments

Spacecraft charging affects telemetry and electronics on spacecraft. Logical circuits and computer chips are becoming smaller and less power consuming but are more delicate. Delicate electronics are susceptible to charging, anomalies, damage, or catastrophes. Undesirable currents entering circuits by conduction, through pinholes, or via electromagnetic waves through inadequate shielding may cause disturbances. Such disturbances often cause anomalies in the telemetry of the data of the measurements.

If neighboring surfaces are at very different potentials, there is a tendency for a sudden discharge to occur. A discharge may be small, large, frequent, or rare. The size of a discharge depends on the amount of charge built up in the electrostatic capacitances, and on the amount of neutral materials, such as gas, which may be ionized when a discharge is initiated. An avalanche ionization can lead to a large current. A small discharge may be simply a spark, generating electromagnetic waves that may disturb telemetry signals. A large discharge may cause damage. Damage to electronics may, in turn, affect operations, navigation, or even survivability of a spacecraft.

---

occur. For dielectrics (insulators), both surface charging and deep dielectric charging can occur. Surface charging can occur in dielectrics if the incoming electrons are below about 70 to 100 keV in energy; deep dielectric charging can occur if they are of higher energy. There is no sharp demarcation line between the two charging regimes. In general, MeV electrons are responsible for deep dielectric charging, while electrons of energy in the keV range are responsible for surface charging of dielectrics. The penetration depth depends on the electron energy and the material density. For kapton polymide, for example, an electron of 100 keV penetrates to about 0.007 cm. The ability to hold charges inside depends on the conductivity of the material. More will be discussed in chapter 16.

Two remarks: (1) A discharge on a spacecraft is often called an *electrostatic discharge* (ESD), because magnetic fields are almost not involved. In this context, *discharge* refers a harmful discharge. (2) The word *discharge* sometimes means "reduce the charging level." For example, suppose that a spacecraft charges to −10 kV, and the person in control suggests discharging the spacecraft to a lower voltage. In this context, to discharge means to mitigate.

If the incoming electrons or ions are of high energies (MeV or higher), they may be able to penetrate, pass through, or deposit inside materials. These high-energy electrons may stay inside nonconductors—i.e., dielectrics—for a long time. After a prolonged period of high-energy electron bombardment, the electrons inside may build up a high electric field. If the field is high enough, it may be sufficient to cause a local dielectric breakdown. When a local breakdown occurs, ionization channels develop extremely rapidly inside the dielectric, allowing currents to flow, which in turn generate more ionization and heat. As a result, internal instruments may be damaged. Fortunately, the densities of high-energy (MeV or higher) electrons and ions in space are low. Internal damage events are rare. However, when they occur, they may, in extreme cases, cause the loss of spacecraft.

1.2.2 Interference with Scientific Measurements

Spacecraft charging may affect scientific measurements on spacecraft. For example, when scientific measurements of space plasma properties such as the plasma density, mean energy, plasma distribution function, and electric fields are needed onboard, the measurements may be affected. The effects on each of these measurements are explained here.

We first examine the basic mechanism of how a charged object disturbs the ambient plasma. A charged spacecraft repels the plasma charges of the same sign and attracts those of the opposite sign (figure 1.1). As a result, a plasma sheath is formed in which the density of the repelled species is lower than that of the attracted species. The plasma density inside the sheath is different from that outside. The plasma in a sheath is nonneutral. Sheath formation occurs not only in space but also for charged objects in laboratory plasma.

Since the mean energy of the charged particles is shifted by repulsion or attraction, the mean energy of the electrons and that of the ions inside the sheath are different from their respective values outside. The amount of shift depends on the magnitude of repulsion or attraction.

The electron and ion energies of a plasma in equilibrium are in Maxwellian distributions:

$$f(E) = n(m/2\pi kT)^{3/2}\exp(-E/kT) \tag{1.3}$$

A graph of the logarithm of the distribution, $f(E)$, versus E would be a straight line with a slope equal to $-1/kT$, if $f(E)$ is Maxwellian (figure 1.2).

$$\log f(E) = \log n + \frac{3}{2}\log\left(\frac{m}{2\pi kT}\right) - \frac{1}{kT}E \tag{1.4}$$

If the distribution $f(E)$ is measured on the surface of a spacecraft charged to a potential, ϕ_s, the distribution measured would be shifted from that of the ambient plasma by an amount of energy $e\phi_s$. If the distribution is not Maxwellian, the graph $f(E)$ will not be a straight line. No matter what the distribution is, the energy shift will be $e\phi_s$. In the following, we will examine Maxwellian distribution only.

For the attracted species, the energy shift is $e\phi_s$, forming a gap from 0 to $e\phi_s$ in the distribution (figure 1.2). Physically, a charged particle initially at rest is attracted and would gain an energy $e\phi_s$ when it arrives at the spacecraft surface. This size of the gap, which can be clearly identified, gives a measure of the spacecraft potential, ϕ_s. The historical discovery[1] of

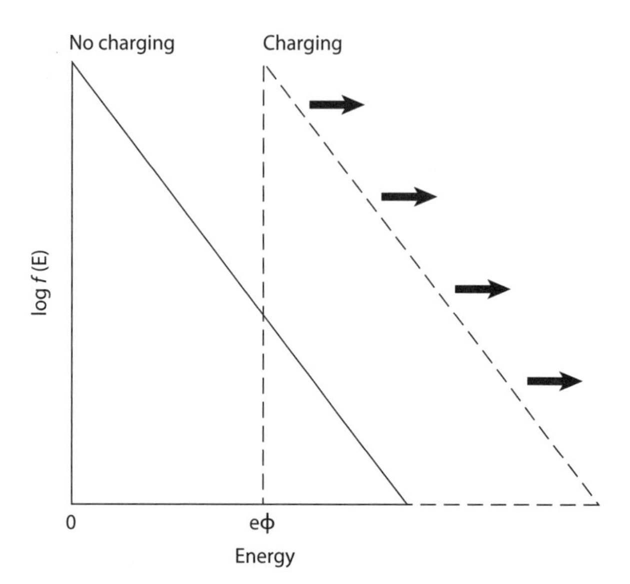

Figure 1.2 Shift of Maxwellian distribution plotted against energy. The amount of shift indicates the spacecraft charging potential. (Adapted from reference 4.)

kilovolt-level charging of a spacecraft (ATS-5) at geosynchronous altitudes at night was made by using a method related to the idea described in this paragraph.

For the repelled species, the shift is $-e\phi_s$. This forms no gap but results in the loss of the negative energy portion of the distribution (figure 1.2). Physically, the repelled ambient species of energy between 0 and $e\phi_s$ are repelled by the charged spacecraft. They cannot reach the spacecraft surface and the instrument on the surface and therefore cannot be measured.

Electric fields are important in governing the flow of electrons and ions in space plasmas. Measurements of electric fields in space are commonly carried out by means of the double probe method. The addition of an artificial potential gradient by a charged spacecraft may affect the measurements. Typical electric fields[2] measured in the ionosphere are of the order of mV/m, which is easily overwhelmed by strong electric fields near the spacecraft surfaces charged to, for example, hundreds of volts.

Spacecraft charging may also affect measurements of magnetic pitch angles of incoming charged particles since charged particles drift in the presence of both electric and magnetic fields. The trajectories of the charged particles are disturbed and therefore are different from those without spacecraft charging.

1.3 How Does Spacecraft Charging Occur?

The cause of surface charging is due to the difference between ambient electron and ion fluxes. Electrons are faster than (all kinds of) ions because of their mass difference. As a result, we have the following theorem:

Theorem: The ambient electron flux is much greater than that of the ambient ions.

To illustrate this point, let us consider hydrogen ions whose mass m_i is about $1837m_e$, where m_e is the electron mass. The electron and ion number densities are equal. Equipartition of energy gives the equality:

$$\frac{1}{2}m_i v_i^2 = \frac{1}{2}m_e v_e^2 \tag{1.5}$$

where the ion energy kT_i equals the electron energy kT_e, where k is the Boltzmann constant. Equation (1.5) yields a ratio of the electron to ion velocities: $v_e \approx 43v_i$.

The equality, equation (1.5), is only approximately valid. If the ion energy kT_i fluctuates and deviates, equation (1.5) would change accordingly. As long as the energies, kT_i and kT_e, are of the same order of magnitude, the preceding equality remains valid as a good approximation, viz, the ambient electron flux is much greater than that of the ambient ions. If some of the ions are heavier than H^+, the velocity difference would be greater.

Therefore, the electron flux, $n_e e v_e$, is greater than the ion flux, $n_e e v_i$. As a corollary of this property of relative velocities, we conclude that surface charging is usually negative, because the surface intercepts more electrons than ions.

It takes a finite time to charge a surface because its capacitance is finite. For typical surfaces at geosynchronous altitudes, it takes a few milliseconds to come to a charging equilibrium. At equilibrium, Kirchhoff's circuital law applies because the surface is a node in a circuit in space.

Kirchhoff's law states that at every node in equilibrium, the sum of all currents coming in equals the sum of all currents going out. Therefore, the surface potential, ϕ, must be such that the sum of all currents must add up to zero. These currents, $I_1, I_2, \ldots, I_k$, account for incoming electrons, incoming ions, outgoing secondary electrons, outgoing backscattered electrons, and other currents if present. The current balance equation, equation (1.6), determines the surface potential ϕ at equilibrium:

$$\sum_k J_k(\phi) = 0 \tag{1.6}$$

1.4 Capacitance Charging

With a steady ambient current, I, the time, τ, for charging a surface is given by

$$\tau I = C\phi \tag{1.7}$$

where C is the capacitance, and ϕ is the surface potential. For a simple example, the surface capacitance C of a spherical object of radius $R = 1$ m is given by $C = \varepsilon_o 4\pi R \approx 10^{-10}$ farad/m. Let us take the ambient current $I = J\pi R^2$, where J is the ambient flux. The object in a space environment of flux density $J = 0.5$ nA/cm^2 would charge from 0 to 1 kV in $\tau \approx 2$ ms, which is a short time for many applications.[3] The charging time, τ, increases directly with the charging level, ϕ, and inversely with the radius, R, and the current density, J:

$$\tau \propto \frac{\phi}{RJ} \tag{1.8}$$

Capacitance charging is analogous to the filling of a tub with water; the water flow rate and the hose size (the cross section of flow) control the time of filling (figure 1.3). During filling, the water level is rising, and therefore the incoming current exceeds the outgoing current. This means that Kirchhoff's law, viz, steady state current balance, is not applicable during capacitance charging; one needs to include a time-dependent term. Once the tub is filled, the water level remains constant, and therefore the incoming and outgoing currents balance each other. The current balance equation, equation (1.6), is not applicable for our simple example during the first 2 ms but is a valid approximation thereafter. (Note that it takes infinite time to charge a capacitance asymptotically, but, for our purpose, we do not need exact values because the space plasma is not measured with high accuracy and varies very much in space and time.)

Note that coupled capacitances take a longer time to charge and thin dielectric layers have larger capacitances than surfaces. For simplicity, we will not consider these complications.

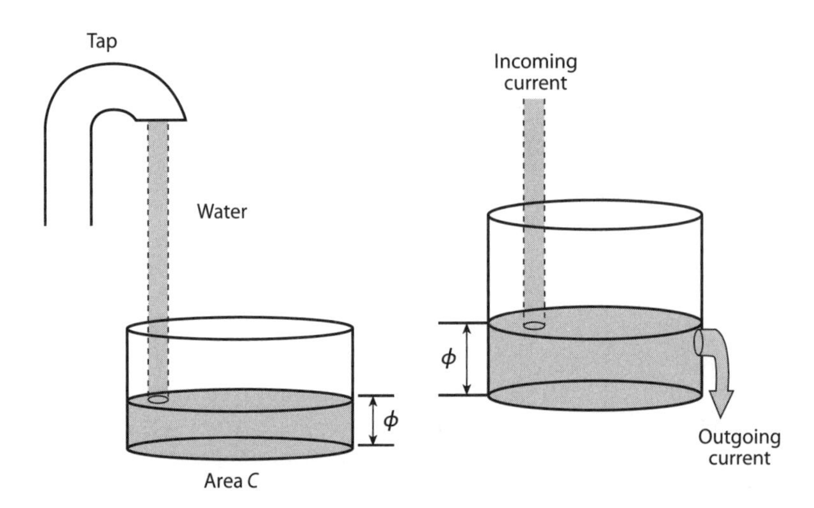

Figure 1.3 Charging of a capacitance. (Left) The water flux J received in a period τ equals the volume of water given by $C\phi$. The level ϕ is changing in time. (Right) At equilibrium, the incoming flux equals the outgoing flux. The level ϕ is constant.

1.5 Other Currents

In sunlight, the photoemission current emitted from a spacecraft surface has to be taken into account. In quiet periods (without severe magnetic storms), the photoemission current often exceeds the ambient currents, thus charging a typical spacecraft surface positively. Since photoelectrons generated by sunlight in the geosynchronous environment have typically a few eV in energy, they cannot leave if the surface potential exceeds a few volts positive. Thus, sunlight charging is typically at only a few volts positive.

Secondary and backscattered electrons are of central importance in spacecraft charging. They will be discussed in detail in chapter 4.

For spacecraft charging induced by charged beam emission, the beam current must be included in the current balance equation. Depending on the properties of the beam and the charging condition of the spacecraft, a fraction f of the beam may leave the spacecraft, while the rest of the beam current returns to the spacecraft. The net current leaving the spacecraft may be very different from that leaving the exit point of the beam,—i.e., the fraction f may be $\ll 1$. If the net beam current leaving the spacecraft exceeds the ambient current arriving at the spacecraft, the beam controls the spacecraft potential.

1.6 Where Does Spacecraft Charging Occur?

Natural surface charging depends on the location of the spacecraft, the material of the surface, the local time, and the space weather. It is customary to delineate four types of locations: geosynchronous altitudes, low Earth orbit altitudes, the auroral latitudes, and the radiation belts.

1.6.1 Geosynchronous Altitudes

The most important region of surface charging is at or near geosynchronous altitudes. This region is important for two reasons: (1) Even though the plasma density is often low, the energy is sometimes high. (2) There are many communication satellites in this region.

The geosynchronous region is sometimes inside the plasmasphere (figure 1.4). During very quiet days, the entire region can be inside the plasmasphere, while during extremely disturbed days, the entire region can be outside it. Most often, the dusk side is inside while the rest of the region is outside. Although the plasmasphere corotates to some extent with the

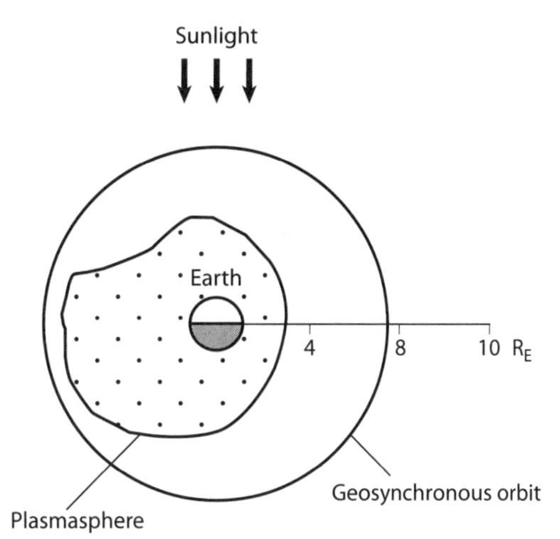

Figure 1.4 The plasmasphere is a dense region of low-energy plasma (about 100/cc, 10 to 30 eV). It sometimes expands into the geosynchronous region on the dusk side. (Adapted from reference 4.)

Earth, the protruding part on the dusk side often persists. The shape of the plasmasphere is not corotating. The plasmasphere usually has relatively high plasma density (> 1 cm^{-3}) and low plasma energy (< 100 eV). Within the plasmasphere, charging of spacecraft surfaces is at zero or low level (usually a few eV negative without sunlight) and is not of concern. In sunlight, the level is at most a few eV positive, which is also not of concern. Occasionally, the spacecraft is outside the plasmasphere and in a low-density (< 1 cm^{-3}), high-energy (keV) plasma region. There, high-level spacecraft surface charging may occur if there is a surge of high-energy electrons and the surface is in eclipse. The high-energy (many keVs and higher) electron cloud typically arrives sunward from the geomagnetic tail.

The initial disturbance usually comes from the Sun in the form of solar wind, high-energy electron and ion clouds, and also x rays. The electrons and ions, upon arrival, compress the dayside magnetosphere and then elongate the nightside magnetosphere to hundreds of Earth radii, forming a long magnetospheric tail (figure 1.5). The elongation is analogous to the stretching of a rubber band. An elongated rubber band eventually snaps back. After hours of stretching, magnetic reconnection occurs somewhere in the geomagnetic tail followed by a snap-back. As a result, an energized electron and ion cloud travels toward the Earth from the tail. This describes the process of a geomagnetic substorm, or simply substorm. It can occur more than once in a night—that is, a storm may consist of a series of substorms.

The energetic electrons and ions enter the Earth's geosynchronous altitudes at about midnight. There, the energetic electrons travel eastward due to the Earth's magnetic field curvature, while the energetic ions travel westward (appendix 1). Since the high-energy electrons are often the cause of spacecraft charging, spacecraft charging at or near the geosynchronous altitude region occurs most probably near midnight and the morning hours. Typically, the charging levels in this region reach hundreds of volts or even several kV, if the spacecraft surface is not in sunlight.

The exact charging level, of course, is determined by current balance. The currents of ambient electrons, ambient ions, and secondary and backscattered electrons have to be taken into account. Photoelectron current is important in sunlight. If there are different neighboring surfaces, currents flowing from one surface to another have to be considered. If electron or ion beams are emitted, the net beam currents leaving the spacecraft have to be taken into account also. The higher the beam-induced potentials are, the more significant are the current flows from surfaces to surfaces. The net beam currents can increase or decrease the surface potentials depending on the nature of the beam flows and the signs of the beam charges leaving or returning.

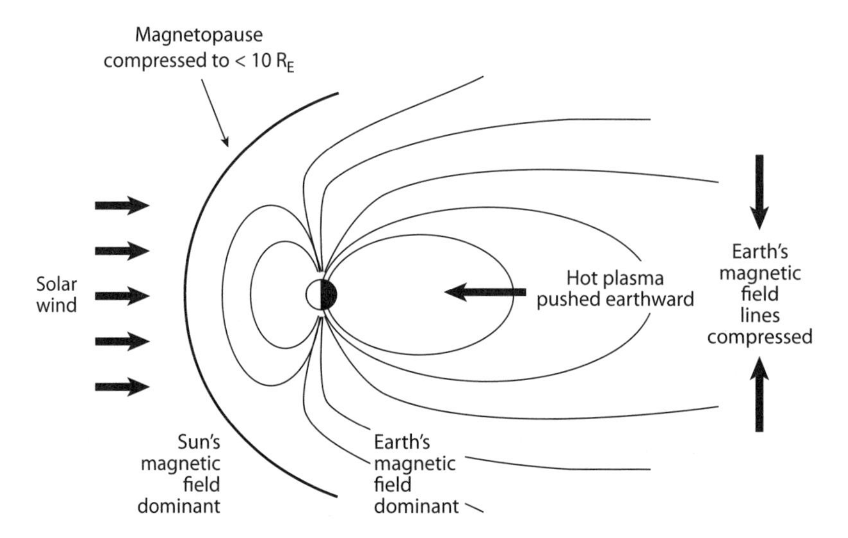

Figure 1.5 Solar plasma disturbance to the Earth's magnetosphere. Solar wind and solar plasma clouds energize the magnetosphere and pull the magnetic field lines to form a long tail. Hot plasma is injected from the tail during substorms and storms. Spacecraft charging is often due to the hot plasma injected around midnight.

1.6.2 Low Earth Orbits

The charging level at low Earth orbits (up to a few hundred km) is usually not of concern. At these altitudes, the space plasma is typically of low energy (about 0.1 eV) and high density (10^5 cm^{-3} or higher). Electrons of 0.1 eV can charge surfaces to 0.1 V at most. If a spacecraft surface is charged to one sign, abundant charges (10^5 cm^{-3}) of the opposite sign can quickly arrive from the vicinity of the charged surface to neutralize it. Charging at low altitudes is not of concern, except (1) when a high-current charged beam is being emitted from the spacecraft, (2) if differentially charged surfaces of solar, or nuclear, batteries are exposed, (3) when a long tether attached to the spacecraft is moving across the Earth's magnetic field lines, (4) in the wake behind a spacecraft moving with an orbiting velocity faster than the ambient ion velocity, or (5) when the spacecraft is at the auroral latitudes during auroral activities. High-current charged beams[4,5] are outside the scope of the introductory part of this course. Exposed battery surfaces are artificial high-voltage devices. Tethers generate electric fields by means of the $\mathbf{V} \times \mathbf{B}$ effect. Natural charging at auroral latitudes will be described briefly in the next section.

1.6.3 Auroral Latitudes

At about 60° to 70° (auroral) latitudes, there are occasionally "inverted V events" during which high-energy (keV) electrons precipitate downward in a beam-like fashion. The events are so called because of the shape of the energy flux plotted as a function of time. During the events, the density of the low-energy plasma component often becomes low, while the the energy distribution of the high-energy electrons (5 to 10 keV typically at 1000 to 1500 km) become beam-like. High-level (hundreds of kV typically) surface charging can occur at these latitudes, and even at fairly low altitudes (approximately 300 km or above). Low-latitude charging in this region is more common than previously thought.

1.6.4 Radiation Belts

Very high energy (MeV or higher) and low flux electrons and ions are in the radiation belts. Spacecraft normally avoid this region because the very high energy radiation (i.e., electrons and ions) may cause internal damage to electronic instruments onboard. The Combined Release and Radiation Effects Satellite (CRRES) flew through this region and collected some interesting data before it suddenly ceased functioning. In this region, spacecraft surface charging is not an important issue, but deep dielectric charging is. Surface charging occurs, but not to very high levels. Deep dielectric charging is due to the very high energy electrons and ions. They can penetrate into and deposit inside nonconducting materials, or even pass through thin insulations. If they enter the electrical wires, they can disturb the circuits, causing anomalies in the electronics, telemetry, and computers. Charge accumulation inside dielectrics may build up a very high electric field, which, if high enough, may lead to a sudden big discharge or dielectric breakdown, damaging instruments onboard.

Very high energy electrons and ions also exist in the geosynchronous region, but with much less intensity and lower fluxes. During the passage of solar coronal mass ejection clouds, very high energy electrons and ions appear in the geosynchronous region and, of course, in the radiation belts.

1.7 Exercises

1. What is spacecraft charging? What are absolute charging, differential charging, spacecraft ground, frame charging, deep dielectric charging, and bulk charging? Is it wrong to think of a spacecraft as at zero potential while the ambient plasma is at a nonzero potential?

2. Spacecraft charging is due to accumulation of charges on spacecraft surfaces. The cause of electron accumulation is often due to the higher flux of electrons than that of ions. In a neutral plasma of density 100 cm^{-3} and average energy 1 eV, calculate the ion and electron velocities and fluxes.

3. Consider a sphere in a vacuum. What is the surface capacitance of a sphere with respect to infinity? How long does it take for a spherical spacecraft of 1-m radius to charge to −1 kV, assuming that the dielectric constant of the ambient plasma is almost that of a vacuum? Calculate the charging time for the sphere as a function of radius, plasma density, and plasma energy.

4. Spacecraft charging may affect some measurements on the spacecraft. How does it affect the plasma density in the vicinity of the spacecraft? How does it affect the plasma electron distribution $f(E)$ measured on the surface of the spacecraft?

5. How does one use the electron or ion energy distribution to measure the level of spacecraft charging? In practice, the energy gap is sometimes partially filled. Why?

6. In the following ion distribution (figure 1.6) plotted as a function of time, the energy scale (left) is in keV, and the flux scale (right) is in #/(cm^2 sec eV). What is the charging voltage?

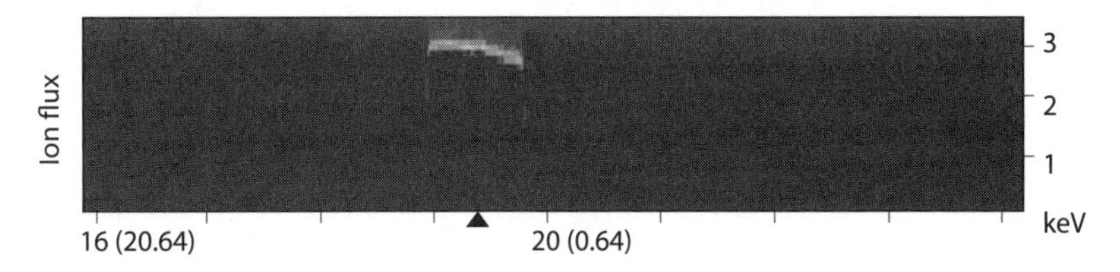

Figure 1.6 A plot of the ion flux measured on a spacecraft surface as a function of time. This figure shows the shift of ion flux during a typical spacecraft charging event. The time given is in universal time (UT), and the time in parentheses is in local time (LT). The triangle indicates midnight in local time.

7. Suggest your own example of measurement that can be affected by spacecraft charging.
8. Why does spacecraft charging not occur at low altitudes—for example, below 5 km?
9. A spacecraft orbit is governed by the balance of the centrifugal force and the gravitational attraction force. Calculate the radius of the geosynchronous orbit.
10. Why is the geosynchronous region important for spacecraft charging? What is the plasmasphere? Do you expect high-level charging when a spacecraft is inside the plasmasphere?
11. At what local time sector in the geosynchronous region is spacecraft charging most likely?
12. At about what latitude at 800 km is spacecraft charging important?
13. Where does deep dielectric charging occur?

1.8 References

For a general reference on the space environment, see, for example, reference 6. For a tutorial on geomagnetic storms and substorms, see, for example, reference 7. For a comprehensive text on space weather and its effects, see, for example, reference 8.

For chapters on spacecraft charging in textbooks, see references 9 and 10. Research papers on spacecraft charging are usually published in *J. Geophys. Res.*, *IEEE Trans. Plasma Sci.*, *IEEE Trans. Nucl. Sci.*, *J. Spacecraft and Rockets*, and *J. Appl. Phys.* There have been Spacecraft Charging Technology Conferences every few years in the past three decades (see, for example, *IEEE Trans. Plasma Sci.,* October 2006 and November 2008).

1. DeForest, S. E., "Spacecraft charging at synchronous orbit," *J. Geophys. Res.* 77: 3587–3611 (1972).
2. Kelley, M. C., *The Earth's Ionosphere*, Academic Press, San Diego, CA (1989).
3. Garrett, H. B., "The charging of spacecraft surfaces," *Rev. Geophys. Space Phys.* 19: 577–616 (1981).
4. Lai, S. T., "An overview of electron and ion beam effects in charging and discharging of spacecraft," *IEEE Trans. Nucl. Sci.* 36, no. 6: 2027–2032 (1989).
5. Wang, J., and S. T. Lai, "Numerical simulations on virtual anodes in ion beam emissions in space," *J. Spacecraft and Rockets* 34, no. 6: 829–836 (1997).
6. Kivelson, M. G., and C. T. Russell, *Introduction to Space Physics*, Cambridge University Press, Cambridge, UK (1995).
7. Lui, A.T.Y., "Tutorial on geomagnetic storms and substorms," *IEEE Trans. Plasma Sci.* 28, no. 6: 1854–1866 (2000).
8. Bothmer, V., and I. A. Daglis, *Space Weather: Physics and Effects*, Springer-Verlag, Berlin (2007).
9. Hastings, D. E., and H. B. Garrett, *Spacecraft-Environment Interactions*, Cambridge University Press, Cambridge, UK (1996).
10. Tribble, A. C., *The Space Environment*, rev. ed., Princeton University Press, Princeton, NJ (2003).

2

The Spacecraft as a Langmuir Probe

In the laboratory, Langmuir probes are commonly used for measuring plasma properties. One applies a controlled potential to the probe and then measures the current collected by the probe. Varying the potential will cause the current to vary, and the resulting current-voltage graph is useful for inferring the plasma density and temperature. The applied potential is the known driving force, while the current is the response to be measured.

In space plasmas, a spacecraft behaves as a Langmuir probe.[1,2] The current-voltage equation[3] of a spacecraft is the same as that of a Langmuir probe. Unlike in the laboratory, one does not apply a potential to a spacecraft. Instead, the spacecraft potential responds to the currents collected from its environment. If one knows all the currents, one can calculate the potential. The currents are the driving forces, while the spacecraft potential is the response, in contrast to the laboratory case.

When the space plasma density is high (although such a high-density regime is rare in space situations), the current density is limited by the Child-Langmuir law.[4,5] Current density saturation by space charge is rare and therefore of little concern for current collection in space. Since the space plasma density is low, the current density impacting on a spacecraft surface is limited by the orbital angular momenta of the incoming charged particles. This is what we refer to as the *orbit-limited regime* (figure 2.1). Spacecraft charging in the geosynchronous environment often occurs in this regime. Because of the low plasma density and often high temperature in the geosynchronous environment, the Debye distance λ_D usually exceeds the spacecraft radius. We will now derive the Langmuir probe equation for a sphere in the orbit-limited regime. The equation is useful for describing a simple current balance model of spacecraft charging for situations where the geometry is isotropic.

2.1 Orbit-Limited Attraction

Consider a particle traveling at velocity v from infinite distance ($r = \infty$) toward the vicinity of a charged spherical spacecraft, which attracts the particle. The sum of the kinetic and potential energies of the particle is constant and is given by

$$mv^2(r)/2 + q\phi(r) = mv^2/2 \tag{2.1}$$

In equation (2.1), q is the charge of the particle being attracted. For electrons, $q_e = -e$, where e is the elementary charge. For positive and singly charged ions, $q_i = e$. For a negative potential ϕ, the sign of $q_e\phi$ is positive, while that of $q_i\phi$ is negative.

As the particle comes near the probe, the velocity $v(r)$ increases with increasing magnitude of the potential $\phi(r)$. The kinetic energy at infinity equals approximately the mean thermal velocity for an equilibrium plasma:

$$mv^2/2 = kT \tag{2.2}$$

The angular momentum is also constant:

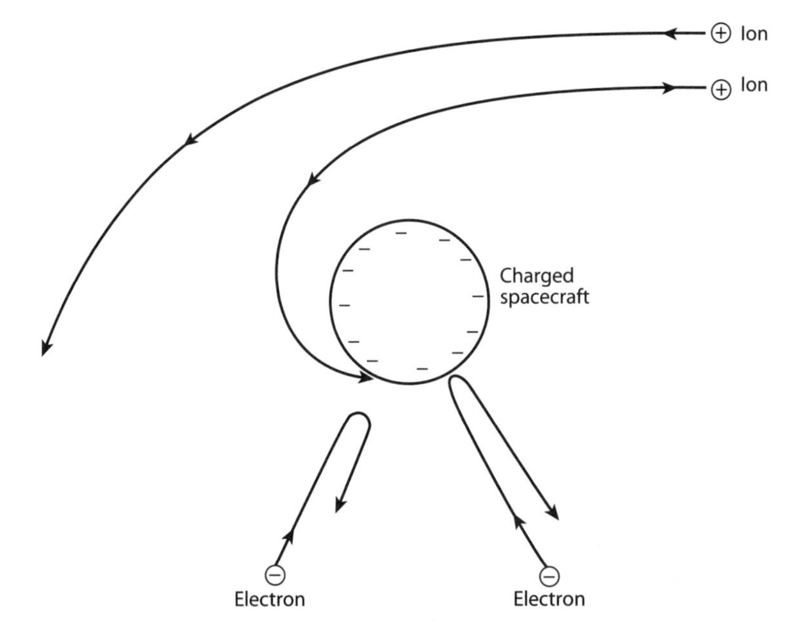

Figure 2.1 The spacecraft as a Langmuir probe. Electrons are repelled by a negative potential. Ions are attracted by a central force, their angular momenta remaining constant.

$$mvh = mv_a a \tag{2.3}$$

where h is the impact distance measured from the center of the sphere to the straight line of travel, a is the spacecraft radius, and v_a is the velocity $v(a)$ of the particle when it touches tangentially the spacecraft at $r = a$. Substituting equations (2.2) and (2.3) into equation (2.1), one obtains the impact distance h:

$$h = a\left(1 - \frac{q\phi}{kT}\right)^{1/2} \tag{2.4}$$

and the velocityv_a at $r = a$:

$$v_a = v\left(1 - \frac{q\phi}{kT}\right)^{1/2} \tag{2.5}$$

2.2 Current Collection in Spherical Geometry

The current, $I(\phi)$, collected by a spherical spacecraft is given by

$$I(\phi) = 4\pi h^2 nqv \tag{2.6}$$

where n is the plasma density. Using equation (2.4), equation (2.6) becomes

$$I(\phi) = 4\pi a^2 nqv\left(1 - \frac{q\phi}{kT}\right) \tag{2.7}$$

which can be written in the form

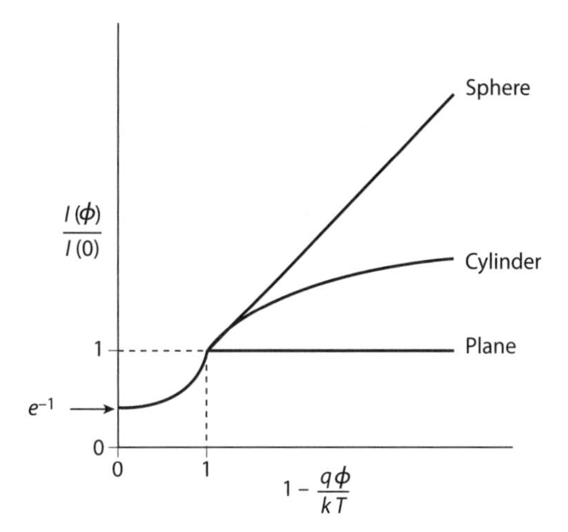

Figure 2.2 Orbit-limited Langmuir probe attraction in one, two, and three dimensions.

$$I(\phi) = I(0)\left(1 - \frac{q\phi}{kT}\right) \tag{2.8}$$

where $I(0) = 4\pi a^2 nqv$ and is the current collected if the spacecraft is not charged—i.e., if $\phi = 0$. Equation (2.8) is the orbit-limited current collection formula for a spherical spacecraft (figure 2.2). It is commonly called the Langmuir probe formula for current collection in the orbit-limited regime; the formula was discovered by Mott-Smith and Langmuir.[6]

2.3 Current Collection in Cylindrical Geometry

In the case of cylindrical geometry, one would be looking in figure 2.1 along the axis of symmetry—i.e., the z axis of the cylinder. The current collected by a long cylinder charged to an attractive potential ϕ is given by

$$I(\phi) = 2\pi hnqv\mathrm{L} \tag{2.9}$$

where L is the length of the cylinder, $2\pi hL$ is the collection surface area, and h is given in equation (2.4). Equation (2.9) can be written as

$$I(\phi) = I(0)\left(1 - \frac{q\phi}{kT}\right)^{1/2} \tag{2.10}$$

where the current collected by the cylinder, if uncharged, is given by $I(0) = 2\pi anqvL$ (figure 2.2).

2.4 Current Collection in Plane Geometry

For an infinite plane, there is no central force and therefore no angular momentum. Symmetry requires the trajectories to be one-dimensional. The attraction term is of the form

$$I(\phi) = I(0) \tag{2.11}$$

This result is shown as a horizontal line in figure 2.2.

The current-balance equation in one dimension is of the form

$$I_i(0) - I_e(0)\exp\left(-\frac{q_e\phi}{kT_e}\right) = 0 \tag{2.12}$$

2.5 Remarks

Remark 1: A more accurate result, which can be derived more rigorously, is of the form

$$I(\phi) = I(0)\mu\left(1 - \frac{q\phi}{kT}\right)^{1/2} \tag{2.13}$$

where $\mu = 1.1$ approximately. Equation (2.13) replaces equation (2.10) for better accuracy. For most practical purposes, however, equation (2.11) is good enough.

Remark 2: The orbit-limited current collection terms, equations (2.8) and (2.10), were first obtained by Mott-Smith and Langmuir.[6] They are usually called the *Langmuir orbit-limited formulas.*

Remark 3: For arbitrary geometry, there is no guarantee that the attractive term is of the form

$$I(\phi) = I(0)\mu\left(1 - \frac{q\phi}{kT}\right)^{\alpha} \tag{2.14}$$

where α and μ are an empirical exponent. Incidentally, the exponent α for the Spacecraft Charging at High Altitude (SCATHA) satellite has been determined to be $\alpha \approx 0.77$ and $\mu \approx 1.1$ by least-square fitting the current-voltage data[7] obtained during electron beam emissions from the spacecraft, the shape of the SCATHA spacecraft being a short cylinder with its length equal to its diameter approximately.

We have examined the current of the attracted species attracted to a Langmuir probe. Next, we will look at the current of the repelled species to the probe.

2.6 Boltzmann's Repulsion Factor

Consider a spacecraft charged to a negative potential ϕ. The incoming ambient electrons are repelled by the negative potential. Suppose that the ambient electrons form a Maxwellian distribution $f(E)$. Because of repulsion, the distribution of the electrons arriving at the surface is shifted in energy.

Accordingly, one must use the shifted distribution $f(E + q_e\phi)$ instead of the ambient distribution $f(E)$. The flux $J(\phi)$ of the ambient electrons arriving at the surface is given by the velocity integral:

$$J(\phi) = \int d^3\mathbf{v}\mathbf{v} f(E + q_e\phi) \tag{2.15}$$

where $q_e = -e$, e is the elementary charge, and $\phi < 0$. The distribution function $f(E)$ in equation (2.15) is a velocity distribution with the velocity expressed in terms of energy E.

For a Maxwellian distribution,

$$f(E + q_e\phi) = f(E)\exp(-q_e\phi/kT_e) \tag{2.16}$$

Therefore, the arriving electron flux, equation (2.15) becomes

$$J(\phi) = \int d^3\mathbf{v}\mathbf{v}f(E)\exp(-q_e\phi/kT_e) \tag{2.17}$$

The current $I(\phi)$ of the electrons arriving at the surface is given by

$$I(\phi) = J(\phi)A = J(0)\exp(-q_e\phi/kT_e)A \tag{2.18}$$

where A is the surface area. Therefore,

$$I(\phi) = I(0)\exp(-q_e\phi/kT_e) \tag{2.19}$$

The exponential term, equation (2.19), is called the Boltzmann factor. It is valid for spherical, cylindrical, or plane Langmuir probes.

For a negatively charged spacecraft, the repelled species is the electrons (figure 2.1). Including the repelled species, the Langmuir equation becomes:

$$I(\phi) = I_i(0)\left(1 - \frac{q_i\phi}{kT}\right) - I_e(0)\exp\left(1 - \frac{q_e\phi}{kT}\right) \tag{2.20}$$

where $I_e(\phi)$ and $I_i(\phi)$ are the collected currents of the attracted and repelled species, respectively. In equation (2.20), the electrostatic energy, $q_i\phi$, is negative. The repulsion term is obtained by assuming that the plasma distribution is Maxwellian. The electrostatic energy, $q_e\phi$, is positive. If $I_e(0)$ and $I_i(0)$ are the only significant currents, equation (2.20) can be used to calculate the spacecraft potential, ϕ, by solving $I(\phi) = 0$. In equation (2.20), we have not included secondary electron and backscattered electron currents, which will be introduced in the next chapter.

Note that $h > a$ in equation (2.4) because $q\phi$ in equations (2.7) and (2.8) is negative and therefore $I(\phi) > I(0)$ in equation (2.8),. If the absolute value $|q\phi|$ is used instead of $q\phi$, the sign preceding $|q\phi|$ in equations (2.7) to (2.9) would be positive. In practice, many users prefer $|q\phi|$, because both $|q\phi|$ and kT have the same unit (eV) and sign, which is positive.

Equation (2.9) is called the *orbit-limited Langmuir equation.*[6] In spacecraft charging, some authors simply call it the *Langmuir probe equation* or the *Langmuir equation*. In spacecraft charging, this equation is the most useful one for describing the balance of currents for a spherical spacecraft in the geosynchronous environment. Since most spacecraft are often considered as spherical for simplicity, equation (2.9) is often used as the current-balance equation. The Langmuir equation, equation (2.20), is a simple model useful for analytical calculations of spacecraft charging. For more detailed models, one has to solve for Poisson equations and calculate particle trajectories on large computers.

2.7 Child-Langmuir Saturation Current

In the Earth's magnetosphere and ionosphere, the space plasma currents are of such low densities that they are far from saturation. The Debye distance λ_D at the geosynchronous environment is larger than, or comparable to, the spacecraft radius. Therefore, "orbit-limited" is the appropriate regime for spacecraft charging.

For completeness, the Child-Langmuir current formula[4,5] is given as follows. This formula is applicable to the current saturation regime only but not the orbit-limited regime. The charge distributed in a local region is commonly called the *space charge*. The current saturation regime is also called the *space charge regime*. In this regime, the space charge potential built up by the charged particles dominates over the angular momentum exerted by the central attractive force of the probe or the charged spacecraft. Therefore, this formula is rarely used in spacecraft charging calculations unless in exceptionally high space charge situations.

In a space charge region, the potential profile $\phi(x)$ at a point x is given by Poisson's equation:

$$\frac{d^2\phi(x)}{dx^2} = 4\pi q(n_e - n_i) \tag{2.21}$$

where q is the elementary charge, n_e is the electron density, and n_i is the ion density. In a plasma sheath, there are more charges of one sign than those of the opposite sign, and therefore the net charge is nonneutral. In the simplest approximation, we assume that there are charges of one sign only. The kinetic energy of a charged particle moving along the potential profile $\phi(x)$ is given by

$$\frac{1}{2}Mv^2 = q\phi \tag{2.22}$$

where v is the charged particle velocity. The current density $j(x)$ is of the form

$$j(x) = qn(x)v(x) \tag{2.23}$$

Substituting equations (2.22) and (2.23) into equation (2.21), one obtains

$$\frac{d^2\phi}{dx^2} = 4\pi\left(\frac{M}{2q}\right)^{1/2}\frac{j(x)}{\phi^{1/2}} \tag{2.24}$$

Integrating equation (2.24) twice, one obtains

$$j(x) = \frac{1}{9\pi}\left(\frac{2q}{M}\right)^{1/2}\frac{\phi^{3/2}(x)}{d^2} \tag{2.25}$$

where d is the sheath thickness. Equation (2.25) is the Child-Langmuir equation for space charge flow. Some authors call it *Child's equation*. Because of the $\phi^{3/2}$ term in the denominator, some authors also call the Child-Langmuir equation (2.25) the *three-halves power law*. Although ions are slower (with smaller velocity v) and therefore have higher charge density n (equation 2.2), in space charge flows the Child-Langmuir equation is also appropriate for electron space charge flows. The equations of space charge flow [equations (2.21) to (2.24)] have often been used for calculating space charge limited currents in electron diodes and triodes with given boundary conditions.

For more on sheaths and space charge flows, refer to standard plasma texts such as references 8 to 10.

2.8 Exercises

1. To get some feeling of the properties of the Langmuir equation, plot the graphs of the two terms in the Langmuir equation, equation (2.9), for $\phi = 0$ to -60 V. You may assume that the electron current is 100 times larger than the ion current when the spacecraft potential is 0. The average plasma thermal energy kT can be taken as 10 to 50 eV.

2. Modify the Langmuir equation for (a) different electron and ion temperatures (b) various ion charge species, and (c) positive charging. You may assume that the different ion species do not affect each other.

3. Assuming that the argument in the exponential of equation (2.9) is much less than unity, expand the exponential in the Langmuir equation in a Taylor series and take the first two terms only. How does the potential behave as a function of the plasma temperature?

4. Plot the current-voltage curves for the charged species attracted to a Langmuir probe of spherical, cylindrical, and plane geometries.

2.9 References

1. Langmuir, I., *Collected Works of Irving Langmuir,* ed. C. G. Suits, Pergamon Press, New York (1960).
2. Chen, F. F., "Electric Probes," chapter 4 in *Plasma Diagnostic Techniques*, eds. R. H. Huddlestone and S. L. Leonid, pp. 113–200, Academic Press, New York (1965).
3. Lai, S. T., "The spacecraft as a Langmuir probe," abstract, in *IEEE International Conference on Plasma Science*, p. 128, Tampa, FL (1992).
4. Child, C. D., "Discharge from hot CaO," *Phys. Rev.* 32: 492–511 (1911).
5. Langmuir, I., "The effect of space charge and residual gases on thermionic currents in high vacuum," *Phys. Rev.* 2: 450–486 (1913).
6. Mott-Smith, H. M., and I. Langmuir, "The theory of collectors in gaseous discharges," *Phys. Rev.* 28: 727–763 (1926).
7. Lai, S. T., "An improved Langmuir probe formula for modeling satellite interactions with near geostationary environment," *J. Geophys. Res.* 99: 459–468 (1994).
8. Birdsall, C. K., and W. B. Bridges, *Electron Dynamics of Diode Regions*, Academic Press, New York (1966).
9. Kirstein, P. T., G. S. Kino, and W. E. Waters, *Space-Charge Flow,* McGraw-Hill, New York (1967).
10. Chen, F. F., *An Introduction to Plasma Physics*, 2nd ed., Plenum, New York (1984).

3

Secondary and Backscattered Electrons

Secondary and backscattered electrons[1-3] are very important in spacecraft charging. They provide outgoing electrons from surfaces. They depend on the surface material properties and on the energy and angle of incidence of the primary electrons.

3.1 Secondary Electron Emission

When an electron impacts a surface, emissions may occur, depending on the electron energy. If the energy is below about 10 eV, the electron may be reflected from the surface. At higher energies, the electron interacts, and thereby shares its energy, with the neighboring electrons in the surface material. One (or more) of the neighboring electrons may gain enough energy to leave the surface as secondary electrons. In other words, the incoming (i.e., primary) electron knocks out one or more electrons, the secondary electrons,[1,2] from the surface (figure 3.1). The energy of a secondary electron is typically a few eV.

At high energies, the primary electron may penetrate so deeply into the material that the probability of escape (i.e., emission) for the energized electrons is low. At very low energies, the primary electron may not be energetic enough to generate secondary electrons, and therefore the probability of emission is also low. The maximum probability (also called *coefficient* or *yield*) of secondary electron emission must lie in an intermediate energy region.

For some materials, the probability $\delta(E)$ of secondary electron emission in an intermediate energy range ($E_1 < E < E_2$) of primary electrons may exceed 1:

$$\delta(E) > 1 \qquad (\text{for } E_1 < E < E_2) \tag{3.1}$$

The energies E_1 and E_2 are typically 70 and 1200 eV, respectively (table 3.1). Physically, when the primary electrons are in this energy range, for every primary electron coming in, there is probably more than one secondary electron going out (figure 3.2). In the literature, the probability $\delta(E)$ is also called the *secondary electron coefficient* or the *secondary electron yield* (SEY).

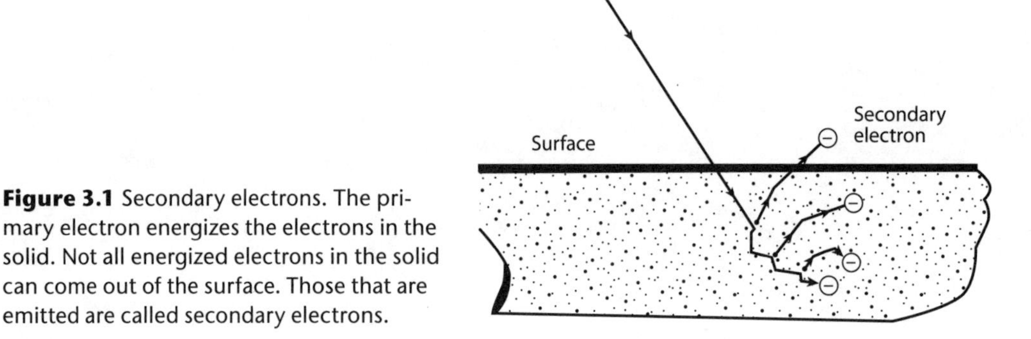

Figure 3.1 Secondary electrons. The primary electron energizes the electrons in the solid. Not all energized electrons in the solid can come out of the surface. Those that are emitted are called secondary electrons.

TABLE 3.1
Secondary and Backscattered Electron Emission Properties of Materials

Material	E_{max} *(keV)*	δ_{max}	*Z*	*A*	*B*	*C*
Magnesium	0.25	0.92	12.0	0.1460	0.0250	0.3440
Aluminum	0.30	0.97	13.0	0.1568	0.0303	0.3431
Kapton	0.15	2.10	5.3	0.07	0.0	0.0
Kapton 2	0.25	1.80	5.3	0.07	0.0	0.0
Aluminium oxide	0.30	2.60	10.0	0.1238	0.01721	0.3435
Teflon	0.30	3.00	8.0	0.09	0.0	0.0
Copper-beryllium	0.30	2.20	29.0	0.3136	0.0692	0.6207
Glass	0.35	2.35	16.9	0.20	0.0420	0.4100
Silver	0.80	1.00	47.0	0.39	0.2890	0.6320
Magnesium oxide	0.40	4.00	10.0	0.1238	0.0172	0.3435
Indium oxide	0.80	1.40	24.4	0.2750	0.0600	0.5400
Gold	0.80	1.45	79.0	0.4802	0.3566	0.6103
Cu-Be (activated)	0.40	5.00	29.0	0.3136	0.0692	0.6207
SiO_2 (quartz)	0.42	2.50	10.0	0.1238	0.0172	0.3435
Colloidal graphite	0.35	0.75	6.0	0.0800	0.0	0.0
Fused silica	0.33	3.46	10.0	0.1238	0.0172	0.3435
SCATHA gold paint	0.72	1.03	70.1	0.4560	0.3380	0.6120
SCATHA yellow paint	0.48	1.49	42.0	0.3730	0.2760	0.6170
SCATHA ML12	0.30	1.00	6.0	0.08	0.0	0.0
SCATHA boom material	0.59	1.86	63.4	0.4380	0.3250	0.6130
MgF_2	0.85	6.38	10.0	0.1238	0.0172	0.3435

Note: Depending on the manufacturer, the material composition and properties may vary. Prolonged exposure in space environment may gradually change material properties.

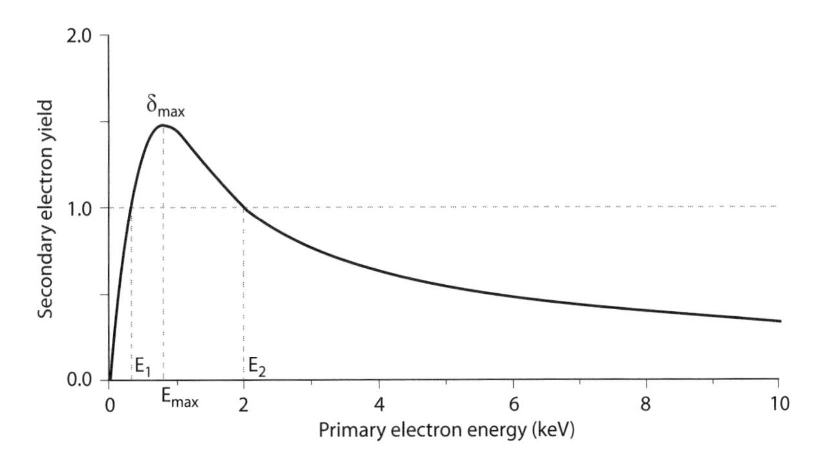

Figure 3.2 The secondary electron emission coefficient. There are two unity crossings at the energies E_1 and E_2 shown in the figure. The maximum δ_{max} of the curve is at energy E_{max}. For some materials, the curve has no crossing, meaning that $\delta(E) < 0$ at any energy E.

3.2 Backscattered Electrons

When a primary electron impacts a surface material, there is a probability η that the electron may backscatter, usually around some ion site. The emitted electron is the same one that came in. The probability η depends on the material, primary electron energy, and angle of incidence. Unlike secondary electrons, the probability η cannot exceed unity, and a backscattered electron may have energy up to nearly the primary electron energy. In the literature, the probability η(E) is also called the *backscattered electron coefficient* or the *backscattered electron yield* (BEY).

3.3 Total Contribution of Electron Emissions

Adding the contributions of the secondary and backscattered electrons, the total probability of electron emission is $\delta + \eta$ (see figure 3.3). For an incoming electron (also called a *primary electron*) of energy E, there are $\delta(E) + \eta(E)$ outgoing electrons. In reality, there are many incoming electrons having different energies. For an incoming electron current I_e, there are two outgoing electron currents, viz, the secondary electron current I_S and the backscattered electron current I_b. The incoming current is defined as the number of particles (or the amount of charge) arriving at an area per unit time. Flux J is defined as the current per unit area. One can use the flux J if it is reasonably homogeneous over the area considered.

Mathematically, one integrates over the electron velocity v, or the electron energy E, to obtain the currents. The primary electron flux J_e is of the form

$$J_e = \int_0^\infty d^3 v v f(E) \tag{3.2}$$

where v is the electron velocity. In practice, it is easier to measure the electron energy E instead of the velocity v. In equation (3.2), $f(E)$ is the electron velocity distribution, with the velocity expressed in terms of energy E by using $E = (1/2)mv^2$. If the electrons are arriving from various directions, one needs to integrate over the angles. For normal incidence on a small plane surface, one can write the primary electron flux, equation (3.2), as follows:

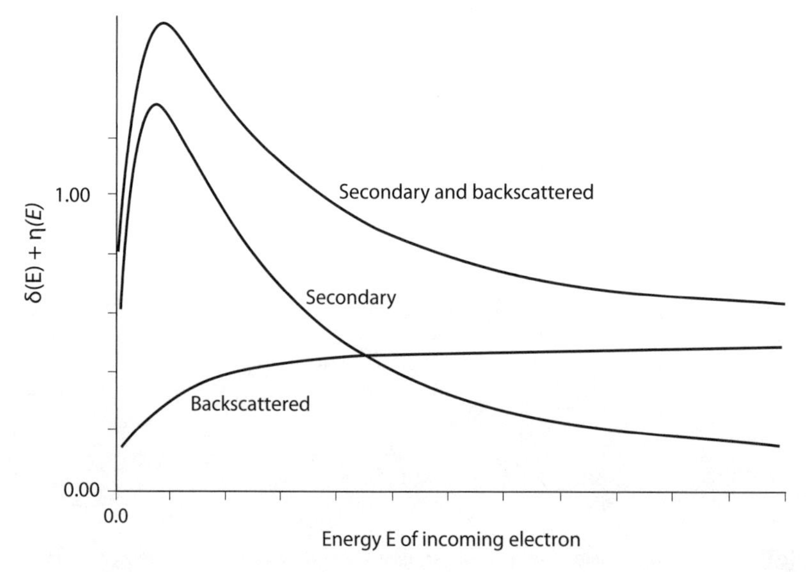

Figure 3.3 Behavior of secondary and backscattered electrons as functions of primary electron energy. There is a range of energy (typically, 60 to 1400 eV depending on the material) in which the total outgoing electron flux exceeds the primary electron flux.

$$J_e = \int_0^\infty dE E f(E) \tag{3.3}$$

A constant multiplicative factor ($4/m^2$ for a normal flow on the plane surface) is neglected in equation (3.3). Equation (3.3) is useful for discussing the basic physics of the incoming and outgoing electron fluxes.

The outgoing secondary electron flux, J_S, is defined as the number of secondary electrons, which are outgoing, leaving the surface of unit area. Since one knows that there are $\delta(E)$ secondary electrons for every incoming electron of energy E, one can use the $\delta(E)$ coefficient and the distribution function $f(E)$ of the primary electrons to obtain the secondary electron flux, J_S, as follows:

$$J_S = \int_0^\infty dE E \delta(E) f(E) \tag{3.4}$$

Similarly, one can write the backscattered electron flux, J_b, as follows:

$$J_b = \int_0^\infty dE E \eta(E) f(E) \tag{3.5}$$

Since equations (3.4) and (3.5) are obtained in the same manner as equation (3.3), the same multiplicative constant is present but not shown.

Note that there is another way to obtain the outgoing electron fluxes. Instead of equations (3.4) and (3.5), one can integrate over the distribution of the outgoing electrons to construct the outgoing currents, but to do so, one needs to know the distributions of the secondary and backscattered electrons. In practice, one does not measure the distributions of the outgoing electrons. We do not need them, since the outgoing fluxes, equations (3.4) and (3.5), can be obtained simply from the incoming electron distribution.

For primary electrons of flux J_e, the total outgoing electron flux is the sum of the secondary electron flux J_s and backscattered electron flux J_b. The net incoming flux J_{net} of electrons is given by subtracting the total outgoing flux from the incoming flux:

$$J_{\text{net}} = J_e - J_s - J_b \tag{3.6}$$

At equilibrium, Kirchhoff's law requires that the currents must balance—that is, the total incoming flux equals the total outgoing fluxes. From equation (3.6), current balance means that the following two equations are satisfied:

$$J_{\text{net}} = 0 \tag{3.7}$$

$$J_e = J_S + J_b \tag{3.8}$$

Using equations (3.2) to (3.5), equation (3.8) can be written in the following form:

$$\int_0^\infty dE E f(E) = \int_0^\infty dE E [\delta(E) + \eta(E)] f(E) \tag{3.9}$$

The same multiplicative constant of equation (3.3) appears on both sides of equation (3.9) and cancels out.

There is a slightly different way to write equation (3.9) in the following manner:

$$\langle \delta + \eta \rangle = \frac{\int_0^\infty dE E f(E) [\delta(E) + \eta(E)]}{\int_0^\infty dE E f(E)} = 1 \tag{3.10}$$

where we have used the notation $\langle \delta + \eta \rangle$ to denote a kind of average with the distributions as the weight. When $\langle \delta + \eta \rangle = 1$, the net incoming or outgoing flux equals zero.

In equations (3.9) and (3.10), we have considered secondary and backscattered electrons only. In sunlight, however, photoelectrons also contribute to the outgoing electron flux. The net incoming electron flux in sunlight becomes

$$J_{\text{net}} = J_e - J_s - J_b - J_{ph} \tag{3.11}$$

Depending on the surface material and the primary electron energies, it is possible that $J_{e,\text{net}} < 0$, shifting the surface potential toward positive, instead of negative. As the surface potential changes, the electron impact energy changes accordingly. If the surface potential becomes positive beyond a few volts, the secondary electrons cannot leave because they have only a few eV in energy.

As the potential ϕ (in negative volts) increases in magnitude, the ions become important because they are attracted toward the spacecraft. When the ions are important, one includes the primary electron current, the ions current, the secondary electron current, the backscattered electron, and the photoelectron current. The current balance equation becomes

$$I_e(\phi) - I_S(\phi) - I_b(\phi) - I_{ph}(\phi) - I_i(\phi) = 0 \tag{3.12}$$

The solution of the current balance equation (3.12) gives the spacecraft potential ϕ. As mentioned earlier, if the current is uniform over the area of consideration, one can replace the current with the flux in the balance.

More on the use of secondary and backscattered electron coefficients for spacecraft charging calculations will be given in ~~the~~ chapter 4.

3.4 Remarks

Chapter 1 gave the statement that "the electron flux is usually greater than the ion flux in space, and therefore the probability of negative charges impacting a spacecraft surface is greater than that of positive charges" as the reason for spacecraft charging. This statement would be valid if the electrons are all either low energy or high energy (e.g., below about $E_1 \approx$ 60 eV or above $E_2 \approx$1200 eV, depending on the material properties). With incoming electrons of a wide range of energy, the preceding statement may be invalid because one needs to take into account the secondary and backscattered electrons. Even with abundant incoming electron flux and negligible incoming ion flux, the surface potential can be positive, because the outgoing secondary and backscattered electron flux can exceed the incoming electron flux. Whether charging occurs depends on the current balance, which in turn depends on the secondary and backscattered electron coefficients and on the primary electron energy distribution.

3.5 Dependence on Incident Angle

Darlington and Cosslett[3] have given the angular-dependent forms of secondary and backscattered electron emission coefficients:

$$\delta(E,\phi) = \delta(E,0)\exp[\beta_S(E) \cdot (1 - \cos\phi)] \tag{3.13}$$

$$\eta(E,\phi) = \eta(E,0)\exp[\beta_b(E) \cdot (1 - \cos\phi)] \tag{3.14}$$

where ϕ is the angle of incidence of the primary electrons. β_S and β_b are empirical factors that can be obtained by fitting the experimental data measured in the laboratory. The forms of β_S and β_b obtained by Laframboise and Kamitsuma.[4] are as follows:

$$\beta_S(E) = \exp(\zeta) \tag{3.15}$$

$$\beta_b(E) = 7.37Z^{-0.56875} \tag{3.16}$$

$$\zeta = 0.2755(\xi - 1.658) - \{[0.2755(\xi - 1.658)]^2 + 0.0228\}^{1/2} \tag{3.17}$$

$$\xi = \ln(E/E_{max}) \tag{3.18}$$

Equation (3.9) is good for normal incidence only. To include angular incidence, equation (3.9) must be modified by including integration over the angles. In general, the integration requires numerical computation. In the case of isotropic incidence, the integration can be done analytically.

3.6 Remarks on Empirical Formulae

The $\delta(E)$ formulae of Sternglass[5] and Sanders and Inouye[6] and the $\eta(E)$ formulae of Sternglass[7] and Prokopenko and Laframboise[8] have been used commonly in the spacecraft charging area in the past decades. For pedagogical purposes, we will use these formulae in chapter 4.

From time to time, new laboratory measurements of secondary electron and backscattered electron coefficients of various materials are published in the literature, and as a result, new empirical formulae based on the new measurements emerge.[9–12] Equation (3.9) is general in the sense that it is not restricted to any particular formula of $\delta(E)$ and $\eta(E)$. One can substitute new empirical formulae of $\delta(E)$ and $\eta(E)$ into the current balance equation, equation (3.9), yielding new results of the equation. If the different formulae used are reasonably accurate, the results yielded are probably not too different from each other.

Last, we remark that surface condition,[13–16] such as smoothness or contamination, is important because it can affect the secondary and backscattered electron coefficients. Since surface condition may change slowly in the harsh space environment, the coefficients may also change slowly in time.

3.7 Exercises

1. Why is the energy of the primary electrons important in spacecraft charging?

2. How does one distinguish a secondary electron from a backscattered electron?

3. Consider a beam of primary electrons of energy $E = 1$ keV at the time of arrival at a spacecraft surface that has a secondary emission coefficient $\delta > 1$ for 1.4 keV $> E >$ 0.2 keV. If the surface is initially uncharged, what would be the resultant surface potential? (For simplicity, let us ignore the backscattered electrons and assume that the secondary electron energy is 2 eV.)

4. Consider a beam of primary electrons of energy $E = 1$ keV at the time of arrival at a spacecraft surface that has a secondary emission coefficient $\delta > 1$ for 1.4 keV $> E >$ 0.2 keV. If the surface is initially charged to −1 kV, what would be the resultant surface potential?

5. Suggest a situation in which secondary electrons may be suppressed.

3.8 References

1. Bruining, H., *Physics and Application of Secondary Electron Emission*, Pergamon Press, New York (1954).
2. Dekker, A. J., "Secondary electron emission," in *Solid State Physics*, vol. 6, eds. F. Seitz and D. Turnbull, Academic Press, New York (1958).
3. Darlington, E. H., and V. E. Cosslett, "Backscattering of 0.5–10 keV electrons from solid targets," *J. Phys. D: Applied Phys.* 5: 1969–1981 (1972).
4. Laframboise, J. G., and M. Kamitsuma, "The threshold temperature effect in high voltage spacecraft charging," in *Proceedings of Air Force Geophysics Workshop on Natural Charging of Large Space Structures in Near Earth Polar Orbit*, eds. R. C. Sagalyn, D. E. Donatelli, and I. Michael, AFRL-TR-83-0046, ADA-134-894, pp. 293–308, Air Force Geophysics Laboratory, Hanscom AFB, MA (1983).
5. Sternglass, E. J., "Theory of secondary electron emission," Scientific Paper 1772, Westinghouse Research Laboratories, Pittsburgh, PA (1954).
6. Sanders, N. L., and G. T. Inouy "Secondary emission effects on spacecraft charging: energy distribution considerations," in *Spacecraft Charging Technology*, eds. R. C. Finke and C. P. Pike, NASA-2071, ADA-084626, pp. 747–755, Air Force Geophysics Laboratory, Hanscom AFB, MA (1978).
7. Sternglass, E. J., "Backscattering of kilovolt electrons from solids," *Phys. Rev.* 95: 345–358 (1954).
8. Prokopenko, S. M., and J.G.L. Laframboise, "High voltage differential charging of geostationary spacecraft," *J. Geophys. Res.* 85, no. A8: 4125–4131 (1980).
9. Suszcynsky, D. M., and J. E. Borovsky, "Modified Sternglass theory for the emission of secondary electrons by fast-electron impact," *Phys. Rev.* 45, no. 9: 6424–6429 (1992).
10. Scholtz, J. J., D. Dijkkamp, and R.W.A. Schmitz, "Secondary electron properties," *Philips J. Res.* 50, nos. 3–4: 375–389 (1996).
11. Cazaux, J., "On some contrast reversals in SEM: applications to metal/insulator systems," *Ultramicroscopy* 108: 1645–1652 (2008).
12. Lin, Y., and D. G. Joy, "A new examination of secondary electron yield data," *Surf. Interface Anal.* 37: 895–900 (2005).
13. Jabonski, A., and P. Jiricek, "Elastic electron backscattering from surfaces at low energies," *Surf. Interface Anal.* 24: 781–785 (1996).
14. Cimino, R., I. R. Collins, M. A. Furman, M. Pivi, F. Ruggerio, G. Rumulo, and F. Zimmermann, "Can low-energy electrons affect high-energy physics accelerators?" *Phys. Rev. Lett.* 93, no. 1: 14801–14804 (2004).
15. Cimino, R., "Surface related properties as an essential ingredient to e-cloud simulations," *Nucl. Instrum. Methods Phys. Res.* 561: 272–275 (2006).
16. Lai, S. T., "The importance of surface conditions for spacecraft charging," *J. Spacecraft and Rockets*, in press (2010).

4

Spacecraft Charging in a Maxwellian Plasma

When a plasma is in thermal equilibrium, its distribution function is a Maxwellian. In the framework of Maxwellian plasma, we will obtain two analytical results from the current balance equation. The results are (1) a critical parameter for the onset of spacecraft charging and (2) a theorem on indefinite charging level. For simplicity, photoemission is omitted in this chapter. The previous chapter introduced the concepts of primary electrons, secondary electrons, and backscattered electrons. In this chapter, we will use an explicit formula of primary electron distribution, an explicit formula of secondary electron yield, and an explicit formula of backscattered electron yield to calculate the flux integrals and provide physical interpretations to the results.

4.1 Velocity Distribution

At zero spacecraft potential (i.e., before the onset of spacecraft charging), the ambient ion currents are about two orders of magnitude smaller than those of electrons. To study the onset of spacecraft charging, it is a good approximation to ignore the ions because the electron currents dominate. In this approximation, the incoming electron current I_{in} is balanced by the outgoing electron currents I_{out}.

$$I_{in} = I_{out} \tag{4.1}$$

Since the current equals the flux multiplied by the area, current balance is equivalent to flux balance if the flux is uniform:

$$J_{in} = J_{out} \tag{4.2}$$

For electrons with a velocity distribution $f(v)$, the flux J is of the form

$$J = \int d^3v f(v) v \tag{4.3}$$

where the velocities are integrated (appendix 2) in all directions of the flux to the surface area. For a Maxwellian plasma, the velocity distribution function $f(v)$ is of the form:

$$f(v) = n\left(\frac{m}{2\pi kT}\right)^{3/2} \exp\left(-\frac{mv^2}{2kT}\right) \tag{4.4}$$

Since the electron energy $E = (1/2)mv^2$, one can write equation (4.4) as follows:

$$f(E) = n\left(\frac{m}{2\pi kT}\right)^{3/2} \exp\left(-\frac{E}{kT}\right) \tag{4.5}$$

For convenience, we will use $f(E)$ instead of $f(v)$. The function $f(E)$ in equation (4.5) is a Maxwellian velocity distribution, with v expressed in terms of E. $f(E)$ is not an energy distribution. The energy distribution function $f_E(E)$ is defined differently. We need not use $f_E(E)$ in the rest of this chapter (see appendixes 3 to 5).

4.2 Critical Temperature for the Onset of Spacecraft Charging: Physical Reasoning

In space and the laboratory, the plasma electrons travel faster than the ions because the ions are heavier. As a result, the electron flux exceeds that of the ions.

For pedagogical purposes, let us consider the average electron velocity v_e and the average ion velocity v_i. At thermal equilibrium, the kinetic energies of the electrons and ions are equal approximately:

$$(1/2)m_e v_e^2 = (1/2)m_i v_i^2 \tag{4.6}$$

Since the ion mass m_i exceeds the electron mass m_e by two orders of magnitude, the electron velocity v_e is much greater than the ion velocity v_i. For charge neutrality, the electron and ion densities are equal, $n_e = n_i$. Therefore, the electron flux, $J_e = q_e n_e v_e$, is higher than the ion flux, $J_i = q_i n_i v_i$—that is, $J_e \gg J_i$. As a result, an uncharged spacecraft intercepts higher electron flux than ion flux—that is, an uncharged spacecraft intercepts more ambient electrons than ions.

However, for every incoming electron impacting a spacecraft surface with an energy E, there are $\delta(E)$ secondary electrons and $\eta(E)$ backscattered electrons going out. The $\delta(E)$ function peaks at a few hundred eV typically and may exceed unity, depending on the surface material. The $\eta(E)$ at $E \neq 0$ is negligibly small typically, depending on the surface material. At the peak, $[\delta(E) + \eta(E)]$ may exceed unity, depending on the surface material. Here, exceeding unity implies that there are more outgoing electrons than incoming ones. Beyond the peak, the number of outgoing electrons $[\delta(E) + \eta(E)]$ decreases as the incoming electron energy E increases. At a sufficiently high energy $E = E^*$, the number of incoming electrons equals the number of the outgoing electrons. Above E^*, the number of incoming electrons exceeds that of the outgoing ones.

The preceding description is for electrons with a single energy E. In reality, the ambient plasma particles travel at various velocities and form a velocity v distribution. Since $\delta(E)$ and $\eta(E)$ are functions of E, it is convenient to express the velocity v in terms of energy E. We denote the distribution as $f(E)$. The velocity distribution $f(E)$ is a function of energy. In statistical mechanics, there is an energy distribution, which is a different function not used here.

At equilibrium, the distribution $f(E)$ is Maxwellian [equation (4.5)]. If one plots log $f(E)$ versus E, the slope is negative, $-1/kT$, which implies that there are more low-energy electrons than high-energy ones (figure 4.1). As the temperature T increases, the slope decreases in magnitude, implying that there are increasingly more high-energy electrons. Since $[\delta(E) + \eta(E)]$ decreases as the incoming electron energy E increases, the high-energy electrons favor negative voltage charging. Above a critical temperature T^*, it is conceivable that the number of incoming electrons exceeds the number of outgoing ones.

The preceding is physics. We will develop the mathematics in the next sections.

4.3 Balance of Currents

To calculate the fluxes or currents, it is convenient to convert the velocity integration to an energy integration, because both the secondary electron emission coefficient $\delta(E)$ and the backscattered electron emission coefficient $\eta(E)$ are measured as functions of energy E. (For more on integrals and distributions, see appendixes 2–5.) At the threshold of spacecraft

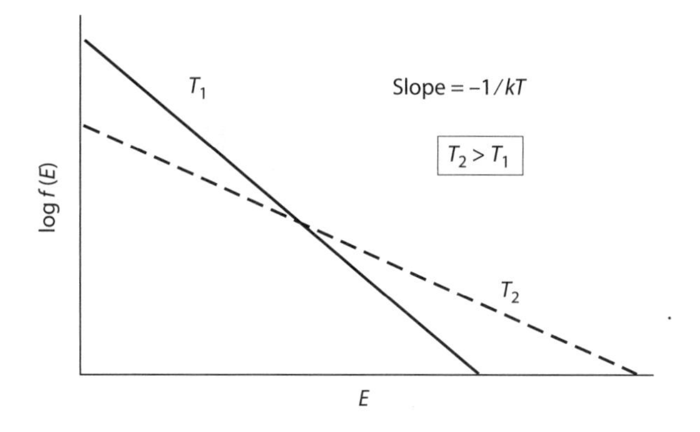

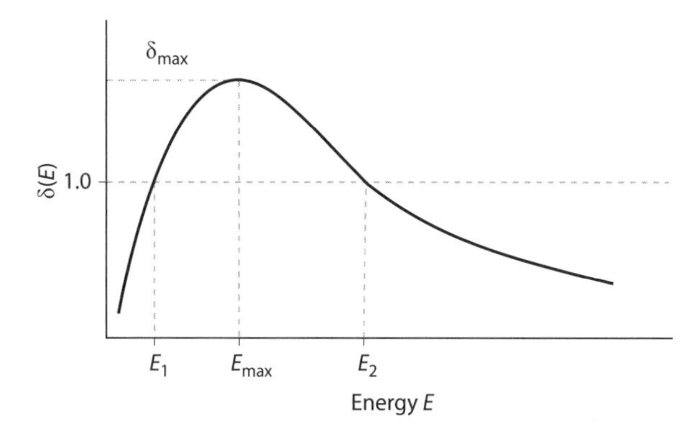

Figure 4.1 Physical reason for the existence of critical temperature for the onset of spacecraft charging.

charging, the spacecraft potential ϕ equals zero. For simplicity, we consider the coefficients $\delta(E)$ and $\eta(E)$ at normal incidence only. The integrations over the angles on both sides cancel each other. The current balance equation is given by

$$\int_0^\infty dEEf(E) = \int_0^\infty dEE[\delta(E) + \eta(E)]f(E) \tag{4.7}$$

Substituting the Maxwellian velocity distribution, $f(E)$ of equation (4.5), into the current balance equation, equation (4.7), earlier, we find that the density n cancels out on both sides. The equation (4.7) is independent of the plasma density, n.

Theorem: In the Maxwellian model, the plasma density plays no role in the onset of spacecraft charging.

The current balance equation, equation (4.7), is a function of plasma temperature, T, only. Let us denote the solution of equation (4.7) as $T = T^*$. Therefore, we have obtained an interesting theorem:

Theorem: In the Maxwellian model, there exists a critical temperature (or threshold temperature), T^*, for the onset of spacecraft charging.

For numerical calculations of the current balance equation, equation (4.7), one needs formulae of $\delta(E)$ and of $\eta(E)$. For example, the Sternglass formulae[1,2] are most popular, the Sanders and Inouye $\delta(E)$ formula[3] is convenient, and the Prokopenko and Laframboise $\eta(E)$ formula[4] is useful. We remark that if one uses angular dependent formulae of δ and η, the preceding two theorems stand because the δ and η formulae are independent of the electron density, and therefore the current balance equation has only one variable, the electron temperature.

The secondary electron formula of Sanders and Inouye[3] is of the form

$$\delta(E) = c\,[\exp(-E/a) - \exp(-E/b)] \tag{4.8}$$

where $a = 4.3\,E_{max}$, $b = 0.367E_{max}$, and $c = 1.37\,\delta_{max}$. The two parameters, E_{max} and δ_{max}, characterize the energy dependent behavior of $\delta(E)$ for a surface material.

The backscattered electron formula of Prokopenko and Laframboise[4] is of the form

$$\eta(E) = A - B\exp(-CE) \tag{4.9}$$

where A, B, and C depend on the surface material.

Using equations (4.8) and (4.9), the current balance equation of equation (4.7) becomes[5]

$$c[(1 + kT/a)^{-2} - (1 + kT/b)^{-2})] + A - B(CkT + 1)^{-2} = 1 \tag{4.10}$$

which can be solved for given values, a, b, c, A, B, and C, of the surface material (Table 4.1).

For primary electrons arriving at various angle of incidence, θ, one needs to use formulae of $\delta(E,\theta)$ and $\eta(E, \theta)$ and integrate over energy and angles in the current balance equation. The following results were first obtained in references 6–9 and reference 5, which is cited in reference 10.

TABLE 4.1
Critical Temperatures (eV)

Material	*Isotropic*	*Normal*
Magnesium	400	—
Aluminum	600	—
Kapton	800	500
Aluminum oxide	2000	1200
Teflon	2100	1400
Copper-beryllium	2100	1400
Glass	2200	1400
Silver	2700	1200
Magnesium oxide	3600	2500
Indium oxide	3600	2000
Gold	4900	2900
Copper-beryllium (activated)	5300	3700
Silicon oxide	2600	1700
Magnesium fluoride	10,900	7800

From reference 5.

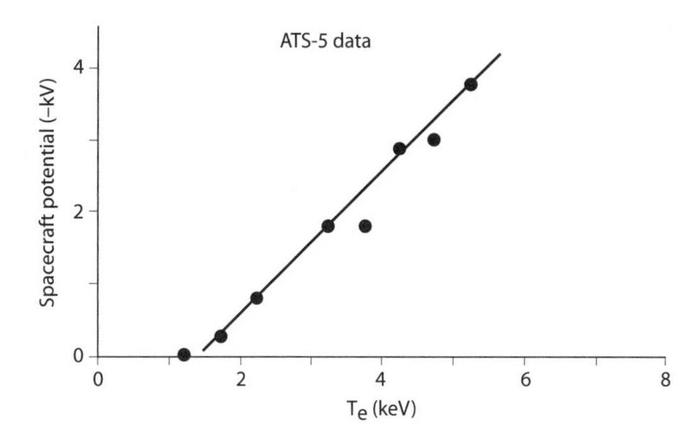

Figure 4.2 ATS-5 satellite data. The existence of a critical temperature is clearly visible. (Adapted from reference 11.)

In figure 4.2, the existence of a critical temperature, T^*, is already apparent in the data from ATS-5[11] in 1978. At that time, a theoretical model of critical temperature, T^*, was not yet available. The model was not available until 1982 (references 6 and 8).

New empirical $\delta(E)$ and $\eta(E)$ formulae[12–16] appear from time to time in journal papers and conference publications. A comparison of some of them is available.[17] For ease of deriving analytical results enabling simple illustration of ideas, we have used the Sanders and Inouye formula[3] in this chapter. One can substitute in any updated $\delta(E)$ and $\eta(E)$ formula in the current balance equation for numerical or analytical results. The results depend not only on the composition of the surface material but also on the surface condition, such as smoothness, contamination, degradation, etc. For pedagogical purposes, we consider simplicity here. Equation (4.7) can be written in the following form:

$$\langle \delta + \eta \rangle = \frac{\int_0^\infty dE\, E f(E)[\delta(E) + \eta(E)]}{\int_0^\infty dE\, E f(E)} = 1 \tag{4.11}$$

As a shorthand notation, one can also write the current balance equation (4.6) as follows:

$$\langle \delta + \eta \rangle = 1 \tag{4.12}$$

The solution of $\langle \delta + \eta \rangle = 1$ [equation (4.12)] gives the critical temperature T^*. Since $\eta \ll \delta$, one can approximate equation (4.12) further as $\langle \delta \rangle = 1$, the solution of which gives an approximate T^*. In the preceding equations, the $\langle\ \rangle$ symbol denotes a special kind of averaging in the sense given by equation (4.10).

4.4 Charging Level

In the preceding section, we have derived a condition for charging onset on an initially uncharged spacecraft. In this model, the incoming electrons are balanced by the outgoing secondary and backscattered electrons. It is natural to ask: "We have found the criterion for the charging onset. Can we derive the charging level?" We now search for an answer as follows.

With charging, the distribution function $f(E)$ measured on the spacecraft surface is shifted (see chapter 2). For the repulsed species, $f(E)$ becomes $f(E + e\phi)$. The current balance equation is approximately as follows:

$$\int_0^\infty dEEf(E + e\phi)\,[\delta(E) + \eta(E)] = \int_0^\infty dEEf(E + e\phi) \tag{4.13}$$

For a Maxwellian distribution, there is a special property that $f(E + e\phi)$ is separable:

$$f(E + e\phi) = f(E) \cdot g(e\phi) \tag{4.14}$$

where the g factor is independent of E, viz,

$$g(e\phi) = \exp(-e\phi/kT) \tag{4.15}$$

Thus, the g factors in equation (4.13) cancel out, resulting in

$$\int_0^\infty dEEf(E)\,[\delta(E) + \eta(E)] = \int_0^\infty dEEf(E) \tag{4.16}$$

Equation (4.16) is independent of the potential ϕ. Therefore, we have the following theorem:

Theorem: In the Maxwellian electron model, the ratio of the outgoing flux to the incoming flux is independent of the spacecraft potential.

Equivalently, the theorem can be written as follows:

Theorem: In the Maxwellian electron model of current balance, the spacecraft charging level is indefinite.

The preceding theorem is surprising. It is due to the special property of the Maxwellian distribution, equation (4.5). Nature cannot be indefinite. Following charging onset, the ion current increases, because of attraction. The charging level has to be determined by the current balance, including electrons and ions.

4.5 Equation of Current Balance in the Orbit-Limited Regime

The ambient electron flux in a plasma exceeds that of the ambient ions by two orders of magnitude. Therefore, a spacecraft intercepts more electrons than ions. Taking into account the emission of secondary and backscattered electrons, a spacecraft may charge to negative voltages. The onset of spacecraft charging is governed by the balance of the incoming and outgoing electron fluxes. One can neglect the ion fluxes in considering the onset of spacecraft charging.

Once negative voltage charging occurs, the voltage will repel the ambient electrons and attract the ambient ions. Thus, the ambient electron current is reduced, but the ion current is enhanced. The spacecraft potential will be governed by the balance of currents, including that of the incoming ambient ions.

The current-balance equation (4.16) is often a good description of the Boltzmann repulsion of the incoming electron current and Langmuir's orbit-limited ion current collection[18] at geosynchronous altitudes:

$$I_e(0)[1-\alpha]\exp\left(-\frac{q_e\phi}{kT_e}\right)-I_i(0)\left(1-\frac{q_i\phi}{kT_i}\right)=0 \tag{4.17}$$

where $I_e(0)$ is the ambient electron current for an uncharged spacecraft (at $\phi=0$), q_e is the negative charge of electron ($q_e=-e$), k is the Boltzmann constant, $q_i=e$ (which is positive), T_e is the electron temperature, T_i is the ion temperature, and

$$\alpha=\langle\delta+\eta\rangle=\frac{\int_0^\infty dEE[\delta(E)+\eta(E)]f(E)}{\int_0^\infty dEEf(E)} \tag{4.18}$$

is the ratio of outgoing to incoming electron currents. Note that the ion collection factor in the last parentheses in equation (4.17) is greater than unity. We have assumed a spherical spacecraft in equation (4.17). For a long cylindrical collector, the last parentheses in equation (4.17) should have a square root sign.

For a simple estimate, we assume $T_e \approx T_i$ and $e\phi/kT \ll 1$ so that $\exp(-e\phi/kT)$ is approximated as a sum of the first two terms only in equation (4.17). With this assumption, the distribution is slightly different from a Maxwellian, but an approximated Maxwellian only. Let the primary electron current, I_e, be modified to $I_e(1-\alpha)$, where α is given by equation (4.18), to account for the secondary and backscattered electrons. With this approximation, the Langmuir equation (4.17) yields a result that, indeed, the charging level, ϕ, is proportional to the plasma temperature, T:

$$I_e(1-\alpha)(1-e\phi/kT+\ldots)=I_i(1+e\phi/kT) \tag{4.19}$$

$$e\phi\approx\frac{I_e(1-\alpha)-I_i}{I_e(1-\alpha)+I_i}kT \tag{4.20}$$

Equation (4.19) gives an interesting proportionality, viz, $e\phi \propto kT$. It is interesting to observe in equation (4.17) that in the limit $I_e/I_i \to \infty$, $e\phi = kT$. In the other limit, viz, $I_i \gg I_e$, which is unlikely because electrons are much faster than ions, the result, $e\phi=-kT$, is obtained. The experimental data[11] obtained on ATS-5 (figure 4.2) shows that indeed the charging level, ϕ, plotted against the plasma temperature, T, is almost a straight line. Thus, the experimental and theoretical results agree with each other[19] in the sense of the following approximate theorem:

Theorem: The spacecraft charging level is approximately proportional to the ambient plasma temperature T.

When the distribution is non-Maxwellian, the concept of temperature T is defined differently. There is no guarantee that the preceding approximate theorems are valid for an arbitrary distribution.

Since better measurements of secondary and backscattered electron coefficients of various materials are published from time to time, substituting them into equation (4.7) will enable updating the critical temperatures of the materials.

4.6 Comparison with Real Satellite Data

In recent years, new evidence[19,20] of the existence of critical temperature was obtained from data obtained on the Los Alamos National Laboratory (LANL) geosynchronous satellites. The phenomenon—onset of spacecraft charging at a critical temperature for a given surface

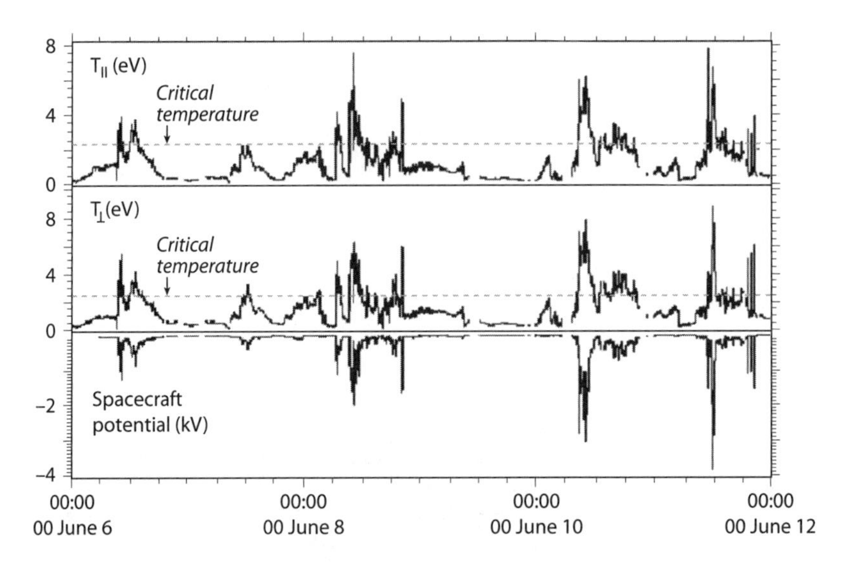

Figure 4.3 Evidence of the existence of critical temperature for the onset of spacecraft charging. The ambient electron temperatures are measured parallel and perpendicular to the ambient magnetic field lines. Below the critical temperature, there is no spacecraft charging; above it, the charging level is proportional to the temperature approximately. (Adapted from reference 19.)

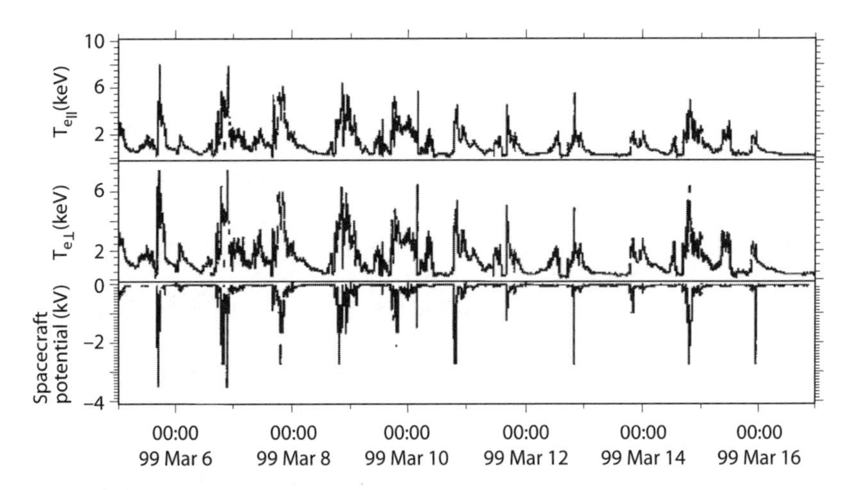

Figure 4.4 Another example of critical temperature for the onset of spacecraft charging. (Adapted from reference 19.)

material—persists, regardless of which day, which season, which year, which satellite. It persists even in daylight[21] and in eclipse,[20] but this aspect will be explained in chapter 7, on charging in sunlight. Figures 4.3 and 4.4 show some typical data taken on a LANL satellite.

4.7 Exercises

1. Suppose that the ambient plasma distribution is a sum of Maxwellians, each having a different density and a different temperature. Is the ratio of the outgoing electron flux to the incoming flux independent of the densities?

2. Suppose that the ambient plasma distribution is a sum of Maxwellians, each having a different density and a different temperature. Is the theorem on indefinite charging level valid?

3. Modify equation (4.19) for the case that the exponential of the Langmuir equation can be approximated as a sum of the first three terms in a Taylor expansion for small $e\phi/kT$. Instead of a straight line in a $e\phi$ (kT) graph, what would one get?

4. Let the electron and ion temperatures be different but greater than the spacecraft potential. Derive an equation of $e\phi$ to replace equation (4.20).

5. In chapter 1, negative voltage charging shifts the peaks of the ion distribution, log $f(E)$, as a function of ion energy E. Suppose that an instrument on a spacecraft measures the differential flux instead of the distribution. The differential flux is proportional to $Ef(E)$ [see equation (Ap.4.9)]. Describe the properties of the charging shift and the magnitude of the peak for the case of using log[$E\,f(E)$] instead of log $f(E)$.

4.8 References

1. Sternglass, E. J., "Theory of secondary electron emission," Scientific Paper 1772, Westinghouse Research Laboratories, Pittsburgh, PA (1954).
2. Sternglass, E. J., "Backscattering of kilovolt electrons from solids," *Phys. Rev.* 95: 345–358 (1954).
3. Sanders, N. L., and G. T. Inouye, "Secondary emission effects on spacecraft charging: energy distribution consideration," in *Spacecraft Charging Technology 1978*, eds. R. C. Finke and C. P. Pike, NASA-2071, ADA-084626, pp. 747–755, Air Force Geophysics Laboratory, Hanscom AFB, MA (1978).
4. Prokopenko, S. M., and J.G.L. Laframboise, "High voltage differential charging of geostationary spacecraft," *J. Geophys. Res.* 85, no. A8: 4125–4131 (1980).
5. Lai, S. T., "Spacecraft charging thresholds in single and double Maxwellian space environments," *IEEE Trans. Nucl. Sci.* 19: 1629–1634 (1991).
6. Lai, S. T., M. S. Gussenhoven, and H. A. Cohen, "Range of electron energy spectrum responsible for spacecraft charging," presented at AGU Spring Meeting, Philadelphia, PA, May 1982; abstract in *EOS* 63, no. 18: 421 (1982).
7. Lai, S. T., M. S. Gussenhoven, and H. A. Cohen, "The concepts of critical temperature and energy cutoff of ambient electrons in high-voltage charging of spacecrafts," in *Spacecraft/Plasma Interactions and Their Influence on Field and Particle Measurements*, eds. A. Pedersen, D. Guyenne, and J. Hunt, pp. 169–175, European Space Agency (ESA) Scientific and Technical Publications Branch (ESTEC), Noordwijk, The Netherlands (1983).
8. Laframboise, J. G., R. Godard, and M. Kamitsuma, "Multiple floating potentials, threshold temperature effects, and barrier effects in high voltage charging of exposed surfaces on spacecraft," in *Proceedings of International Symposium on Spacecraft Materials in Space Environment*, Toulouse, France, ESA Report ESA-178, pp. 269–275, European Space Agency, Noordwijk, The Netherlands (1982).
9. Laframboise, J. G., and M. Kamitsuma, "The threshold temperature effect in high voltage spacecraft charging," in *Proceedings of Air Force Geophysics Workshop on Natural Charging of Large Space Structures in Near Earth Polar Orbit*, eds. R. C. Sagalyn, D. E. Donatelli, and I. Michael, AFGL-TR-83-0046, ADA-134-894, pp. 293–308, Air Force Geophysics Laboratory, Hanscom AFB, MA (1983).
10. Hastings, D. E. and H. B. Garrett, *Spacecraft-Environment Interactions,* Cambridge University Press, Cambridge, UK (1996).

11. Rubin, A., H. B. Garrett, and A. H. Wendel, "Spacecraft charging on ATS-5," AFGL-TR-80-0168, ADA-090-508, Air Force Geophysics Laboratory, Hanscom AFB, MA (1980).
12. Katz, I., M. Mandell, G. Jongeward, and M. S. Gussenhoven, "The importance of accurate secondary electron yields in modeling spacecraft charging," *J. Geophys. Res.* 91, no. A12: 13739–13744 (1986).
13. Suszcynsky, D. M., and J. E. Borovsky, "Modified Sternglass theory for the emission of secondary electrons by fast-electron impact," *Phys. Rev.* 45, no. 9: 6424–6429 (1992).
14. Scholtz, J. J., D. Dijkkamp, and R.W.A. Schmitz, "Secondary electron properties," *Philips J. Res.* 50, nos. 3–4: 375–389 (1996).
15. Furman, M., "Comments on the electron-cloud effect in the LHC dipole bending magnet," *KEK Proc.* 97-17, 234 (1997).
16. Lin, Y., and D. G. Joy, "A new examination of secondary electron yield data," *Surf. Interface Anal.* 37: 895–900 (2005).
17. Lai, S. T., "The importance of surface conditions in spacecraft charging," *J. Spacecraft and Rockets* 47, no. 4, 634–638 (2010).
18. Mott-Smith, H. M., and I. Langmuir, "The theory of collectors in gaseous discharges," *Phys. Rev.* 28: 727–763 (1926).
19. Lai, S. T., and D. Della-Rose, "Spacecraft charging at geosynchronous altitudes: new evidence of the existence of critical temperature," *J. Spacecraft and Rockets* 38, no. 6: 922–928 (2001).
20. Lai, S. T., and M. Tautz, "High-level spacecraft charging in eclipse at geosynchronous altitudes: a statistical study," *J. Geophys. Res.* 111: A09201 (2006).
21. Lai, S. T., and M. Tautz, "Aspects of spacecraft charging in sunlight," *IEEE Trans. Plasma Sci.* 34, no. 5: 2053–2061 (2006).

5

Spacecraft Charging in a Double Maxwellian Plasma

In general, the current balance equation is of the form $J_T(\phi) = 0$, where J_T is the total (or net) flux. It is possible that the equation has multiple roots,[1-6] i.e., solutions. If it has three roots, the $J_T(\phi)$ curve as a function of ϕ is a *triple-root curve*. The spacecraft potential is at one of the roots. As the ambient plasma condition changes in time, a *triple-root jump* may occur—that is, the spacecraft potential may jump from one root to another. This behavior will be explained in the double Maxwellian plasma model.

5.1 A General Theorem on Multiple Roots

Our sign convention is that incoming flux of positive ion is positive, and so is outgoing electron flux. In a general curve of flux-voltage (figure 5.1), there exists at least one root, $J_T(\phi) = 0$, where J_T is the total flux. This is because at high positive potential $\phi > 0$, incoming electron flux must dominate, and therefore $J_T < 0$. At high negative potential ϕ (<0), incoming ion flux must dominate, and therefore $J_T > 0$. Between these two extremes, there must exist at least one or an odd number of zero crossings, $J_T(\phi) = 0$.

Therefore, we have a general theorem:

Theorem: The number of roots, $J_T(\phi) = 0$, must be odd.

The even roots are unstable, because their slopes, $dJ_T/d\phi$, have the wrong sign, corresponding to negative resistance. Only the odd roots are stable. Since the spacecraft potential cannot be multiple valued at the same time, it is at one of the odd roots only. The theorem in this section is general; it applies to not only double Maxwellian but also arbitrary distributions.

5.2 Double Maxwellian Space Plasma

The space plasma environment varies in time. In the outer region of the geosynchronous orbit, energetic plasma clouds from the magnetotail may come in at about midnight hours. Due to the curvature of the magnetic field and other factors, the energetic electrons tend to drift eastward and the energetic ions westward. As they move nearer the Earth, the corotation effect tends to move everything eastward. This describes what usually happens during a *substorm injection*, which may occur from once in many days to a few times a night.

In a quiet period, it is often a good approximation to describe the energy distribution of the space plasma at geosynchronous altitudes as a Maxwellian $f_1(E)$. When a plasma cloud arrives, the plasma distribution changes. It is conceivable that when a hot plasma cloud arrives at a region where a cold plasma is located, the resultant plasma is a mixture of the two plasmas having different temperatures. It is often convenient to describe the resultant distribution, f, as a double Maxwellian, which is a sum of a low-temperature, T_1, component and a high-temperature, T_2, component.

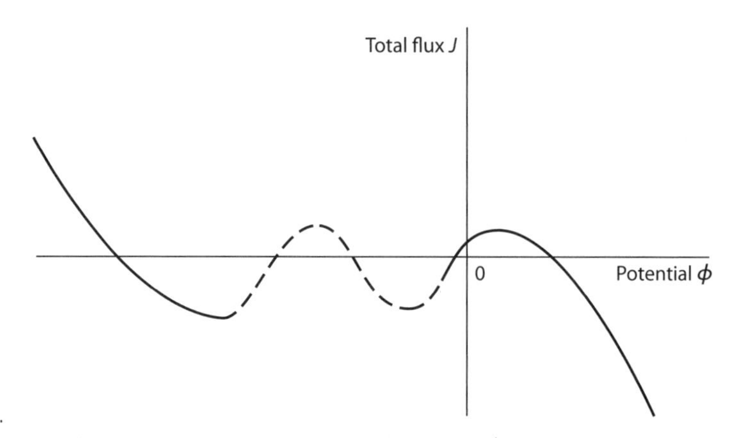

Figure 5.1 Schematic diagram of a general flux-voltage curve. The number of roots (crossings at zero flux) must be odd. The even roots are unstable. (Adapted from reference 6.)

It should be mentioned that any modeling of the space plasma distribution is not exact. The distribution fluctuates and evolves in time. If there is little or no external disturbance, a system would evolve toward equilibrium, according to the second law of thermodynamics. In reality, there are external disturbances, some of which may be transient. Sometimes a kappa distribution describes the space plasma better, especially in the high-energy (tens of keV) region. The mathematics of kappa distribution is more complicated. We will not need to discuss it further here.

We will study the properties of a double Maxwellian distribution in section 5.3. As the space plasma environment changes, it may happen that two neighboring roots, including the spacecraft potential, disappear together (figure 5.1). The spacecraft potential would jump to the next neighboring (the third) root. When the space plasma parameters reverse their course, the two lost roots may appear again. Yet, the spacecraft potential may remain at the new root. A return to its first root may occur but at different values of plasma parameters. This hysteresis behavior will be explained in section 5.6.

5.3 Triple-Root Situation of Spacecraft Potential

If both distributions charge the spacecraft surface potential to the same sign, there is only one root for the spacecraft potential. However, if each of the two distributions charges the surface to potentials of opposite signs, the distributions compete with each other. The one that wins will determine the surface potential sign, and this is the crux of this section. At zero to moderate ϕ (< 0) values, the ambient electron current exceeds that of ambient ions. Therefore, we can neglect the ions for zero to moderate ϕ (< 0) values. We will consider the competition between two Maxwellian electron distributions (figure 5.2).

Consider a double Maxwellian velocity distribution $f(E)$ where we have expressed the velocity in terms of the energy E:

$$f(E) = f_1(E) + f_2(E) \tag{5.1}$$

Each of the two electron distributions is of the form

$$f_1(E) = n_1 (m/2\pi k T_{e,1})^{3/2} \exp(-E/kT_{e,1}) \tag{5.2}$$

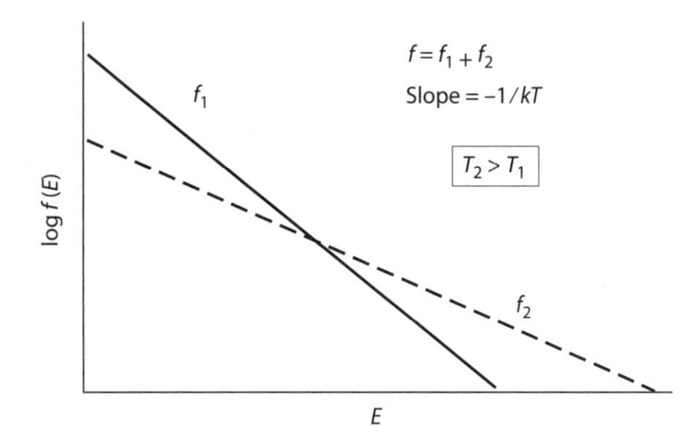

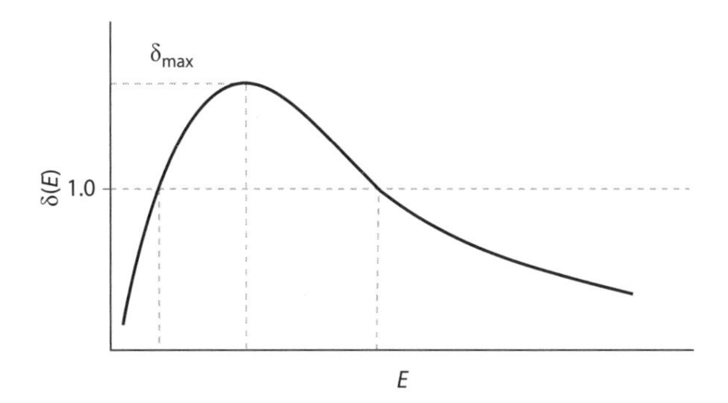

Figure 5.2 A double Maxwellian distribution $f = f_1 + f_2$, with the temperature T_2 of f_2 exceeding the temperature T_1 of f_1. Suppose that T* is the critical temperature for the onset of spacecraft charging. The basic reason for the existence of T* is the secondary electron coefficient δ of the surface material (see chapter 4). The backscattered electron coefficient η contributes less and is not shown here, for simplicity. If $T_2 > T_1 > T^*$, both distributions together will charge the surface to a resultant negative potential. Likewise, if $T^* > T_2 > T_1$, the resultant potential must be positive. However, if $T_2 > T^* > T_1$, there is a competition.

$$f_2(E) = n_2(m/2\pi kT_{e,2})^{3/2}\exp(-E/kT_{e,2}) \tag{5.3}$$

where m is the electron mass, k is the Boltzmann constant, and $T_{e,1}$ and $T_{e,2}$ denote the first and the second Maxwellian electron temperatures, respectively. Three analogous equations can be written for the double Maxwellian ions, with the subscript e replaced by i. The temperature in a Maxwellian distribution is defined as the inverse of the slope of the graph of log $f(E)$ as a function of E. We assume that $T_{e,1} < T_{e,2}$ and $T_{i,1} < T_{i,2}$. For simplicity, we assume charge neutrality, $n = n_e = n_i$.

If both temperatures $T_{e,1}$ and $T_{e,2}$ are below the critical temperature T^* (see chapter 4), each of the two electron distributions would charge the surface to a potential of the same (positive) sign, and therefore the two distributions do not compete with each other. The resultant surface potential is, of course, positive. Likewise, if both $T_{e,1}$ and $T_{e,2}$ are above T^*, both electron distributions would charge the surface to a potential of the same (negative) sign, and again there is no competition. Therefore, it is necessary to satisfy the following inequality (5.4) for the two distributions to compete with each other:

$$T_1 < T^* < T_2 \tag{5.4}$$

To study the competition, let us begin by writing down the full expressions of the fluxes collected by a spacecraft of a negative voltage, ϕ. The total flux J_T is given by the sum of the separate contributions from the two Maxwellian distributions:

$$J_T(\phi) = J_1(\phi) + J_2(\phi) \tag{5.5}$$

$$J_1(\phi) = -j_{e,1}(0)\,[1 - \langle \delta + \eta \rangle] \exp\left(-\frac{q_e \phi}{kT_{e,1}}\right) + j_{i,1}(0)\left[1 - \frac{q_i \phi}{kT_{i,1}}\right] \tag{5.6}$$

$$J_2(\phi) = -j_{e,2}(0)\,[1 - \langle \delta + \eta \rangle] \exp\left(-\frac{q_e \phi}{kT_{e,2}}\right) + j_{i,2}(0)\left[1 - \frac{q_i \phi}{kT_{i,2}}\right] \tag{5.7}$$

where

$$\langle \delta + \eta \rangle = \frac{\int_0^\infty dE E f_h(E)[\delta(E) + \eta(E)]}{\int_0^\infty dE E f_h(E)} \qquad (h = 1,2) \tag{5.8}$$

$$q_e = -e$$

$$q_i = +e$$

In equations (5.6) and (5.7), $j_{e,h}(\phi)$ and $j_{i,h}(\phi)$ are the ambient electron and ion currents, subscript h (= 1,2) labels the Maxwellian, k is the Boltzmann constant, $\delta(E)$ and $\eta(E)$ are the secondary electron[7,8] and backscattered electron[9,10] emission coefficients, respectively, and e is the elementary charge. The exponentials are due to repulsion of electrons by the negative potential; the square bracketed ion term is due to the attraction of positive ions in the orbit-limited regime of Mott-Smith and Langmuir.[11] The regime is usually valid at geosynchronous altitudes. We assume that $j_{e,h} \gg j_{i,h}$ (h = 1,2), which is usually valid because of the difference in electron and ion masses. The sign convention in equations (5.6) and (5.7) is such that the incoming electron flux is negative. The admissible values for the spacecraft potentials are given by the zeros (roots) of the total flux equation:

$$J_T(\phi) = 0 \tag{5.9}$$

Let us now examine the nature of the function $J_T(\phi)$ [equations (5.6) and (5.7)] by varying ϕ as a mathematical parameter. At $\phi = 0$, $J_T(\phi = 0)$ is dominated by the contribution from the electron term of the first Maxwellian distribution:

$$J_T(0) = -j_{e,1}(0)\,[1 - \langle \delta + \eta \rangle] \tag{5.10}$$

Since $T_1 < T^*$ (equation 5.8), there are more outgoing electrons than incoming ones, thereby rendering $\langle \delta + \eta \rangle$ greater than unity, and therefore $J_T(0)$ is positive (figure 5.3, top).

As the magnitude of $\phi(<0)$ increases, the exponentials in equations (5.6) and (5.7) become increasingly influential. Since $T_1 < T_2$, the exponential in equation (5.6) decreases faster than that in equation (5.7). $J_T(\phi < 0)$ decreases. Suppose that at sufficiently large magnitude of $\phi(<0)$, $J_T(\phi < 0)$ is dominated by the contribution from the electron term of the second Maxwellian distribution:

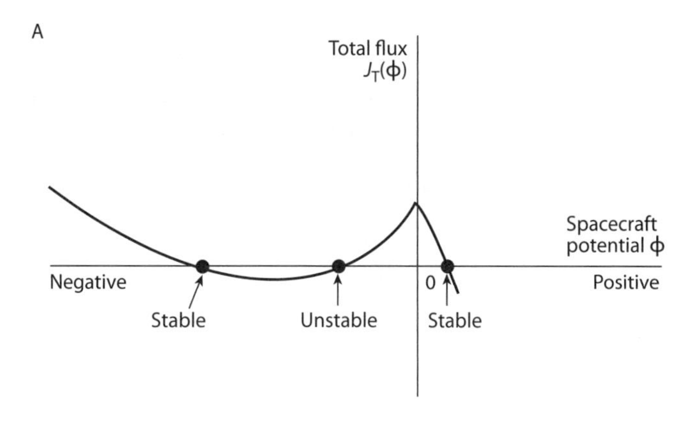

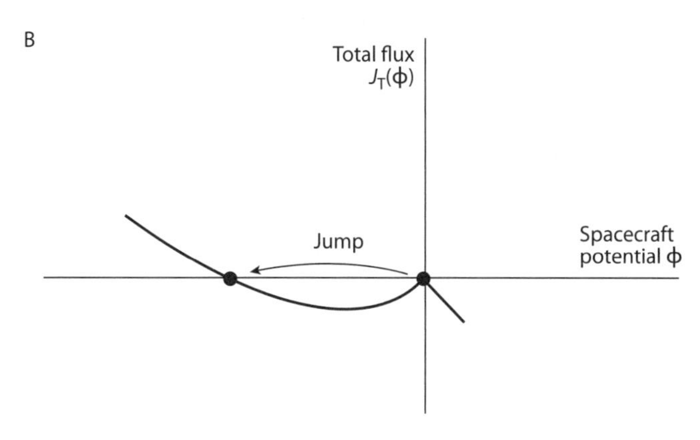

Figure 5.3 (A) A triple-root situation. The flux-potential curve has three roots (zeros). The first and third roots are stable, but the second root is not. (B) A triple-root jump. As the environment changes, it is possible that two of the roots coalesce and disappear, resulting in a jump from the first root to the third. (Adapted from reference 5.)

$$J_T(\phi < 0) = -j_{e,2}(0)\,[1 - \langle \delta + \eta \rangle]\exp\left(-\frac{q_e \phi}{kT_{e,2}}\right) \tag{5.11}$$

Since $T_2 > T^*$, the $\langle \delta + \eta \rangle$ term in equation (5.11) is less than unity, rendering $J_T(\phi < 0)$ negative (figure 5.2, top). As $J_T(\phi)$ changes from being positive to being negative, it must cross the zero value, $J_T(0) = 0$, which is a root.

As the magnitude of $\phi(< 0)$ increases further, eventually the ion terms in equations (5.6) and (5.7) will become increasingly important. This is because more and more positive ions are attracted by the increasingly negative potential $\phi(< 0)$. At sufficiently large negative potential $\phi(< 0)$, the total incoming flux $J_T(\phi < 0)$ rises to positive values, the incoming ion flux dominating over the total electron flux.

There are two possible behaviors of the total flux $J_T(\phi < 0)$ as $\phi(< 0)$ increases in magnitude: In case (a), $J_T(\phi)$ changes from being negative to being positive, it must cross the zero value, $J_T(\phi) = 0$, which is a root. In case (2), it may happen that the flux-voltage curve $J_T(\phi)$ rises so early that the curve $J_T(\phi)$ never reaches zero. This would happen if the second Maxwellian electron flux, with $T_2 > T^*$, is too small to compete with the ion current. In this case, the root $J_T(\phi < 0) = 0$ does not exist in the negative $(\phi < 0)$ regime.

Now we look at the positive potential $\phi(>0)$ regime. The equations are different from equations (5.6) and (5.7) because the signs of attractive and repulsive charges are interchanged and so are the factors of Boltzmann repulsion and Langmuir orbit-limited attraction. Nevertheless, without using elaborate mathematical formulation, it is easy to think as follows. There are two possible behaviors—cases (c) and (d)—of the total flux $J_T(\phi > 0)$ as $\phi(>0)$ increases:

In case (c), suppose that $J_T(\phi = 0)$ is positive at $\phi = 0$. As $\phi(>0)$ increases, the incoming ambient electron flux must increase and ions are repelled, and therefore, according to the sign convention, $J_T(\phi > 0)$ must fall as $\phi(>0)$ increases. As ϕ increases further, $J_T(\phi > 0)$ must cross zero (root). Since the electrons are so much faster than the ions, a high positive potential is unlikely. Even with the emission of photoelectrons from the spacecraft surfaces, the positive surface potential must be low, because photoelectrons emitted from spacecraft at geosynchronous altitudes have typically a few eV in energy only. The photoelectrons must return if the surface potential exceeds a few V. The root on the positive ϕ regime is likely up to a few V only.

In case (d), suppose that $J_T(\phi = 0)$ is negative at $\phi = 0$. As $\phi(>0)$ increases, $J_T(\phi > 0)$ must fall for the same reason as in case (a)—that is, $J_T(\phi > 0)$ remains negative and increases in magnitude as $\phi(>0)$ increases. In other words, $J_T(\phi > 0)$ is negative at $\phi = 0$ and remains negative for all $\phi(>0)$. Therefore, $J_T(\phi > 0)$, being always negative for $\phi > 0$, will never cross zero for any value of $\phi(>0)$. In case (d), there is no root for $J_T(\phi > 0)$ in the positive $\phi(>0)$ domain. However, there must be at least a root, $J_T(\phi > 0)=0$, in the negative $\phi(<0)$ domain. This is because $J_T(\phi < 0)$ must become positive eventually as the negative $\phi(>0)$ increases to large magnitude, attracting abundant positive ions toward the spacecraft.

Counting all of the roots obtained so far, we now have three, if case (a) occurs. In other words, we have a triple-root situation. If case (b) occurs, there is only one root.

5.4 Physical Interpretation of Triple-Root Situation

The previous section, although based on physics, is mathematical in its logical development. It is instructional to explain the same concepts in physics terms despite being seemingly repetitive at places. In a double Maxwellian plasma distribution, it is the competition between the first Maxwellian distribution and the second Maxwellian that determines the fate of charging. If both Maxwellian temperatures are below T^*, the net electron flux is outgoing, and therefore the charging voltage is positive. If both Maxwellian temperatures are above T^*, the charging voltage is negative. If T_1 is below T^* while T_2 is above T^*, we have a competition between the two Maxwellian distributions. Whether $J_{e,1}$ or $J_{e,2}$ is larger depends not only on their number densities n_1 and n_2 together with the velocities v_1 and v_2 but also on the spacecraft potential ϕ. Spacecraft usually charge to negative potentials $\phi(<0)$ because electrons are lighter and faster than ions.

The physics of the total flux $J_T(0)$ at $\phi = 0$ is noteworthy. The spacecraft potential is considered as zero, $\phi = 0$, in this paragraph. The temperature T_1 of the first Maxwellian distribution $f_1(E)$ is lower than that, T_2, of the second Maxwellian distribution. Normally, the density n_1 of the first Maxwellian distribution is much higher than that of the second Maxwellian distribution. In rare events, the density n_1 may decrease, as time progresses, so much that the second Maxwellian distribution becomes more significant.*

* Such natural events are rare. Artificial events are not necessarily rare. For example, when a high-current electron beam is emitted from a spacecraft that charges to the beam energy, the incoming electrons arriving at the spacecraft surface would consist substantially of the returning electrons, which may have impacting energies near the beam energy. More on this topic of returning electrons will be given in chapter 11, on supercharging.) If the density n_2 is significant compared with n_1, and since T_2 exceeds the critical temperature T^*, charging to negative voltages occurs. That is, the total flux becomes negative, because the total incoming electron flux exceeds that of the total outgoing one. When this happens, $J_T(0)$ becomes negative.

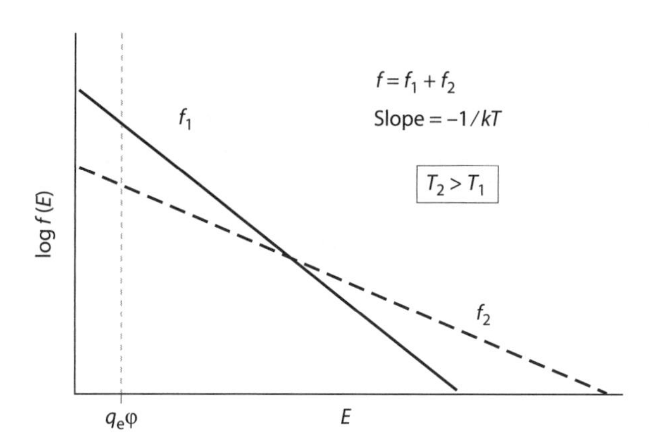

Figure 5.4 Competition between two distributions f_1 and f_2 at various spacecraft potential ϕ. By convention, the flux of f_1 exceeds that of f_2 when there is no charging. Let $T_2 > T^* > T_1$. That is, at $\phi = 0$, f_1 wins, and therefore the total flux is positive. Consider ϕ varying as a mathematical parameter. At higher magnitudes of $q_e\phi$, where both the electron charge q_e and ϕ are negative, the low-energy electrons (with energy E below $q_e\phi$) of both distributions cannot reach the surface because of repulsion. That is, both distributions shown are partially blocked by a curtain from E = 0 to $q_e\phi$. As the parameter $q_e\phi$ increases, the curtain pulls to the right, blocking more of both distributions. Eventually, f_1 loses its dominance to f_2. Since f_2 favors charging to negative voltages, the total flux becomes negative. Eventually, at very high magnitude of ϕ, abundant ions will arrive at the surface because of attraction by ϕ. Therefore, the total flux rises to positive values at very high values of negative potential ϕ.

Next, we discuss the sign of the net flux $J_T(\phi)$ for finite values of potential $\phi(> 0)$ as a varying parameter. At positive potentials, no secondary electrons (or photoelectrons) emitted from the surface can leave if the potential $\phi(> 0)$ is higher than a few volts, because these electrons have a few eV in energy only. If they cannot leave, while the incoming ambient electrons are attracted toward the spacecraft, the net flux must be negative $J_T(\phi) < 0$. At low-to-moderate negative potentials $\phi(< 0)$, both Maxwellian distributions of electrons are important, because both can come in, despite the small potential $\phi(< 0)$ repelling some low-energy electrons. If the first Maxwellian distribution ($T^* > T_1$) dominates, we have a net outgoing electron flux, i.e., $J_T(\phi) > 0$. If the second Maxwellian distribution ($T_2 > T^*$) dominates, we have $J_T(\phi) < 0$ (figure 5.4).

At high negative surface potentials, the first Maxwellian electrons are repelled more than the second Maxwellian electrons because the first Maxwellian distribution is of lower temperature and therefore lower mean energy. As a result, most of the first Maxwellian electrons cannot arrive at the highly negatively charged surface, resulting in the second Maxwellian distribution dominating the incoming electron flux. Since the second Maxwellian distribution has a temperature T_2 exceeding the critical temperature T^*, the incoming electron flux exceeds the outgoing electron flux, resulting in $J_T(\phi) < 0$ according to the sign convention. At very high negative potentials, practically both Maxwellian distributions of electrons cannot come in, but positive ions are attracted toward the spacecraft. Therefore, in this regime, $J_T(\phi) > 0$ again.

5.5 Triple-Root Jump in Spacecraft Potential

A triple-root curve (or situation), does not mean that there are multiple spacecraft potentials simultaneously. The spacecraft potential ϕ is always single-valued and can take up the value of only one of the roots $J_T(\phi) = 0$. Which one is taken up depends on the history of the roots, as will be discussed in section 5.6, on hysterisis. As the space environment changes in time, the roots change. They may disappear or appear in pairs, enabling jumps in spacecraft potential to occur (see figure 5.3, bottom). Although the time scale appears instant, in reality, it takes more time because of surface capacitance and capacitance coupling between surfaces. For typical spacecraft surfaces, it takes milliseconds to charge to kilovolts; for large or coupled capacitances, it may take much longer (see chapter 1).

Although a triple-root jump in spacecraft potential from one root to another can occur, it must be pointed out that a jump is never from one root to the adjacent one. Mathematically, since the roots of $J(\phi) = 0$ always appear and disappear in pairs, the jump must be from an odd root to another odd root. For example, the first root jumps to the third, and never to the second. Physically, the even roots are unstable because they behave as negative Ohm's law, as discussed in section 5.1.

5.6 Hysteresis

A reverse triple-root jump in spacecraft potential can occur. For example, after jumping from the first root to the third, a reverse jump from the third to the first can occur later as the ambient plasma condition (n_1, T_1, n_2, T_2) changes.

Suppose that after a jump from the first root to the third, the ambient space plasma condition is changing. A reverse jump back to the first root is unlikely to occur when the plasma returns to its initial condition, i.e., the condition just before the jump. In this manner, the spacecraft potential does not retrace its track as a function of the space plasma condition. This irreversibility property is called *hysteresis*.

To understand the hysteresis, one is reminded that the jump from the first root to the third occurs when the first and second roots coalesce and disappear together. If the first and second roots reemerge later, the spacecraft potential-flux curve $J_T(\phi)$ becomes a triple-root curve again. However, a jump back to the first root would not occur until the second and the third roots coalesce and disappear together (figure 5.3, bottom).

An illustrative example of hysteresis is shown in figure 5.5. In this figure (top), as the first Maxwellian electron density n_1 decreases from 100 to 0.5 cm^{-3}, the potential root (zero of the total flux) remains positive until about 0.7 cm^{-3}, when the adjacent roots C and D coalesce and disappear together. At that density, the potential root jumps to about -6 kV at F. If the density n_1 decreases, the potential root remains at about -6 kV until the adjacent roots E and F coalesce and disappear together. At that point, the root jumps to a positive value again. The hysteresis of $\phi(n_1)$ manifests in figure 5.5 (bottom). The time evolution of $\phi(n_1)$ does not trace back its original path. In this simple example, the potential of the point F in figure 5.5 (top) is nearly constant, resulting in a square-shaped hysteresis diagram (figure 5.5, bottom). In general, the point F may vary as the density n_1 varies. That is, the time-dependent potential at F may vary, causing the hysteresis diagram to deviate from being a simple square.

5.7 Triple-Root Spacecraft Charging Domains

In a single Maxwellian space plasma (see chapter 4), the plasma temperature is the only parameter controlling the onset of spacecraft charging, while the plasma density plays no role. In a double Maxwellian plasma, both the temperatures and densities participate in controlling the charging onset. To calculate the flux-voltage curve $J_T(\phi)$ as a function of the four plasma electron Maxwellian distribution parameters (n_1, n_2, T_1, T_2) and the four plasma ion Maxwellian distribution parameters, one formulates the $J_T(\phi)$ function in full, in terms of outgoing electron flux minus incoming electron flux plus incoming ion flux, as follows:

A

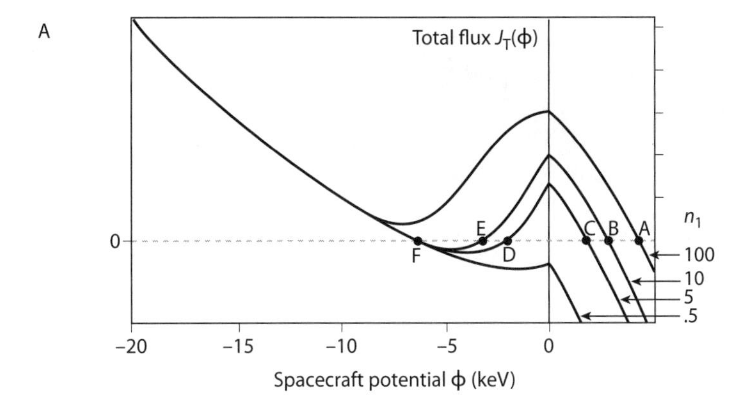

B

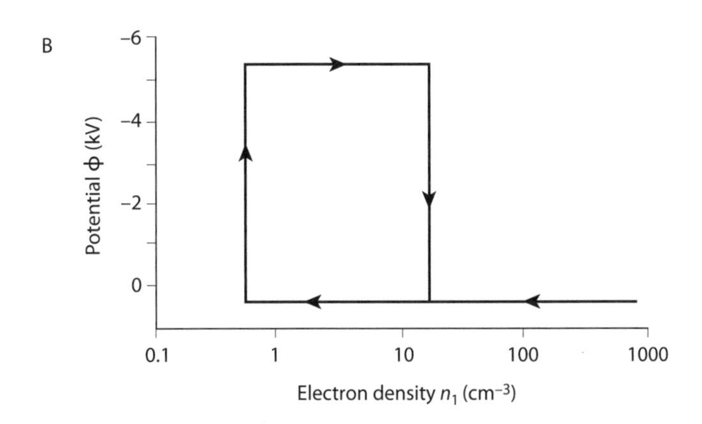

Figure 5.5 An illustrative example of hysteresis. (A) As the first Maxwellian electron density n_1 decreases from 100 to 0.5 cm^{-3}, the potential root (zero of the total flux) remains positive until about 0.7 cm^{-3}, when the adjacent roots C and D coalesce and disappear together. At that density, the potential root jumps to about −6 kV at F. If the density n_1 decreases, the potential root remains at about −6 kV until the adjacent roots E and F coalesce and disappear together. At that point, the root jumps to a positive value again. (B) The hysteresis of $\phi(n_1)$ is manifest. The time evolution of $\phi(n_1)$ does not trace back its original path.

$$J_T(\phi) = J_1(\phi) + J_2(\phi)$$

which is equation (5.5). In this equation, one can write the two fluxes J_1 and J_2 in terms of the electron distributions $f_{e,1}$ and $f_{e,2}$ and the ion distributions $f_{i,1}$ and $f_{i,2}$.

$$\begin{aligned} J_1(\phi) = &-n_{e,1}\left(\frac{m_e}{2\pi kT_{e,1}}\right)\int_0^{\infty} d^3\text{v}.\text{v}[1-\delta(E)-\eta(E)]f_{e,1}(E)\exp\left(-\frac{q_e\phi}{kT_{e,1}}\right) \\ &+ n_{i,1}\left(\frac{m_i}{2\pi kT_{i,1}}\right)\mu\int_0^{\infty} d^3\text{v}.\text{v}f_{i,1}(E)\left[1-\frac{q_i\phi}{kT_{i,1}}\right]^{\alpha} \end{aligned} \tag{5.12}$$

$$\begin{aligned} J_2(\phi) = &-n_{e,2}\left(\frac{m_e}{2\pi kT_{e,1}}\right)\int_0^{\infty} d^3\text{v}.\text{v}[1-\delta(E)-\eta(E)]f_{e,2}(E)\exp\left(-\frac{q_e\phi}{kT_{e,2}}\right) \\ &+ n_{i,2}\left(\frac{m_i}{2\pi kT_{i,2}}\right)\mu\int_0^{\infty} d^3\text{v}.\text{v}f_{i,2}(E)\left[1-\frac{q_i\phi}{kT_{i,2}}\right]^{\alpha} \end{aligned} \tag{5.13}$$

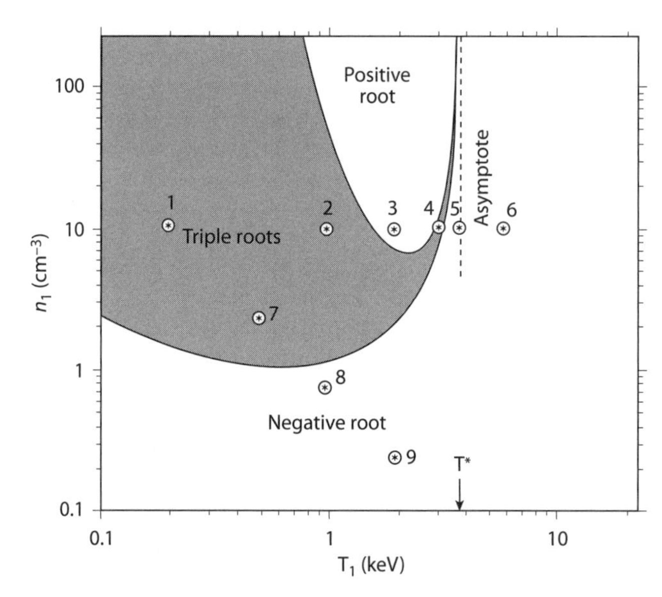

Figure 5.6 A domain diagram. The domain boundaries delineate the negative root, positive root, and triple-root domains. The points 1 through 6 refer to the parametric states (n_1, T_1), to be discussed in figure 5.7, and the points 1, 7, 8, and 9 refer to the parametric states in figure 5.8. (Adapted from reference 5.)

where the symbols are as usual, $\alpha = \mu = 1$ for a sphere, and $\alpha = ½$, $\mu \approx 1.1$ for an infinite cylinder. Furthermore, if the spatial distributions of the incoming electrons and ions are anisotropic, one needs to integrate equations (5.12) and (5.13) over the incoming angles.

Unlike the single Maxwellian case, all electron and ion parameters (densities and temperatures) are involved in determining the spacecraft potential ϕ in the double Maxwellian case. A graph of the function ϕ in an eight-dimensional space would be impossible to visualize. It is instructional to plot a two-dimensional domain diagram delineating the properties of the roots, $J_T(\phi) = 0$, if all the parameters, except two, are fairly constant within the period of interest.

For example, a domain diagram[5] for a period of interest on day 114 of the SCATHA satellite is shown in figure 5.6. The domain boundaries delineate those of positive, negative, and triple-root charging. Interestingly, the double Maxwellian parameters during 5:30–7:30 UT on day 114 are n_1 decreasing from about 2 cm^{-3} to nearly 0.01 cm^{-3}, T_1 staying around 0.4 to 0.5 keV, $n_2 \approx 0.9$ cm^{-3}, and $T_2 \approx 24.8$ keV. Noting that T_1 is varying only slightly, and, n_2, and T_2 are fairly constant while n_1 is decreasing steadily by more than one decade, one can therefore plot a two-dimensional approximate domain diagram, figure 5.6, by calculating the fluxes $J_T(\phi)$ as a function of two parameters n_1 and T_1 [equations (5.12) and (5.13)]. In figure 5.7, the root jumps from the root J to the root H shortly after the moment 4, when the Max[$J_T(0)$] at $\phi = 0$V drops below zero (i.e., Max[$J_T(0)$] < 0). In figure 5.8, a jump occurs at the moment E. Using the domain diagram figure 5.6, one is able to predict that a triple-root jump would occur as n_1 decreases through the range 2 cm^{-3} to nearly 0.01 cm^{-3}, while the other parameters remain approximately constant.

Such a domain graph may be useful for predicting single and triple-root charging onset as the ambient plasma parameters changes in changing space weather. Weather forecasts always tell us barometer pressure, temperature, winds, cloud cover, cold fronts, etc. They are parameters, or features, characterizing the daily weather shown on a weather map. For example, one

Figure 5.7 Time evolution of the approximate flux-voltage curve $J_T(\phi)$ for day 114 on SCATHA. The six panels correspond to the six fictituous moments 1 to 6 labeled horizontally in the domain diagram shown in figure 5.6, the density n_1 being constant. This figure shows the effects as the temperature T_1 alone increases. As the flux at $J_T(0)$ decreases below the origin, the two roots (I and J) coalesce, and therefore a jump to the third root (K) occurs. (Adapted from reference 5.)

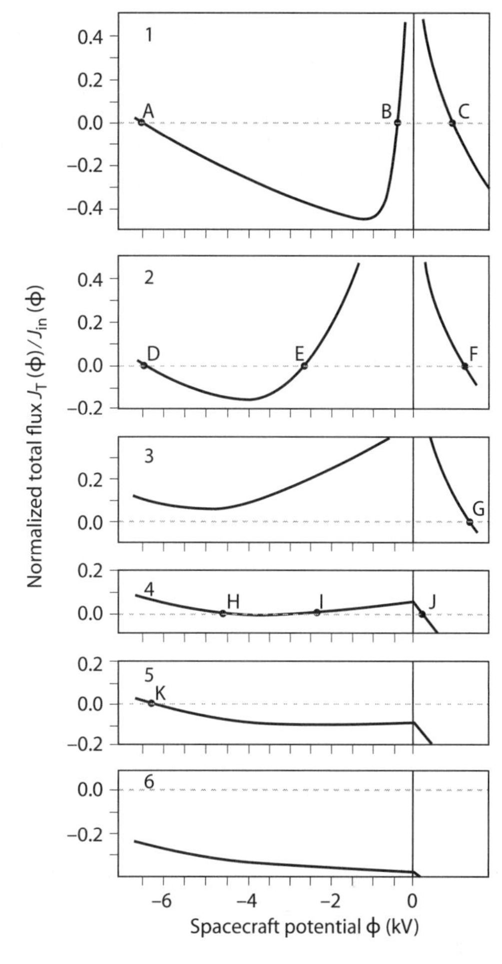

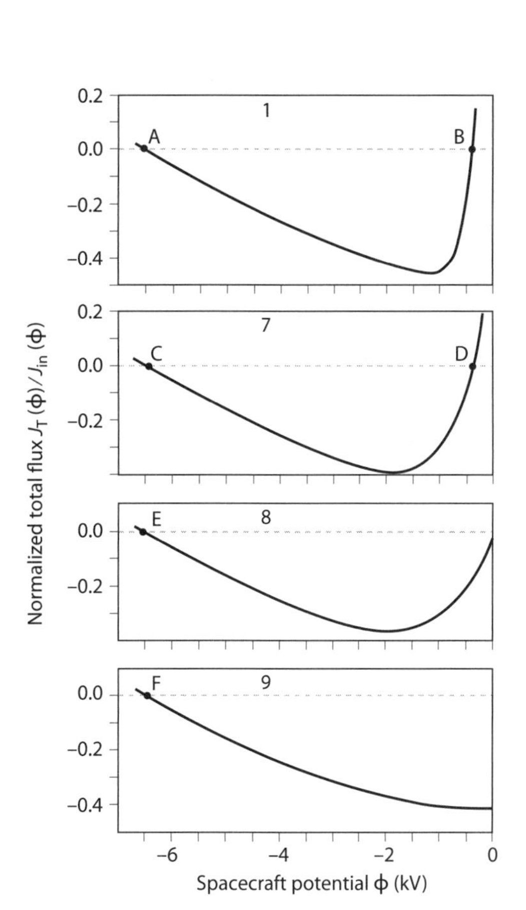

Figure 5.8 Time evolution of the approximate flux-voltage curve $J_T(\phi)$ for day 114 on SCATHA. Four panels are shown, corresponding to the four states labeled 1, 7, 8, and 9 in the domain diagram shown in figure 5.6. These states are chosen because both the density n_1 and the time T are changing as the state (n_1, T) evolves from 1 to 9. As the flux at $J_T(0)$ decreases below the origin, two roots coalesce, and therefore a jump to the third root occurs. (Adapted from reference 5.)

never hears from the radio or TV the velocity distribution of every atmospheric molecule at every spatial grid point in the Atlantic. This is because one cannot digest and use effectively too much data. By analogy, it may be important to observe the changes in a few space environment parameters for forecasting. If a parametric map were available, it would be helpful. The example of SCATHA day 114 shows that if one's spacecraft is at a certain spatial location where the parameter n_1 is decreasing toward the lower boundary between the triple-root domain and the negative root domain (figure 5.3), one should issue a "weather forecast" warning that a finite jump in potential, via the triple-root mechanism, to a negative potential is imminent. In this example, the lower boundary is analogous to a cold front, across which the weather is expected to change abruptly.

This chapter has revealed the physical mechanism of the triple-root situation and triple-root jumps in spacecraft potential. The observation of day 114 on SCATHA gives evidence of the existence of triple-root jumps in potential. Last, we have pointed out that observing the changes in space environment parameters may be helpful in forecasting "space weather."

5.8 Exercises

1. Explain in your own words why there can be three roots in a double Maxwellian model.
2. The features in figure 5.3 can be explained. Why is the asymptote located at the critical temperature T^*? Why is the positive root domain located at high n_1 and temperature T_1 below T^* but not at $T = 0$? Why does negative charging occur when n_1 becomes very small?
3. In a single Maxwellian plasma, the spacecraft potential is proportional to the electron temperature approximately. In a double Maxwellian plasma, there are two temperatures. How would you modify the theorem to relate the spacecraft potential to the two temperatures?
4. If a spacecraft is emerging from eclipse to sunlight, the photoemission from the spacecraft surface can change the total flux $J_T(\phi)$. Which way will the flux move? Does eclipse exit make it more or less likely for a triple-root jump in spacecraft potential to occur?

5.9 References

1. Whipple E. C., Jr., "Potential of surface in space," *Rep. Prog. Phys.* 44, 1197–1250 (1981).
2. Besse, A. L., "Unstable potential of geosynchronous spacecraft, *J. Geophys. Res.* 86, no. A4: 2443–2446 (1981).
3. Laframboise, J. G., R. Godard, and M. Kamitsuma, "Multiple floating potentials, threshold temperature effects, and barrier effects in high voltage charging of exposed surfaces on spacecraft," in *Proceedings of International Symposium on Spacecraft Materials in Space Environment*, ESA SP-178, pp. 269–275, ESA, Paris (1982).
4. Meyer-Vernet, N., "Flip-flop of electric potential of dust grains in space," *Astron. and Astrophys.* 105: 98–106 (1982).
5. Lai, S. T., "Theory and observation of triple-root jump in spacecraft charging," *J. Geophys. Res.* 96, no. A11: 19269–19282 (1991).
6. Lai, S. T., "Spacecraft charging thresholds in single and double Maxwellian space environments," *IEEE Trans. Nucl. Sci.* 19: 1629–1634 (1991).
7. Sternglass, E. J., "Theory of secondary electron emission," Paper 1772, Westinghouse Research Laboratories, Pittsburgh, PA (1954).
8. Sanders, N. L., and G. T. Inouye, "Secondary emission effects on spacecraft charging: energy distribution consideration," in *Spacecraft Charging Technology 1978*, eds. R. C.

Finke and C. P. Pike, NASA-2071,ADA-084626, pp. 747–755, U.S. Air Force Geophysics Laboratory, Hanscom AFB, MA (1978).

9. Sternglass, E. J., "Backscattering of kilovolt electrons from solids," *Phys. Rev.* 95, no. 2: 345–358 (1954).
10. Prokopenko, S. M., and J.G.L. Laframboise, "High voltage differential charging of geostationary spacecraft," *J. Geophys. Res.* 85, no. A8: 4125–4131 (1980).
11. Mott-Smith, H. M., and I. Langmuir, "The theory of collectors in gaseous discharges," *Phys. Rev.* 28, 727–763 (1926).

6

Potential Wells and Barriers

This chapter considers potential wells and potential barriers as mutual interactions between surfaces on spacecraft. Such a complexity is beyond the Langmuir current-balance model, which accounts for the interaction between a spacecraft and its ambient space plasma only. Another concept of spacecraft interaction with objects nearby is also introduced—the charging of objects in the wake of spacecraft.

6.1 Introduction

Charging of spacecraft surfaces depends not only on the ambient plasma conditions but also on the secondary and backscattered properties of the surfaces. Since the properties of different surfaces may be different, the potentials of the surfaces may therefore be different. This we refer to as *differential charging*. In the Langmuir probe approximation, one calculates the potential of each surface individually by considering the current balance between the surface and the ambient plasma. In this approximation, one ignores the interaction between surfaces.

To include the interactions between surfaces is an enormous task. To appreciate the difficulty, imagine that when the potential of one surface changes, the potentials of all other surfaces have to change. The changes of the other surfaces would, in turn, affect the potentials of the first surface and all others. One needs to calculate the potentials in a self-consistent manner. It is a challenge to future computer programs to include self-consistency for better approximation.

In addition, changes of surface potentials may affect the trajectories of incoming and outgoing charged particles. The Langmuir probe model in its original form does not apply to differential charging. Some other examples of complexity are geometrical effects and spacecraft velocity effects. These are challenging details. In this chapter, we will examine two very common mutual interactions between surfaces: potential wells and potential barriers. We will also briefly examine charging in spacecraft wakes.

6.2 Formation of Potential Wells and Barriers

An important effect of differential charging is the formation of local potential wells. Surface materials of different properties have different critical temperatures for the onset of surface charging (see chapter 4). Above the critical temperature, the magnitude of the surface potential increases with the temperature, but the rate of increase depends on the surface properties. Therefore, different surface materials charge to different potentials in space plasma environments (see chapter 4). This phenomenon is called *differential charging*.

Consider, for simplicity, a flat, circular, small surface (figure 6.1) with a surface potential, ϕ_s, of +5 V, say. Suppose that the surface is surrounded by a larger concentric circular surface of −200 V. The smaller surface would have little effect on the larger one, but the large one may have a big effect on the small one. To ambient electrons, the small inner surface is attractive, but the big outer surface is repulsive. The contours of the repulsive potential of the outer

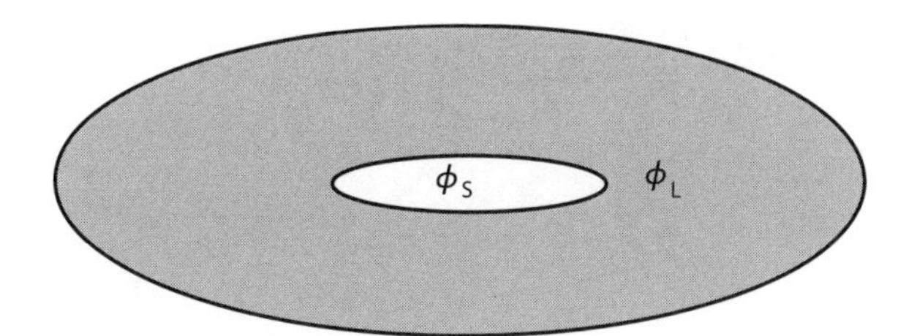

Figure 6.1 A circular double disk of different potentials. The potential ϕ_L of the larger disk is repulsive to electrons, while the potential ϕ_S of the much smaller disk is attractive. The potential contours above the disks may form an overhang over the smaller one, forming a potential barrier and a potential well above it.

(larger) disk may form an overhang above the inner (smaller) one, thus blocking the escape of some of the electrons from the region between the barrier and the smaller surface. In this way, a potential barrier and a potential well are formed.

The potential $\phi(r,z)$, where r, z are the cylindrical coordinates, of a double disk can be obtained by solving the Laplace equation. Analytical results of $\phi(r,z)$ are available in reference 1. In particular, the potential along the z axis of symmetry is of the form

$$\phi(0,z) = \frac{2}{\pi}\left\{\gamma \tan^{-1}\left(\frac{1}{z}\right) + (1-\gamma)\left[\frac{z}{(z^2+\alpha^2)^{1/2}}\right]\tan^{-1}\left[\frac{(1-\alpha^2)^{1/2}}{(z^2+\alpha^2)^{1/2}}\right]\right\} \tag{6.1}$$

In equation (6.1), the potential ratio $\gamma = \phi_S/\phi_L$, where S and L denote the inner and outer disks. The radius ratio $\alpha = r_a/r_b \leq 1$ where r_a and r_b are the radii of the inner and outer disks. The function $\phi(r,z)$ has a barrier located near the surface along the z axis (reference 2).

Potentials are relative. It is interesting that even if the potentials of both inner and outer disks are negative relative to that of the space plasma, a potential barrier can still be formed provided that the inner disk potential is positive relative to that of the outer one. Figure 6.2 shows the calculated potential contours above a double disk. It also shows the trajectories of electrons, each with its given initial energy and angle. Note that the electrons bounce off the potential contours. Whether an electron can escape depends on its initial energy and angle.

A very common occurrence of potential barrier formation is in the charging of a nonconducting satellite in sunlight. Consider such a satellite with photoemission from the sunlit surfaces. The sunlit side charges to low (positive or negative) potentials, depending on the balance of all the currents. The dark side, without photoemission, charges to high negative potentials. The high negative potential contours wrap around and may form barriers on the sunlit side.

A simple analytical model (figure 6.3) shows the monopole-dipole[3] potential $\phi(\theta,R)$:

$$\phi(\theta,R) = K\left(\frac{1}{R} - \frac{A\cos\theta}{R^2}\right) \tag{6.2}$$

In equation (6.2), K is the monopole potential, $A = \alpha/R$ where α is dipole strength, R is the radial distance from the center of the spherical satellite, $R = 1$ is the satellite radius, and θ is the angle subtended by sunlight, where $\theta = 0°$ is the noon direction, and $\theta = 180°$ is the midnight direction. The potential barrier and its saddle point are located at a short distance from the surface at noon. As an exercise, calculate the loci of the barrier and its saddle point. The monopole-dipole model will be discussed in more detail in chapter 7, on charging in sunlight.

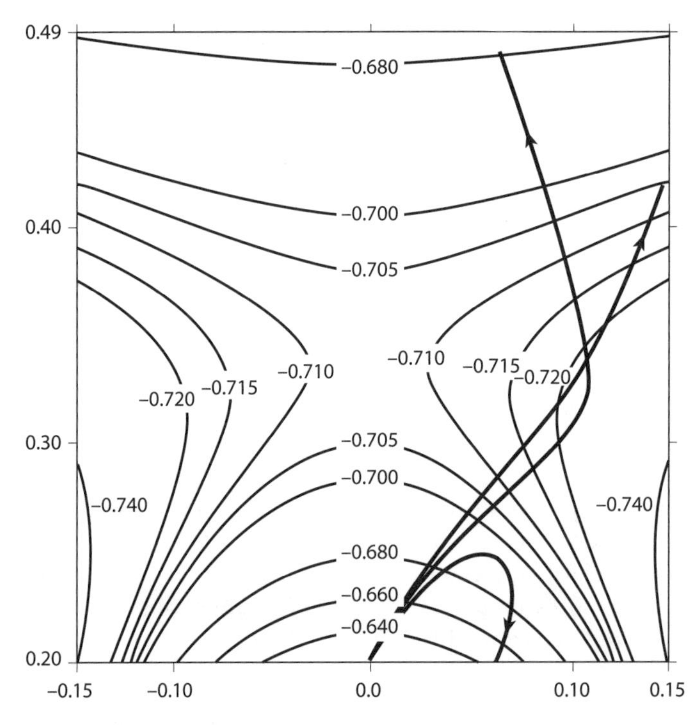

Figure 6.2 Example of potential contours above a double disk (two concentric, circular, and coplanar surfaces) located at $y = 0$ (see figure 6.1). A potential barrier and a potential well are formed. The potential contours form a saddle point where the potential barrier is at a minimum. The low-energy electron of 7 eV bounces back and cannot escape. The electrons with higher energies can escape. (Adapted from reference 2.)

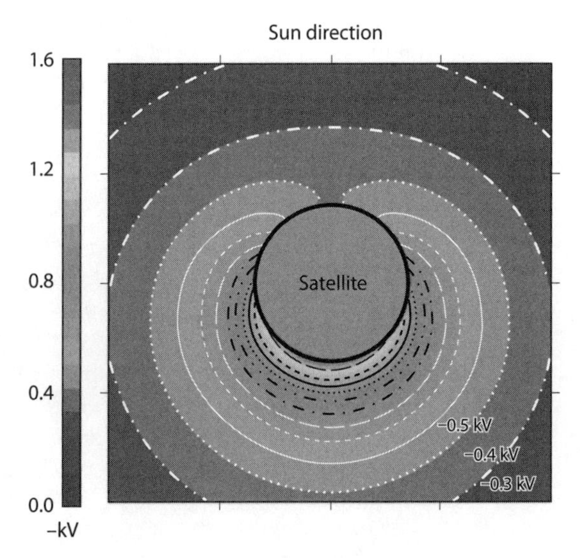

Figure 6.3 Charging of a nonconducting spacecraft in sunlight. The dark side charges to high negative voltages, while the sunlit side charges to low voltages. The potential contours wrap around the sunlit side. A potential barrier with a saddle point is formed very near the surface at noon. The barrier and the saddle point are so near the surface that they are invisible, but nearly visible, in this figure. Charging in sunlight will be described in more detail in chapter 7. (Adapted from reference 9.)

6.3 Effects of Potential Barriers on Electron or Ion Distribution Functions

Suppose that the potential energy on the attractive surface is $e\phi_s$, which is 5 eV for electrons in our example. Suppose that the barrier potential energy is $e\phi_m$ at a radial distance, x_m, from the surface (figure 6.4). And suppose, for example, that $e\phi_m = 60$ eV. All potential energies are measured with respect to that of the ambient plasma. The electron distribution, $f(E)$, measured on the surface ($x = 0$) would be shifted by an amount, $e\phi_m - e\phi_s$, which equals 65 eV. Without a barrier, the distribution would peak to the original value $f(E = 0)$. With a barrier, the distribution would peak to a lower value $f(e\phi_m - e\phi_s)$.

Theorem: If a small surface charged to an attractive ϕ_s V is blocked by a potential barrier of potential ϕ_m, the electron distribution, $f(E)$, measured on the surface ($x = 0$) would be shifted by an amount, $e\phi_m - e\phi_s$. The peak value of the shifted distribution is less than that without a barrier.

6.4 Interpretation of Experimental Data

Changes of electron and ion distributions measured on spacecraft surfaces may affect measurements of spacecraft surface potentials. A common method to measure spacecraft surface

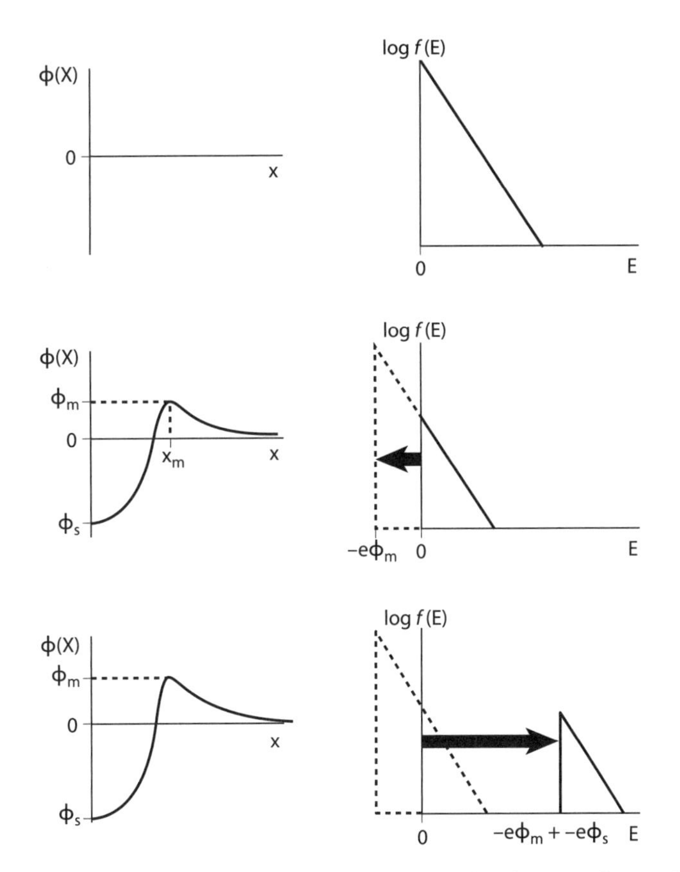

Figure 6.4 Effect of potential wells and barriers on the distribution of incoming ambient electrons. The distribution measured on the surface has an energy cutoff.

potential is to measure the shift of the electron distribution (see chapter 1). A shift in energy from 0 to $q\phi_{shift}$ would infer that the surface potential is ϕ_{shift}. With a potential well, however, this interpretation is erroneous. In the example given in section 6.2, the correct surface potential with respect to the ambient plasma is +5 V, but the usual interpretation, based on the shift amount, would give an erroneous result of +65 V.

6.5 Double Maxwellian Distribution Formed by a Potential Barrier

In figure 6.2, we notice that most of the low-energy electrons generated from the small surface in the potential well cannot escape. For example, the low-energy electrons can be the photoelectrons and secondary electrons generated from the surfaces. The photoelectrons generated from typical spacecraft surfaces in sunlight at geosynchronous altitudes have a characteristic temperature[4] of about 1.5 eV. Secondary electrons[4,5] have typically a few eV in energy. Suppose that the low-energy electrons are trapped in the potential well. They come to equilibrium, which practically takes no appreciable time. Suppose that the equilibrium distribution is Maxwellian with a low temperature T_1. Most of the electrons of the low-temperature Maxwellian distribution cannot escape, except some energetic ones. As a result of the escape of the energetic electrons, the Maxwellian distribution has an upper cutoff at the barrier potential energy B.

A detector on the surface would detect a double Maxwellian distribution $f(E)$, which is the sum of two electron distribution functions, each with a cutoff (figure 6.5). One of them is the Maxwellian distribution function $f_1(E)$, that of the electrons with density n_1 and the lower temperature T_1. The other one is $f_2(E)$ of the incoming ambient electrons of density n_2 and the higher temperature T_2:

$$f(E) = f_1(E) + f_2(E) \tag{6.3}$$

where

$$f_1(E) = n_1\left(\frac{m}{2\pi kT_1}\right)^{3/2} \exp\left(-\frac{E}{kT_1}\right) \tag{6.4}$$

and

$$f_2(E) = n_2\left(\frac{m}{2\pi kT_2}\right)^{3/2} \exp\left(-\frac{E}{kT_2}\right) \tag{6.5}$$

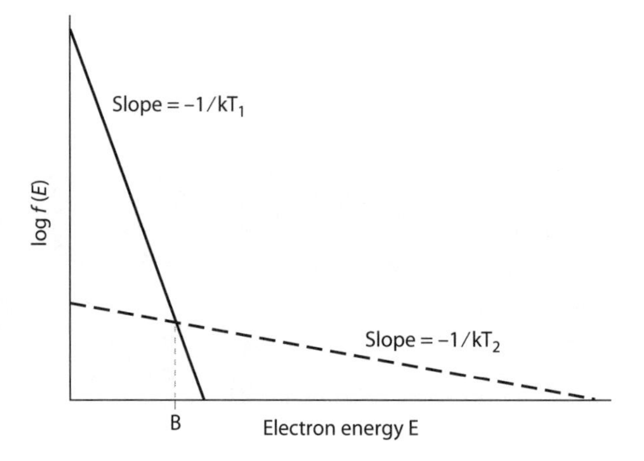

Figure 6.5 Double Maxwellian distribution measured inside a potential well. The low-temperature electron distribution with temperature T_1 has an upper cutoff at the barrier height *B*, while the high-temperature, ambient electrons with temperature T_2 have a lower cutoff at the same barrier height *B*.

The notations of equations (6.3) to (6.5) are given in chapter 5. Let the potential of the surface, where the detector is, be zero. The low-temperature Maxwellian distribution $f_1(E)$ has an upper cutoff, while the high-temperature (T_2) ambient electron Maxwellian distribution has a lower cutoff, both cutoffs being at the same energy, viz, the barrier energy B. The two Maxwellian distributions have no overlap.

In figure 6.5, one notes that the y-intercept of $\log f_1(E)$ gives the density n_1 of the trapped low-density electrons, albeit there is a multiplicative factor $(m/2\pi kT_1)^{3/2}$, which one can evaluate by measuring the inverse of the slope:

$$f_1(0) = n_1\left(\frac{m}{2\pi kT_1}\right)^{3/2} \tag{6.6}$$

Instead, one can use the following method to obtain n_2, the ambient electron density. From equations (6.4) and (6.5), one can write the following:

$$\frac{\log f_1(B)}{\log f_2(B)} = \frac{n_1}{n_2}\left(\frac{T_2}{T_1}\right)^{3/2}\frac{\exp(-B/kT_1)}{\exp(-B/kT_2)} \tag{6.7}$$

Since one knows n_1 from the intercept, the barrier energy B from the graph (figure 6.5), and the temperatures T_1 and T_2 from the slopes, one can determine the ambient electron density n_2. Since the two distribution functions intersect at B, their values are equal at B. Therefore, one can simplify equation (6.7) as follows:

$$\frac{n_1}{n_2}\left(\frac{T_2}{T_1}\right)^{3/3}\frac{\exp(-B/kT_1)}{\exp(-B/kT_2)} = 1 \tag{6.8}$$

It is amazing that one does not need to measure the distribution functions $f_1(E)$ and $f_2(E)$ at the barrier energy B for evaluating equation (6.7). Equation (6.8) is a handy equation for yielding the ambient electron density despite the barrier.

6.6 Bootstrap Charging

Not only can the potential well with a barrier affect measurements of surface potentials, it can also affect the surface potentials. Suppose that a small surface alone charges to its equilibrium potential. As usual, the incoming and outgoing currents balance each other, determining the level of the equilibrium potential. If, however, the surface is surrounded by a potential barrier, the barrier would partially block the currents of the repelled charged species. As a result of this interaction between surfaces, the currents change, thus changing the potential of the surface.

As an illustration (figure 6.6), consider a surface whose potential is determined by the balance of the incoming ambient electrons, incoming ambient ions, outgoing secondary electrons, outgoing backscattered electrons, and outgoing photoelectrons. Suppose that a barrier of −500 V forms because of the surfaces nearby. Such a barrier would block nearly all secondary electrons[4–6] generated from the surface, because they have an average energy of a few eV. Photoelectrons generated from typical spacecraft surfaces at geosynchronous altitudes have a few eV in energy. This is because the photoelectrons are generated by the main solar line, Lyman Alpha, which has an energy $h\nu = 10$ eV approximately, while the surface work function W is usually about 4 to 5 eV. The work function is the departure tax that a photoelectron has to pay when leaving the surface. Therefore, the kinetic energy of a photoelectron is at most $h\nu - W = 5$ eV only. One also has to account for the attenuation as the photoelectrons

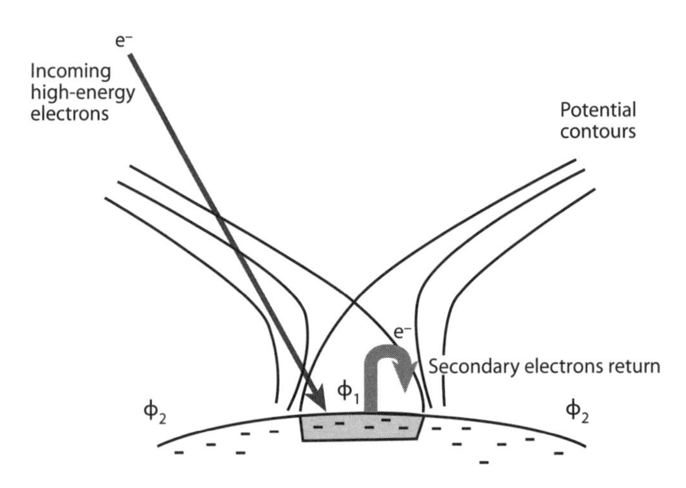

Figure 6.6 Concept of bootstrap charging of a small surface. Let $|\phi_2| \gg |\phi_1|$, and both are negative. Secondary electrons (few eV) cannot leave. High-energy electrons from space can come in easily. The potential barrier blocks the escape of the low-energy electrons, while the high-energy ambient electrons are free to come in. As a result, the small surface charges to a higher negative potential. Depending on the height of the barrier and its saddle point and the angles of the low-energy electron trajectories, some of the low-energy electrons may escape. The final potential of the small surface may or may not be the same as that of the large surface surrounding the former.

interact with the atoms inside the solid layer beneath the surface while the photoelectrons are on their way to the surface. The attenuation reduces the photoelectron energy further. Another contribution to the outgoing electrons is the backscattered electrons. They have energies ranging up to the incoming electron energy, but they constitute a small percentage compared with the photoelectrons and secondary electrons. On the other hand, the incoming ambient electrons may have keV energy.

With the suppression of the outgoing (secondary and photo) electrons, the incoming electrons would keep on accumulating on the surface until a sufficiently high negative potential has built up. Thus, the presence of the barrier changes the current balance, suppressing the low-energy outgoing electrons while allowing high-energy incoming electrons, resulting in a negative surface potential higher in magnitude than without the barrier. This is the idea of *bootstrap charging*[7,8] (figure 6.6).

As a result of bootstrap charging, the potential ϕ_1 of the smaller surface, which is surrounded by the potential barrier, changes. The potential ϕ_1 may change completely—that is, to nearly the same as that of the larger surface—implying the disappearance of differential charging that was present before the bootstrap. As a consequence, the potential barrier also disappears. Or the potential ϕ_1 may change partially—that is, to a potential different from that of the larger surface—but the level of differential charging is reduced nevertheless. In this case, the barrier still exists, but the barrier height is adjusted.

To be rigorous, the calculation of the resultant barrier height has to be done self-consistently. On the one hand, the balance of currents determines the potential of a surface. The surface potentials then determine the potential barrier. On the other hand, the potential barrier determines the net incoming and outgoing currents, thus determining the surface potentials. This is the self-consistent scheme.

The bootstrap mechanism does not necessarily reduce the barrier completely. That is, it does not necessarily bring the small surface to the same potential as that of the large surface.

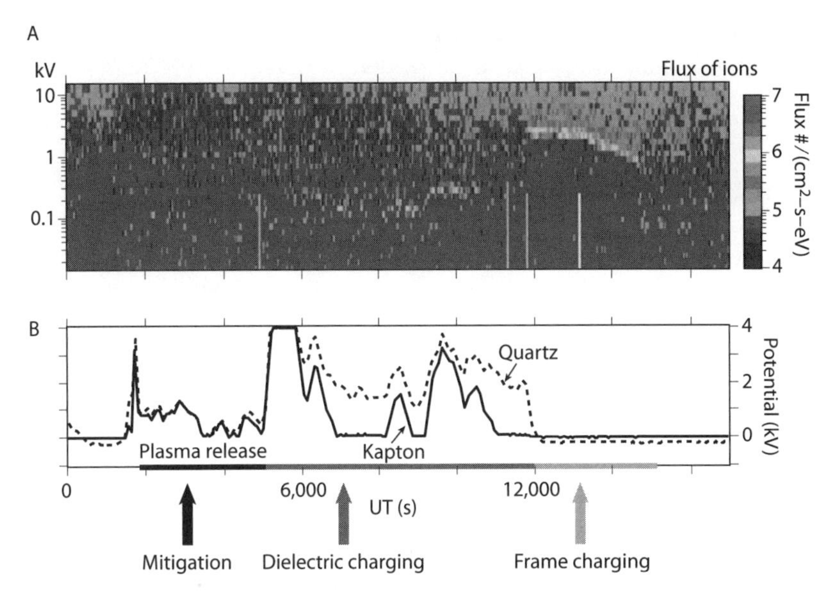

Figure 6.7 Surface charging of a satellite with two small nonconducting (kapton and quartz) surfaces. (A) Shift of the ion spectrum indicates charging of the satellite frame (or ground) to about 4 kV negative at about 12,000 s UT. (B) Charging of the kapton and quartz (both small) surfaces. The potentials are measured relative to the frame. From about 1900 to 5000 s UT, a low-energy plasma is released from the spacecraft for demonstrating mitigation, which is not a topic of discussion in this chapter. As soon as the plasma release has stopped, the small (kapton and quartz) surfaces resume charging relative to the larger surface (frame). Frame charging to high to 4 kV (negative) occurs at about 12,000 s UT (see top panel). At that moment, one can imagine the wrapping of potential contours (not shown) and the formation of a potential barrier (not shown) above each small surface, thus switching on the bootstrap mechanism. As a result, the small surfaces charge to potentials equal to that of the frame (bottom panel). (Adapted from reference 10.)

The key to the outcome lies mainly on the fraction of the low-energy electrons escaping from the barrier and its saddle point (figure 6.2).

Last, we look at some actual data obtained from the DSCS satellite for demonstrating bootstrap charging (figure 6.7). The satellite has two small nonconducting (kapton and quartz) surfaces. The shift of the ion spectrum (top) indicates charging of the satellite frame (or ground) to about 4 kV negative at about 12,000 s UT. Charging of the kapton and quartz (both small) surfaces is shown in the bottom panel. The potentials are measured relative to the frame.

These data require explanation and physical interpretation. From about 1900 to 5000 s UT, a low-energy plasma release takes place from the spacecraft for demonstrating mitigation, which is not a topic of discussion in this chapter. As soon as the plasma release has stopped, the small (kapton and quartz) surfaces resume charging relative to the larger surface (frame). Frame charging to high to 4 kV (negative) occurs at about 12,000 s UT (see top panel). At that moment, one can imagine the wrapping of potential contours (not shown) and the formation of a potential barrier (not shown) above each small surface, thus switching on the bootstrap mechanism. As a result, the small surfaces charge to potentials equal to that of the frame (bottom panel).

The sporadic noisy dots at low energies are not well understood. They represent possibly stray fluxes from neighboring surfaces to the detector. But interpreting the noise is not the main point here.

6.7 Charging in Spacecraft Wakes

If a boat is traveling through water or a spacecraft through the atmospheric neutral particles (figure 6.8), a wake is formed behind the object (boat or spacecraft). From the viewpoint of the object, the neutral particles are traveling backward. Suppose that the neutral particle velocity is V_x and its thermal velocity is $V_{thermal}$. The wake length is given simply by the triangle of V_x and $V_{thermal}$.

Ion depletion in the spacecraft wake has been measured in the early sixties.[11,12] In recent years, there has been interest in charging of an object, such as an astronaut, in the plasma wake of a large spacecraft. Gurevich et al.[13] gave a complicated analytical expression for the current density in the wake as a function of angle. Rubin and Besse[14] gave a simple model of charging in a wake (figure 6.9). The model is based on the orbit-limiting Langmuir probe idea (see chapter 2). In the ionosphere, a spacecraft in orbit is about 3 to 4 km/s faster than ambient ions.

To an observer on a spacecraft, the ions are moving backward with a velocity V. For a spacecraft of radius R, the ion trajectory grazes a small object of radius a in the wake if

$$R^2 = a^2\left(1 - \frac{2e\phi}{mV^2}\right) \tag{6.9}$$

where a is the radius of the satellite, m the ion mass, e the ion charge, and ϕ the potential of the object. That is,

$$-e\phi = \frac{1}{2}mV^2\left(\frac{R^2}{a^2} - 1\right) \tag{6.10}$$

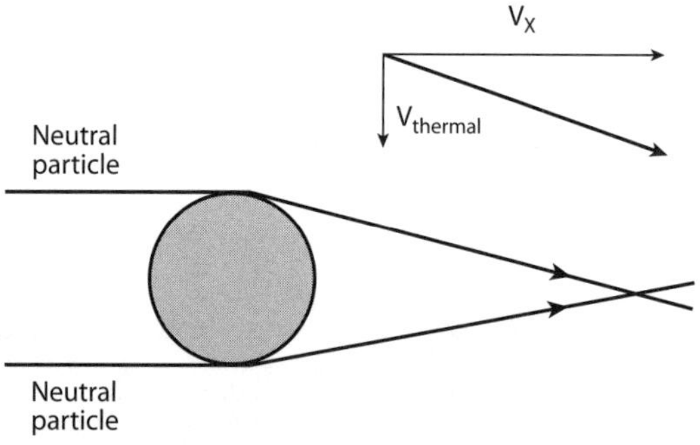

Vx
Vthermal
Neutral particle
Neutral particle

Figure 6.8 A neutral particle wake. Scattering on the ram side is not shown. The length of the wake is given by the triangulation of the stream velocity V_x and thermal velocity $V_{thermal}$ of the particles.

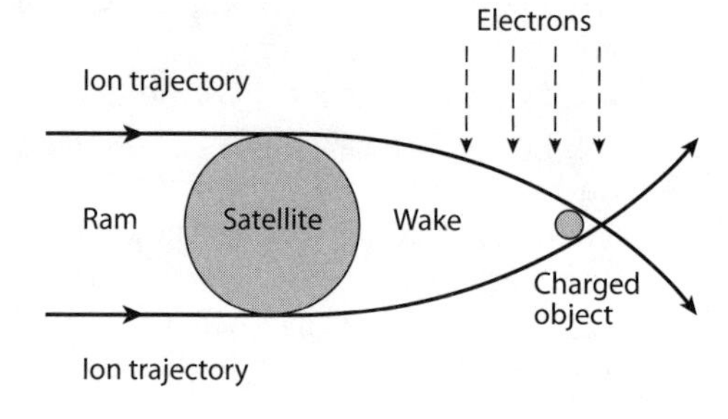

Figure 6.9 A satellite wake. The satellite orbital velocity in the ionosphere is faster than the ions. Ion density is low in the wake.

This ϕ is the voltage to which the object will charge before it is able to attract ions from the inner edge of the wake cone. Taking $R = 15$ m, $a = 1$, and an O^+ ion energy of 6 eV, Rubin and Besse[14] obtained a potential of −1344 V. This is a threshold. If the negative potential rises above this value, a copious current of ionospheric ions is available for neutralization. Therefore, the object will charge almost precisely to the threshold value.

Since a spacecraft is much larger than a typical object, such as a floating instrument or an astronaut, the spacecraft affects a small object. Conversely, the object, because of its smallness, would have little effect on the spacecraft.

Although the Rubin and Besse model of wake charging is not realistic because it neglects collisions in the ionosphere, the model is instructional. Recent observations in space have confirmed the existence of plasma wakes behind spacecraft. The advances in modern computers enable detailed calculations on the wake structures and the potentials of objects at various locations in wakes.[15]

Recently, Engwall et al.[16] observed another type of satellite wake. It can occur in a low-velocity directional ion flow if the spacecraft charges to a positive potential $\phi (> 0)$ such that

$$q_i > \tfrac{1}{2} M V_i^2 > kT_i \tag{6.11}$$

where M is the mass of an ion, V_i the ion velocity, q_i the ion charge, k the Boltzmann constant, and T_i the ion temperature. Normally, a conducting satellite charges to 4 to 5 positive volts only in sunlight at geosynchronous altitudes. There are, however, regions in the magnetospheric tail where the plasma density is sometimes very low. There, spacecraft charging to a few tens of positive volts has been observed. Under such circumstances, the low-energy ions cannot reach the satellite because of the repulsive satellite potential energy so that the ions are scattered off at some angles. Engwall et al.[16] coined a term, "enhanced wake," to describe this phenomenon (figure 6.10).

As a corollary, consider the enhanced wake charging of a nonconducting satellite. The wake side would charge to negative voltages relative to the ram side. This is because the low-energy ions needed for current balance on the surface of the wake side are depleted by the scattering off the ram side.

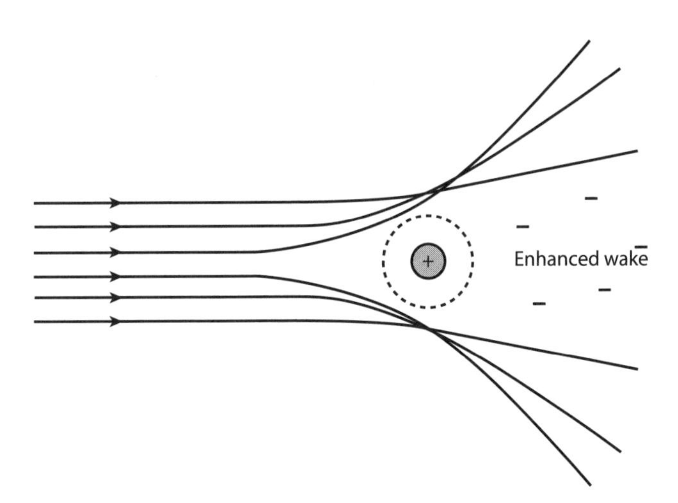

Figure 6.10 An enhanced wake. The ions are scattered off by a positively charged satellite. (Adapted from reference 16.)

6.8 Exercises

1. Calculate the double-disk potential $\phi(0,z)$ of equation (6.1) as a function of the axial distance z. Let the ratio of the inner and outer radii be 1:10. Show that a minimum exists.

2. Calculate the monopole-dipole potential $\phi(0,R)$ of equation (6.2) as a function of the radial distance R. Let the value of A vary between 0 and 1. Under what condition of A can you find the existence of a mimimum? Calculate the loci of the potential barrier for various values of Sun angle θ.

3. What would figure 6.6 look like if the detector is located on a surface that has an attractive potential?

4. In the ionosphere, the spacecraft orbital velocity (typically about 7 km/s) is supersonic, meaning that the spacecraft velocity is faster than the ions (0.25 km/s for O^+). An ion wake forms behind the spacecraft in the ionosphere.

   In the solar wind, however, the ion velocity is typically much faster than the spacecraft velocity. The solar wind ion (H^+) velocity is typically between about 200 to 700 km/s. The thermal velocity is typically 40 km/s. For an uncharged spacecraft, where is the position where the ion flow converges in the wake behind the spacecraft? For a spacecraft charged to a negative potential ϕ, can a wake form behind the spacecraft?

   Let the potential ϕ be -10 and the solar wind ion velocity be 400 km/s. What would the wake look like?

   Suppose that the spacecraft is in a sunlit region where the ion flow is 10 eV in energy and the ion thermal energy is a few eV, while the spacecraft potential is +50 V. In this case, the electron density is so low that charging to positive voltages is possible. In this case, the spacecraft potential energy exceeds the ion flow energy. Let the ion thermal velocity be small. A wake can form. How would the wake look like qualitatively?

6.9 References

1. Sherman, C., and L. Parker, "Potential due to a circular double disk," *J. Appl. Phys.* 42: 870–872 (1971).
2. Lai, S. T., "Recent advances in spacecraft charging," presented at AIAA Aerospace Science Meeting, Reno, NV, Paper AIAA-94-0329 (1994).
3. A. L. Besse and A. G. Rubin, "A simple analysis of spacecraft charging involving blocked photo-electron currents," *J. Geophys. Res.* 85, no. A5: 2324–2328 (1980).
4. Whipple, E. C., "Potentials of surfaces in space," *Rep. Progr. Phys.* 44: 1197–1250 (1981).
5. Sternglass, E. J., "Theory of secondary electron emission," Westinghouse Research Laboratories, Pittsburgh, PA (1954).
6. Lye, R. G., and A. J. Dekker, "Theory of secondary electrons," *Phys. Rev.* 107, no. 4: 977–982 (1957).
7. Mandell, M., I. Katz, G. Schnuelle, P. Steen, and J. Roche, "The decrease in effective photocurrents due to saddle points in electrostatic potentials near differentially charged spacecraft," *IEEE Trans. Nuc. Sci.* 26, no. 6: 1313–1317 (1978).
8. Higgins, D., "An analytic model of multi-dimensional spacecraft charging fields and potentials," *IEEE Trans. Nucl. Sci.* 26, no. 6: 5162–5167 (1979).
9. Lai, S. T., and M. Tautz, "Why do spacecraft charge in sunlight? Differential charging and surface condition," presented at Ninth Spacecraft Charging Technology Conference, Tsukuba, Japan (2005).
10. Lai, S. T., D. Cooke, B. Dichter, K. Ray, A. Smith, and E. Holeman, *Bootstrap Charging on the DSCS Satellite*, European Space Agency, Noordwijk, The Netherlands (2001).

11. Boudreau, R. E. and J. L. Donley, "Explorer VIII Satellite Measurements in the Upper Ionosphere," *Proc. Roy. Soc.* A281: 487–504 (1964).
12. Samir, U., and A. P. Wilmore, "The distribution of charged particles near a moving spacecraft," *Planet. Space Sci.* 13: 285–296 (1965).
13. Gurevich, A. V., L. P. Pitaevsky, and V. V. Smirnova, "Ionospheric aerodynamics," *Space Sci. Rev.* 9: 805–871 (1969).
14. Rubin, A., and A. Besse. "Charging of a manned maneuvering unit in the shuttle wake," *J. Spacecraft and Rockets* 23: 122–124 (1986).
15. Wang, J., P. Leung, H. Garrett, and G. Murphy, "Multi-body plasma interactions—charging in the wake," *J. Spacecraft and Rockets* 31, no. 5: 889-894 (1994).
16. Engwall, E., A. I. Eriksson, and J. Forest, "Wake formation behind positively charged spacecraft in flowing tenuous plasmas," *Phys. Plasmas* 13: 062904-1 (2006).

7

Spacecraft Charging in Sunlight

Photoelectron current emitted from spacecraft surfaces often dominates over all other ambient currents in the magnetosphere. Therefore, according to current balance, it seems impossible for spacecraft charging to negative potentials to occur in sunlight. Yet, spacecraft charging to negative potentials is reported not only in eclipse but also sometimes in sunlight. To explain spacecraft charging in sunlight, we examine two significant causes: the blocking of photoelectrons in the monopole-dipole model and the reduction of photoemission by high surface reflectance.

7.1 Photoelectron Current

In chapter 3, the equations of secondary and backscattered electron fluxes are expressed in terms of the incoming ambient electron flux and the secondary electron yield (SEY) and backscattered electron yield (BEY). Analogously, one writes down an equation of photoelectron flux J_{ph} emitted from a surface in sunlight at normal incidence.

$$J_{ph} = \int_0^{\infty} d\omega f_S(\omega) Y(\omega) \tag{7.1}$$

where $f_S(\omega)$ is the incoming solar photon flux per unit area per second per unit photon energy, ω the photon energy, and $Y(\omega)$ the photoelectron yield per unit incident photon of energy ω at normal incidence.

7.2 Surface Reflectance

Taking the surface reflectance R into account, the photoelectron flux J_{ph} is of the following form:

$$J_{ph}(R) = \int_0^{\infty} d\omega f_S(\omega) Y(\omega, R) \tag{7.2}$$

where R is a function of ω also. In the literature, the word *reflectivity* is used by some authors to mean reflectance.

For a given frequency ω (or energy $h\omega$), Y is the yield of photoelectrons per incident photon. It can be written as follows:

$$Y[\omega, R(\omega)] = Y^*[\omega, R(\omega)][1 - R(\omega)] \tag{7.3}$$

In equation (7.3), $Y^*(\omega, R)$ is the yield per absorbed photon. One can write it as follows:

$$Y^*[\omega, R(\omega)] = \frac{Y[\omega, R(\omega)]}{1 - R(\omega)} \tag{7.4}$$

In practice, Y is the measured quantity, and not Y^*. The latter is useful for understanding the physics. In equations (7.3) and (7.4), when R approaches unity, Y approaches zero.

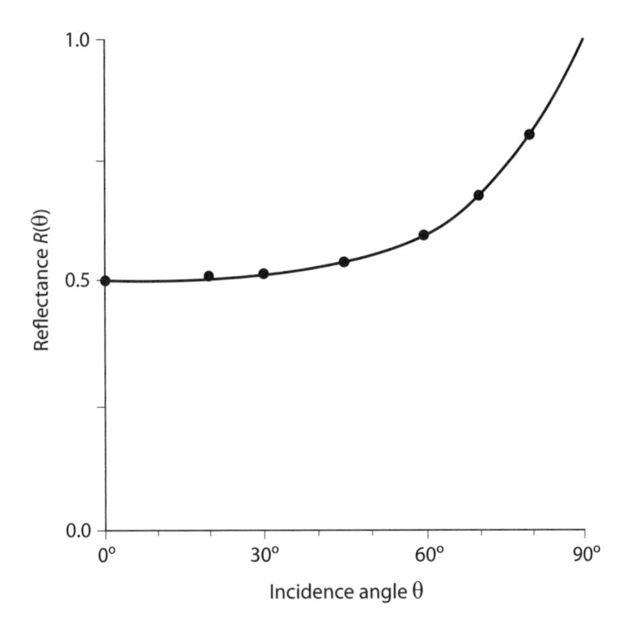

Figure 7.1 Reflectance of beryllium surface. The sunlight incidence angle θ is 0° at normal incidence. The reflectance R approaches 100% as the incidence angle θ approaches grazing. (Adapted from reference 12.)

The photoelectron flux emitted from the surface diminishes as the reflectance R increases. A perfectly reflecting mirror, with $R = 1$, would emit no photoelectron.[1] Aluminum mirror surfaces in space have $R = 0.9$ approximately. With good mirror coating, the reflectance is less than 0.9 but above 0.8.

For finite incidence angle θ (0° for normal incidence), the photoelectron flux is modified because both the yield function and the reflectance are angle dependent (figure 7.1). Furthermore, there is cos θ factor due to the effective surface area receiving sunlight.

$$J_{ph}(\theta) = \int_0^{\infty} d\omega f_S(\omega) Y^* [\omega, R(\omega,\theta)] [1 - R(\omega,\theta)] \cos\theta \tag{7.5}$$

The photon coming in at a shallower angle θ penetrates less deeply into the solid material so that the photoelectron generated can escape more easily than in the normal incidence case. A simple model by Jeanneret[2] has shown that the θ dependence of $Y^*(\omega)$ is of the form

$$Y^* [\omega, R(\omega,\theta)] \approx \frac{Y^* [\omega, R(\omega,0)]}{\cos\theta} \tag{7.6}$$

Laboratory measurements have shown that the factor, $[1 - R(\omega,\theta)]$, also has an approximate cosine dependence[3,6]:

$$1 - R(\omega,\theta) \approx [1 - R(\omega,0)] \cos\theta \tag{7.7}$$

Multiplying equation (7.6) with equation (7.7), the two cosine factors cancel each other. The product $Y^*(\omega,R(\omega,\theta))\,[1 - R(\omega,\theta)]$ in equation (7.5) has no θ dependence. As a result, there is only one $\cos\theta$ in equation (7.5). We remind ourselves that the cos factor is due to the effective surface area receiving sunlight at an angle θ.

From equation (7.5), we can write the photoelectron flux $J_{ph}(\theta)$ as follows:

$$J_{ph}(\theta) = J_{ph}(0)\cos\theta \tag{7.8}$$

where

$$J_{ph}(0) = \int_0^\infty d\omega f_S(\omega) Y(\omega)\,[1 - R(\omega)] \tag{7.9}$$

The photoelectron current $I_{ph}(\theta)$ emitted from a surface area A is obtained from the photoelectron flux multiplied by A:

$$I_{ph}(\theta) = J_{ph}(0) A \cos\theta \tag{7.10}$$

where $J_{ph}(0)$ is given by equation (7.9).

For calculating spacecraft potential ϕ in sunlight, one needs to include the photoelectron current, equation (7.10), in the current balance equation. If the potential ϕ is negative, the photoelectron current leaves entirely. If ϕ is positive, the photoelectron current $I_{ph}(\alpha,\phi)$ cannot leave completely or partially, depending on the surface potential ϕ and the potential contours in the vicinity of the surface.

7.3 The Prominent Solar Spectral Line

It takes energy for a photon to generate a photoelectron. A photoelectron generated at some depth inside a solid has to find (zigzag) its way to the surface. Along the photoelectron's way, it loses some energy to attenuation. When the photoelectron reaching the surface wants to leave, it is required to pay a departure tax, which equals the work function of the surface material. For typical surface materials, the work function is about 4 to 6 eV. Therefore, the low-energy portion of the solar spectrum $f_s(\omega)$ below the work function does not contribute to photoemission.

Unlike the continuous energy distribution of electrons in space, the solar spectrum $f_s(\omega)$ includes many emission lines from the solar atoms and ions. The Lyman Alpha (Lyα) line, which has about 10.2 eV, is the most prominent solar line for photoemission in Earth's magnetosphere. The higher energy solar spectral lines are less intense and are mostly attenuated by the magnetosphere. The photoelectrons generated by the higher energy lines are not abundant and are therefore insignificant in the current balance for controlling the spacecraft potential, unless the spacecraft is in a very low plasma density location, outside the magnetosphere, or near the Sun. For more discussion on photoemission yields and the solar spectrum, see references 4 and 5.

7.4 Can Spacecraft Charging to Negative Voltages Occur in Sunlight?

The photoelectron emission current, or photoemission current for short, can be measured in the laboratory. For typical spacecraft surface materials, such as kapton, teflon, and astroquartz, measurements in vacuum chambers under space-like conditions give the photoemission current for normal photon incidence as about $J_{ph} = 2 \times 10^{-9}$ A/cm^2 (reference 7). Other surface materials have photoelectron yields of the same order of magnitude. In comparison, the average electron current density[8] measured on the SCATHA (Spacecraft Charging at High Altitudes) satellite was $J = 0.115 \times 10^{-9}$ A/cm^2, and the worst case[9] reported was $J_{max} = 0.501 \times 10^{-9}$ A/cm^2.

Thus, the (outgoing) photoemission current density J_{ph} exceeds the (incoming) average ambient electron current J by a factor $F \approx 20$ and the worst stormy current J_{max} by a factor $F \approx 4$:

$$J_{ph} \approx 20J \qquad \text{(in average space condition)} \tag{7.11}$$

$$J_{ph} \approx 4J_{max} \qquad \text{(in worst space condition)} \tag{7.12}$$

If a spacecraft is charged to a negative potential ($\phi < 0$) in steady state, the potential must satisfy the current balance equation. For a spherical spacecraft in sunlight at the geosynchronous environment in the larger orbit limit regime, the current balance equation[10] is of the form

$$I_e(0)\exp(-e_e\phi/kT_e) - I_i(0)\left(1 - \frac{e_i\phi}{kT_i}\right) - I_{ph}(\phi) = 0 \tag{7.13}$$

where the notations are as usual (see previous chapters), and ϕ is negative. If $I_{ph} > I_e(0)$, there is already no solution for the preceding equation. If $I_{ph} > I_e(0)$ by a factor of $F = 4$ to 20, it is impossible to find a solution, no matter how one argues. How, then, can spacecraft charging to negative potentials occur in sunlight? This is the main question in this chapter.

7.5 Spacecraft Charging to Positive Potentials

Before we discuss why spacecraft charging to negative potentials can occur, let us consider in this section a spherical spacecraft of conducting surfaces in sunlight. Since the outgoing photoemission current dominates, charging to positive potentials can occur. With a positive surface potential $\phi(> 0)$, the photoemission current $I_{ph}(\phi)$ emitted from the charged surface is a function of ϕ. Laboratory measurements show that the energy distribution of photoelectrons is approximately Maxwellian with a temperature[11] of about 1.2 to 1.5 eV, depending on the work function of surface material and the ultraviolet radiation condition. When the spacecraft potential ϕ is near 0, the photoelectrons can leave. When ϕ exceeds about a few (positive) volts, some of the photoelectrons cannot leave and have to come back to the spacecraft surface. If they come back, the effective current of photoelectrons leaving the spacecraft is reduced. Therefore, at equilibrium, the spacecraft potential is at a few positive volts, with the photoemission current $I_{ph}(\phi)$ leaving partially. At equilibrium, current-balance is satisfied. Thus, spacecraft charging of conducting spacecraft in sunlight in quiet space conditions is usually at a few (positive) volts only.[12]

7.6 The Photoemission Current at Negative Spacecraft Potentials

In section 7.4, you have seen that the outgoing photoelectron current exceeds the incoming electron current by a factor of F. The factor F is higher than unity. Negative voltage charging cannot occur if F exceeds unity. In this section, let us search for mechanisms that may reduce F to a smaller number (hopefully below unity), so that negative voltage charging may occur in sunlight. Three mechanisms are discussed in this section: spacecraft geometry, surface reflectance, and differential charging between neighboring surfaces in light. The important topic of differential charging between the sunlit and the dark surfaces will be discussed in detail in section 7.7.

7.6.1 Spacecraft Geometry

The F factor can be relaxed further if we notice that only one side of a satellite is in sunlight. This reduces the photoemission current I_{ph} but does not change the ambient electron current $I_e(0)$. With this reduction, the F factor is reduced from "4 to 2" and "20 to 10," respectively. Can we argue that F can be reduced further?

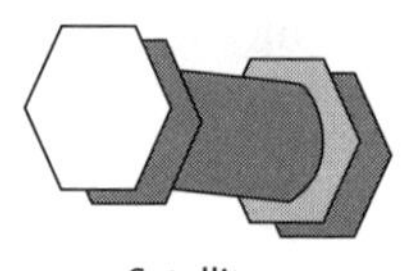

Figure 7.2 Shadows in series. The total shadowed surface area of a spacecraft is equal to or greater than the sunlit area.

Theorem: For any spacecraft geometry, the total area A_S of sunlit surfaces is at most equal to, but not greater than, the total area A_D of the shadowed surfaces.

$$A_S \leq A_D \tag{7.14}$$

For a simple convex geometry, such as a sphere or a spheroid, the equality holds. For complex spacecraft geometries with both convex and concave surfaces, it is possible, and in fact fairly common, to have more shadowed surfaces than sunlit surfaces. An example is shown in figure 7.2. The example is exaggerated for the purpose of illustrating a concept. With the effective ratio of sunlit to shadowed areas reduced, the ratio F of the (outgoing) photoemission current I_{ph} to the (incoming) ambient electron current $I_e(0)$ can be reduced further, depending on how the shadows stack up in series.

7.6.2 Surface Reflectance

We can reduce the value of F further by using the following argument. The laboratory measurements quoted in section 7.4 are for surfaces without specifying the surface condition. Surface condition can make a significant difference in photoemission. The photoemission current I_{ph} is a function of reflectivity R of the surface. For a given frequency, the photoelectron current can be written (for example, reference 13) as follows:

$$I_{ph}(R) = I_{ph}(0)(1 - R) \tag{7.15}$$

The reflectance R depends on the Sun angle θ of the photons. In the literature, the incidence angle χ is sometimes defined as $90^\circ - \theta$. Depending on the surface material and surface geometry, the reflectance may reduce the photoemission current from a spacecraft, perhaps by about 50 to 80%. For some high-reflectance materials, such as smooth aluminum, which has been used for mirrors or reflectors in space, the reflectance is near 90%, reducing the photoemission significantly.[1]

Accounting for all the preceding arguments, the F factor can now be below 1 approximately, depending on the surface reflectance, the sunlight incidence angles, and whether the spacecraft geometry allows for shadows to occur in series.

7.6.3 Differential Charging

In section 7.4, we have assumed that all photoelectrons leave the spacecraft if the spacecraft charging voltage is negative. For a highly conducting spacecraft, charging is uniform. However, for a spacecraft covered by a surface, or some surfaces, of nonconducting materials, charging may not be uniform. If so, differential charging occurs. As a consequence, there may be potential barriers and wells (see chapter 6). In this section, we will first examine a scenario of differential charging between neighboring surfaces of a spacecraft in sunlight. We will devote the whole section 7.7 to the important topic of differential charging between the sunlit and dark surfaces.

On the sunlit side, a potential well due to high negative potentials on neighboring surfaces may form a potential well blocking the photoemission current. The potential well usually

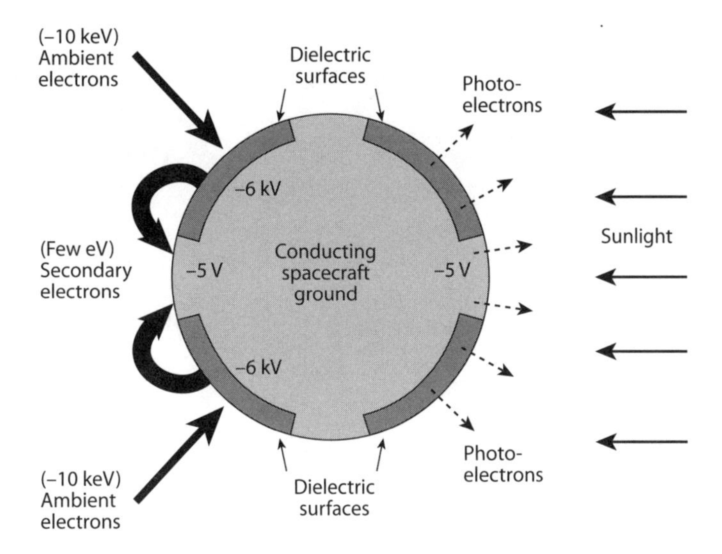

Figure 7.3 Example of secondary electrons returning to the satellite ground. As a result, the net outgoing electron current is reduced.

features a saddle point at the middle of the potential barrier, as discussed in the previous chapter. The potential barrier and well trap the low-energy part of the photoemission current I_{ph}. The reduction of photoelectrons leaving the sunlit side diminishes the level of charging to positive potentials on the sunlit side. On the shadowed side, the differential charging may be more severe than on the sunlit side because, while some dielectric surfaces may charge to high negative potentials, some conductive surfaces may be electrically connected (i.e., grounded) to some others on the sunlit side (figure 7.3). As a result, some secondary electrons from the negatively charged surfaces on the shadowed side cannot leave; they return to the grounded surfaces. The reduction of secondary electron current leaving the satellite enhances the negative potentials of the satellite.

7.7 The Monopole-Dipole Potential

This section discusses an important and interesting topic of differential charging—the charging of a spacecraft to very different voltages on the sunlit and the dark sides. Suppose that a satellite with dielectric surfaces is in sunlight. While the sunlit side receives sunlight and emits photoelectrons, the dark, or shadowed, side can charge to a higher (negative) potential if the incoming ambient electrons are of high temperature (keV or higher). The different potentials of the sunlit and dark sides form a dipole (figure 7.4). The net potential is the sum of a monopole potential plus that of a dipole.[14] The average potential of the spacecraft is the monopole potential, while the deviations in potentials of the shadowed and sunlit sides form a dipole potential. The monopole-dipole potential is of the form[14]

$$\phi(\theta,R) = K\left(\frac{1}{R} - \frac{A\cos\theta}{R^2}\right) \tag{7.16}$$

where $1/2 < A < 1$. The Sun angle $\theta = 0°$ facing the Sun. In equation (7.16), K is the monopole strength, and A is the dipole strength relative to that of the monopole.

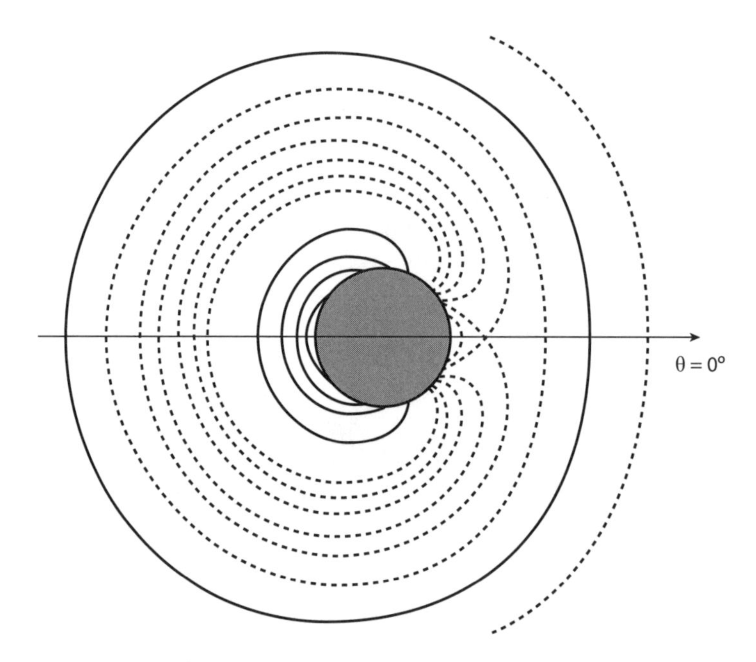

Figure 7.4 Equipotential contours for the monopole-dipole field with parameter $A = 3/4$. $\theta = 0°$ is the Sun direction. The solid curves are for −50, −70, −80, −90, and −110 V. The −80 V curve crosses itself at the saddle point. The spherical spacecraft potentials are −60 V and −420 V at $\theta = 0°$ and 180°, respectively. (Adapted from reference 14.)

The first term K/R on the right-hand side of equation (7.16) is the monopole potential, and the second is the dipole potential. For a unit radii satellite, the radial distance $R = 1$ at the satellite surface, and $K = \phi(90°, 1)$. A in equation (7.16) can be written as

$$A = \frac{\alpha}{K} \tag{7.17}$$

where α is the dipole strength given by

$$\alpha = [\phi(180°, 1) - \phi(0°, 1)]/2 \tag{7.18}$$

It is useful to provide some algebraic properties of the monopole-dipole model. From equation (7.16), one obtains a ratio:

$$\frac{\phi(0°, 1)}{\phi(180°, 1)} = \frac{1 - A}{1 + A} \tag{7.19}$$

From equation (7.19), A can be written explicitly as the ratio

$$A = \frac{\phi(180°, 1) - -\phi(0°, 1)}{\phi(180°, 1) + \phi(0°, 1)} \tag{7.20}$$

Similarly, the monopole potential strength K can be written as

$$K = [\phi(180°,1) + \phi(0°,1)]/2 \tag{7.21}$$

The potential barrier is located at the maximum potential for $\theta = 0°$:

$$\left[\frac{d\phi(0°,R)}{dR}\right]_{R=R_S} = 0 \tag{7.22}$$

which yields

$$R_S = 2A \tag{7.23}$$

Equation (7.23) requires that $R_S > 1$, implying

$$A > 1/2 \tag{7.24}$$

Otherwise, the barrier would be inside the spacecraft.

The barrier height B is defined as $\phi(0°, R_S) - \phi(0°,1)$. From equations (7.16) to (7.20), B is given in the form

$$B = K\frac{(2A-1)^2}{4A} \tag{7.25}$$

Let the dark side potential be E:

$$E = K(1+A) \tag{7.26}$$

Eliminating K in equations (7.25) and (7.26), one obtains the barrier height B:

$$B = \frac{E(2A-1)^2}{4A(1+A)} \tag{7.27}$$

Since the potential barrier B is of the order of a few volts while the monopole potential K can be of the order of kilovolts, B/K is nearly zero. From equation (7.25), $B/K \approx 0$ implies that $A \approx 1/2$.

Substituting this result of $A \approx 1/2$ into equation (7.20), one obtains the ratio of the sunlit side and dark side potentials as follows:

$$\frac{\phi(0°,1)}{\phi(180°,1)} \approx \frac{1}{3} \tag{7.28}$$

This result is simple and important. It enables a convenient estimate of the ratio of charging voltages in sunlit and in darkness. For example, if the dark side of a spacecraft charges to −10 kV, the sunlit side would charge to about −333 kV.

7.8 Fraction of Photoemission Current Trapped

As an example, suppose that $\phi(180°,1) = -4$ kV and $\phi(0°,1) = -1$ kV. Equation (7.21) yields $A = 0.52$, and equation (7.25) yields a potential barrier $B = 2$ V. The fraction f of photoelectron flux trapped by the potential barrier is given by (figure 7.5):

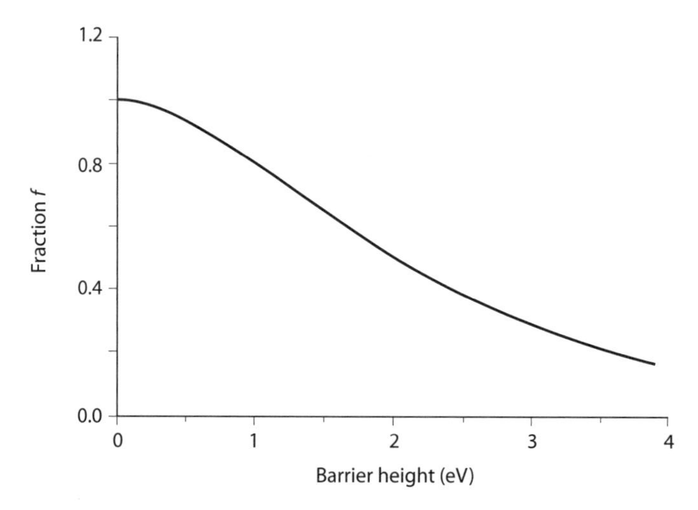

Figure 7.5 Fraction of photoemission current blocked as a function of barrier height. A potential barrier of a few volts is capable of reducing the net photoelectron current to a small fraction.

$$f = \frac{\int_0^B dE E \exp(-E/kT)}{\int_0^\infty dE E \exp(-E/kT)} = \left(\frac{B}{kT} + 1\right) \exp\left(-\frac{B}{kT}\right) \tag{7.29}$$

where the photoelectron distribution used is Maxwellian approximately[14] and has a typical temperature T of 1.2 eV. A potential barrier B of 2 V would trap a substantial fraction of the photoelectron flux.

7.9 Competition between Monopole and Dipole

The potential barrier in the monopole-dipole model diminishes as the spacecraft potential increases to high values (negative volts). A graph of the potential barrier B (normalized by the monopole potential K) as a function of the dipole strength α (normalized by K) is shown in figure 7.6. The figure shows that when the monopole potential K exceeds twice the dipole strength α, the potential barrier B disappears. That is, when $K/\alpha > 2$, the monopole dominates over the dipole. Therefore, at sufficiently high ($K/\alpha > 2$) spacecraft potential (negative voltage), the photoelectrons can leave freely without encountering any potential barrier.

7.10 Measurement of Spacecraft Potential in Sunlight

In sunlight, a spacecraft with dielectric surfaces charges to different potentials on the dark and the sunlit sides. This property often makes single measurements of spacecraft potential misleading. For example, measurements on the dark side of a nonrotating satellite may indicate high negative volts, whereas measurements on the sunlit side may indicate low positive volts. If the photoelectrons from the sunlit surfaces are blocked by the monopole-dipole voltage barrier, even the sunlit side may charge to negative volts (see section 7.7). In this case, the sunlit side charges to about one-third[6,14] of the voltage level of the dark side, as given by equation (7.29).

Some measurements are made at discrete moments of time. Some measurements are time averaged. For example, if the instrument on a rotating satellite measures −4 kV and 0

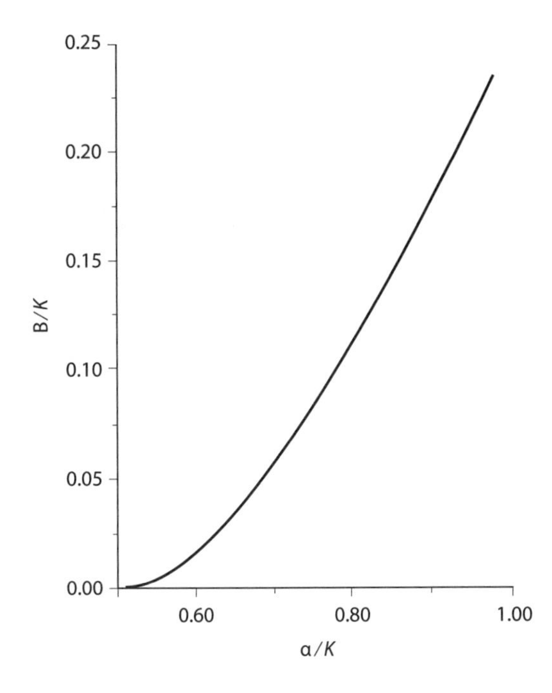

Figure 7.6 Competition between the monopole and the dipole. If the ratio of α over K is less than 0.50, the potential barrier B disappears.

V alternately as the satellite rotates, the time-averaged measurement may be near −2 kV. For modeling spacecraft charging on rotating satellites, see, for example, references 12 and 15.

The ambient electron distributions $f(E)$ measured on the dark and the sunlit sides are also different. The presence of the photoelectrons greatly enhances the electron distribution at low energies on the sunlit side. The energy boundary of the enhancement indicates the potential barrier formed by the potential contours wrapped from the highly negative charged dark side. Such a distribution deviates significantly from being Maxwellian. In such a situation, to calculate the onset of spacecraft surface charging by using only the electron distribution (which is affected by the photoelectrons at low energies) measured on the sunlit side would give a wrong result. As the temperature, or the average energy, of the ambient electron distribution rises, the onset of (negative voltage) charging starts on the dark side. One should use the electron distribution measured on the dark side to calculate the charging (to negative volts) there. The potential contours of the dark side often wrap to the sunlit side and affect the potential on that side in a monopole-dipole manner.

7.11 Exercises

1. If a mirror is sanded, changing the reflectance R from 90% to 10%, how would the photoelectron yield per absorbed photon, Y^*, change accordingly? What is the ratio of the new value (for $R = 90\%$) of the photoelectron yield per incident photon to its old value (for $R = 10\%$)?

2. If the sunlight incidence angle θ changes from 0° (normal) to 90°, what is the ratio of the new value of the photoelectron yield per absorbed photon to the old value? What is the ratio of the new value of the photoelectron yield per incident photon to the old value?

3. Suppose that the photoelectron current per unit area emitted from a conducting spacecraft exceeds that of the ambient electrons per unit area by 10 times. Suppose that the ion

current can be ignored. What is the ratio of the shadow area to the sunlit area so that the spacecraft does not charge?

4. Consider a one-dimensional photoelectron flow from a spacecraft charged to 3 V (positive). What is the fraction of the photoelectron current leaving the spacecraft if the photoelectron temperature is (a) 1 eV, (b) 1.5 eV, or (c) 3 eV ?

5. Plot the potential contours for a spherical spacecraft, if $\phi(0^\circ, 1) = -10$ V, $\phi(180^\circ, 1) = -300$ V.

7.12 References

1. Lai, S. T., "Charging of mirrors in space," *J. Geophys. Res.* 110, no. A01: 204–215 (2005).
2. Jeanneret, J. B., "Photoemission at LHC—a simple model," CERN/Note 97-48 (AP) (1997).
3. Powell, C. J., "Analysis of optical and inelastic electron scattering data, III. Reflectance data for beryllium, germanium, antimony, and bismuth," *J. Opt. Soc. Am.* 60, no. 2: 214–220 (1970).
4. Feuerbacher, B. and B. Fitton, "Experimental investigation of photoemission from satellite surface materials," *J. Appl. Phys.*, 43, no. 4, 1563–1572 (1972).
5. Grard, R. J. L., "Properties of the satellite photoelectron sheath derived from photoemission laboratory measurements," *J. Geophys. Res.*, 78, no. 16, 2885–2906 (1973).
6. Lai, S. T., and M. Tautz, "Aspects of spacecraft charging in sunlight," *IEEE Trans. Plasma Sci.* 34, no. 5: 2053–2061 (2006).
7. Stannard, P. R., et al., "Analysis of the charging of the SCATHA (P78-20) satellite," NASA CR-165348 (1981).
8. Purvis, C. K., H. B. Garrett, A. C. Whittlesey, and N. J. Stevens, "Design guidelines for assessing and controlling spacecraft charging effects," NASA Technical Paper 2361, NASA (1984).
9. Gussenhoven, M. S., and E. G. Mullen, "Geosynchronous environment for severe spacecraft charging, *J. Spacecraft and Rockets* 20, no. 1: 26–34 (1982).
10. Mott-Smith, H. M., and I. Langmuir, "The theory of collectors in gaseous discharges," *Phys. Rev.* 28: 727–763 (1926).
11. Whipple, E. C., Jr., "Potential of surface in space," *Rep. Prog. Phys.* 44: 1197–1250 (1981).
12. Lai, S. T., H. A. Cohen, T. L. Aggson, and W. J. McNeil, "Boom potential on a rotating satellite in sunlight," *J. Geophys. Res.* 91, no. A11: 12137–12141 (1986).
13. Samson, J.A.R., *Techniques of Ultraviolet Spectroscopy*, John Wiley, Hoboken, NJ (1967).
14. Besse, A. L., and A. G. Rubin, "A simple analysis of spacecraft charging involving blocked photoelectron currents," *J. Geophys. Res.* 85, no. A5: 2324–2328 (1980).
15. Tautz, M., and S. T. Lai, "Analytic models for a rapidly spinning spherical satellite charging in sunlight," *J. Geophys. Res.* 110, no. A07: 220–229 (2005).

8

Space Tethers, Plasma Contactors, and Sheath Ionization

8.1 Lorentz Force

If a system K is moving with a nonrelativistic velocity **v** relative to a system K_0, the electric field E in the system K is related to the electric field E_0 in K_0 by[1]

$$\mathbf{E} = \mathbf{E}_0 + \mathbf{v} \times \mathbf{B} \tag{8.1}$$

Consider a tether (K system) moving relative to the space plasma (K_0 system) in the Earth's ionosphere. The plasma (K_0) is a system even though the individual electrons and ions in it are moving in various directions. The plasma is neutral because a snapshot of it shows approximately equal numbers of electrons and ions in the region of view. If the systems K and K_0 are moving with velocities $\mathbf{v}_t$ and $\mathbf{v}_0$ respectively in some stationary system (such as Earth), the relative velocity **v** is given by

$$\mathbf{v} = \mathbf{v}_t - \mathbf{v}_0 \tag{8.2}$$

The magnetic field is approximately the same in both systems K and K_0 in the nonrelativistic regime. From equation 8.1, the electric field **E** in the tether system is given by

$$\mathbf{E} = \mathbf{E}_0 + (\mathbf{v}_t - \mathbf{v}_0) \times \mathbf{B} \tag{8.3}$$

Since the plasma velocity $\mathbf{v}_0$ is negligibly slow compared with the tether velocity $\mathbf{v}_t$ in the Earth's ionosphere, the relative velocity **v** equals $\mathbf{v}_t$ approximately. Since $\mathbf{E}_0 = 0$ for a neutral plasma, equation 8.3 gives the induced electric field $\mathbf{E} = \mathbf{v} \times \mathbf{B}$ on the tether.

As an example, let us consider a tether in the ionosphere. Let the tether velocity $\mathbf{v}_t$ be 7.5 km/s and the average magnetic field **B** be 0.3 G. From equation 8.1, we obtain an electric field **E** of about 0.22 V/m and, if the tether is of 1 km in length, its induced potential difference is about 220 volts.

Note that the plasma velocity $\mathbf{v}_0$ in some planetary ionospheres may not be negligible. In such situations, both $\mathbf{v}_t$ and $\mathbf{v}_0$ need to be included in equation 8.3.

8.2 Tether Moving across Ambient Magnetic Field

Consider a vertical and straight tether attached to a satellite traveling across the Earth's magnetic field. Suppose that the tether is traveling eastward in the equatorial region of the ionosphere, the Earth's magnetic field being horizontal and pointing from south to north (figure 8.1). We will consider three cases: (1) the tether is completely insulated, (2) the tether wire is insulated but the tether ends are not, and (3) the tether is completely bare (not insulated).

8.2.1 Insulated Tether

Suppose that the tether is conducting but completely insulated; the tether has no electrical contact with the ionosphere, but the ambient magnetic field is felt. As the tether is traveling

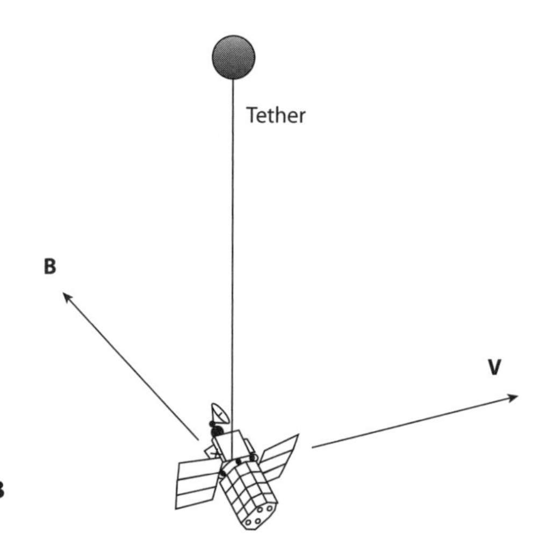

Figure 8.1 A tether moving eastward across **B** pointing northward.

eastward, the electrons in the tether move downward because they feel the electric field E_i, while the positive ions or holes in the tether material move upward. The movements are transient. They continue until an equilibrium field is reached. It is equal and opposite to the Lorentz electric field and therefore halts further movements. The transient movement takes practically no time, depending on the wire resistivity, to reach electrostatic equilibrium. As equilibrium, the top end of the tether develops a positive potential and the lower end a negative potential, both relative to that of the ambient plasma—i.e., the surrounding ionosphere. The net electric field E_{net} along the insulated tether wire is zero at equilibrium:

$$E_{net} = E_i - \frac{\phi_2 - \phi_1}{L} = 0 \tag{8.4}$$

The different potentials ϕ_2 and ϕ_1 at the upper and lower ends of the tether, respectively, are maintained by the external force driven by the motion of the tether across the magnetic field. In this case, even though the tether ends are at different potentials from their surrounding ionosphere, there is no conduction between the tether and the ambient plasma outside because the tether is insulated. If the motion stops, a current would flow in the reverse direction along the wire transiently.

The same process would happen if the tether is not insulated but the environment is a vacuum that is assumed nonconducting. (We are not in the quantum field regime, in which virtual pair creation can occur with finite probability in vacuum according to the uncertainty principle.)

8.2.2 Tethers with Plasma Contactors

Suppose that the ends of the tethers have conducting contacts with the ionosphere. The positive potential at the upper end attracts ambient electrons, and the negative potential at the lower end repels electrons. The current flow completes a circuit by flowing through the ambient plasma and the tether.

The plasma contact at the end of a tether behaves like a plasma probe. It is called a plasma contactor. Currents flow to or from the probe (figure 8.2). The balance of currents determines the probe's floating potential. The floating potential of a plasma contactor (which can be modeled as a Langmuir probe in the ionosphere) will be studied later in this chapter. When a current $\boldsymbol{I}$ flows through a tether of length L, the current gives a force[2]:

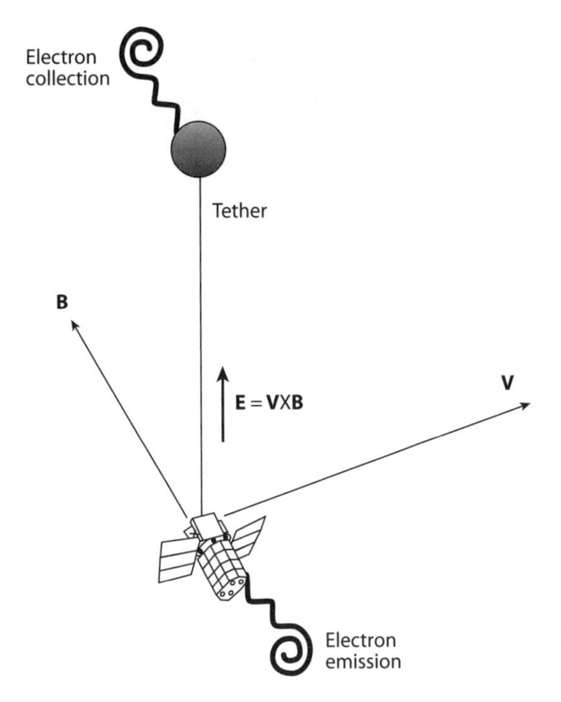

Figure 8.2 Electron emission and tether electron collection in the ionosphere.

$$\mathbf{F} = \int_0^L ds\,(\mathbf{I} \times \mathbf{B}) \tag{8.5}$$

This force can be opposite to, or in the same direction as, the satellite velocity **v**, relative to the space plasma frame depending on the direction of the current ***I*** and **B**. Deceleration requires the inequality condition:

$$\mathbf{F} \cdot \mathbf{v}_t < 0 \tag{8.6}$$

and acceleration requires the positive inequality. To control the current direction, one can use a battery to impose an electric field in a desired direction (figures 8.3 and 8.4). If the force is pushing the satellite forward, the tether behaves like a plasma sail. If the force is backward, the tether retards the satellite. If the current ***I***, the ambient magnetic field **B**, and the tether length L are large, the induced force can be significant. In this manner, one can think of various ways of applications of tethers in space.

8.3 Bare and Conducting Tether

If a tether is bare, it can collect current from the ambient plasma. Figure 8.5 shows a bare tether moving across Earth's magnetic field. The Lorentz (**v**×**B**) force **F** drives a current ***I*** along the wire[3]:

$$\mathbf{F} = \int_0^L ds\,(\mathbf{u} \times \mathbf{B})\,I \tag{8.7}$$

where **u** is the tangential unit vector in the direction of the current ***I*** along the tether. The magnitude of the electric field **E**(s) along the tether is given by

$$E = \mathbf{u} \cdot (\mathbf{v} \times \mathbf{B}) \tag{8.8}$$

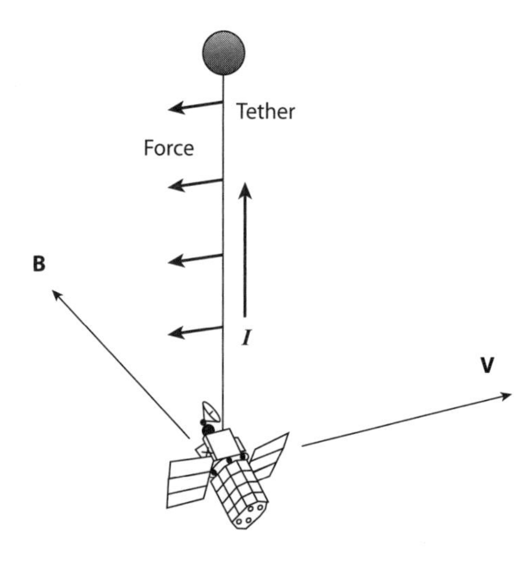

Figure 8.3 Current applied in tether generating retarding force.

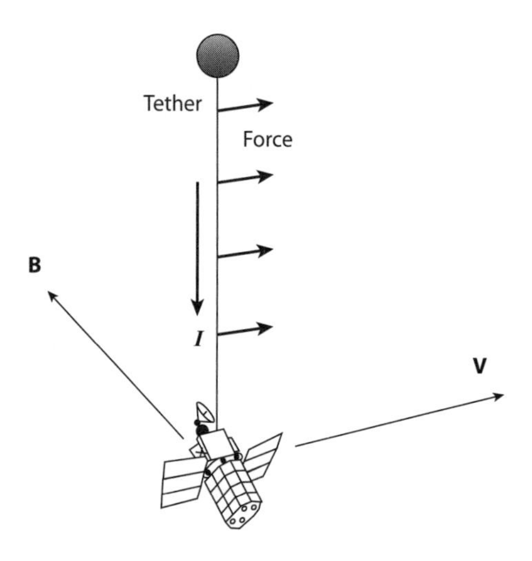

Figure 8.4 Current applied in tether generating accelerating force.

where **v** is relative velocity between the tether and the space plasma, which can be regarded as stationary approximately in Earth's ionosphere.

As a consequence of the driving force, the positive and negative charges pile up at the ends A and C, resulting in potentials ϕ_A and ϕ_C, respectively (figure 8.5). Unlike an insulated tether, there is ambient current I_e collected at every point along the tether. The gradient of the potentials is not equal to the electric field **E** driving the current because one needs to take into account the current collected along the tether. The actual electric field along the tether is not simply the Lorentz electric field **E** but includes the effect of the current collected:

$$E - \frac{I_e}{\sigma A} = -\frac{d\phi}{dx} \tag{8.9}$$

where σ is the conductivity of the tether wire, A is the wire's cross-sectional area, and x is the distance element along the tether wire.

To calculate current collection in the magnetosphere where spacecraft charging is most important, it is often a good approximation to use Mott-Smith and Langmuir's orbit-limited

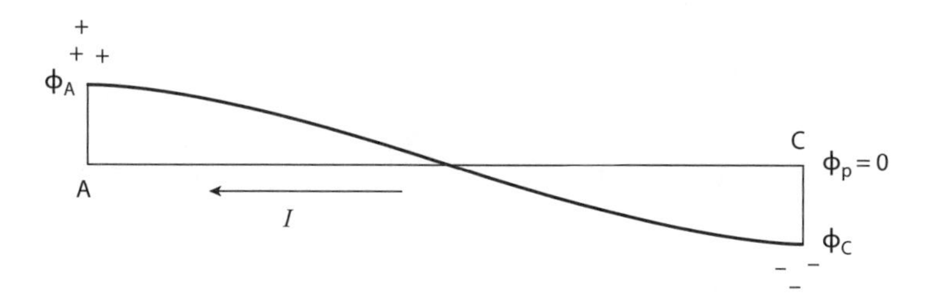

Figure 8.5 Potential developed along an uninsulated (bare) tether as a result of a current I driven by the **V**×**B** force.

formula[3] in cylindrical coordinates. Reference 4 has given careful justification on the use of orbit-limited current collection formulae for tethers in the ionosphere. With collisions and magnetic fields in the ionosphere, there are analytical formulae for current collection (for example, references 6, 7) but they are all approximations. For a heated presheath model, see reference 8. For more on tethers, see a brief review (reference 9), the fundamentals in details (reference 10), some recent applications (reference 11), a recent review (reference 12), and the special issue in *Geophys. Res. Lett.* (reference 13).

The effect of ionization is important for current collection by objects charged artificially to high voltages in the ionosphere.[14,15] In the next sections, we examine the effect of ionization on the potential of a charged body.

8.4 Floating Potential of Plasma Contactor

Langmuir probes are often used in the laboratory and in space. Usually, one applies a sweeping voltage to a Langmuir probe for a range of a few volts and measures the current response. The usual Langmuir probe model has no problem for such applications. However, if one applies a high voltage (for example, +100 V) in the ionosphere, one enters a different physics regime in which the physical processes are different from the usual one. With high voltage in the ionosphere, electron impact ionization of the atmospheric neutrals can occur. Atmospheric neutrals are abundant in the ionosphere. Therefore, one needs a new model.

Models are imperfect description of reality. Good research can improve existing models and advance our present knowledge of reality. The textbook of Hastings and Garrett[16] mentions a few models of plasma contactors but offers no complete description of any. Here, we will look at a fairly complete description of a model—an ionization sheath probe model of plasma contactors.[14,15] This theory can be applied to spacecraft with electron beam emission in the ionosphere or with high current drawn by tether wires (figure 8.6).

8.5 Sheath Model

Consider an electron current *I* going away from a spherical probe in the ionosphere. The probe charges to positive voltage relative to ambient plasma. A sheath forms around the probe. In the sheath, ambient ions are repelled and ambient electrons are attracted. As mentioned in reference 15, there are various ways to define the radius of a sheath, such as matching the densities at the sheath edge or matching the plasma frequencies at the sheath edge. Perhaps none is simpler and more natural than the one using current balance, as described in reference 14, as follows.

For simplicity, we assume spherical symmetry because (1) the tether wire is insulated and much smaller than the plasma contactor or probe, and (2) the electron beam is of high energy

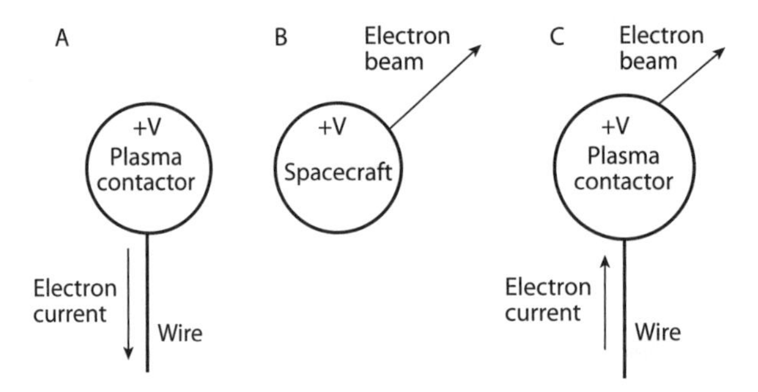

Figure 8.6 (A) Plasma contactor, (B) spacecraft with electron beam emission, and (C) plasma contactor with electron beam emission.

and therefore not ionizing in the vicinity of the probe. For cylindrical probes, one should use the Poisson equation for cylindrical geometry.

The spherical sheath radius r_o is given by equating the outgoing current I with the incoming current from the ambient plasma (figure 8.7):

$$I = 4\pi r_o^2 n_e e v_{th} \tag{8.10}$$

where n_e is the density and v_{th} the thermal velocity of the ambient plasma. Some typical values of sheath radius as calculated by using equation (8.10) are shown in figure 8.8.

The potential ϕ at any point inside the sheath is governed by the Poisson equation:

$$\nabla^2 \phi = -\frac{\rho}{\varepsilon_o} \tag{8.11}$$

where ρ is the charge density, and ε_o is permittivity of empty space.

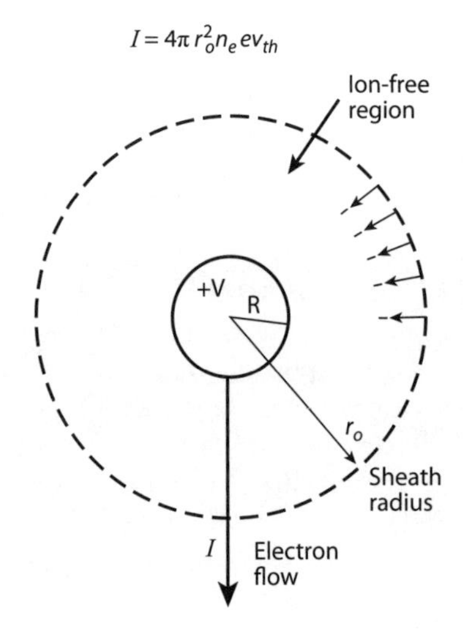

Figure 8.7 Sheath region defined by current balance.

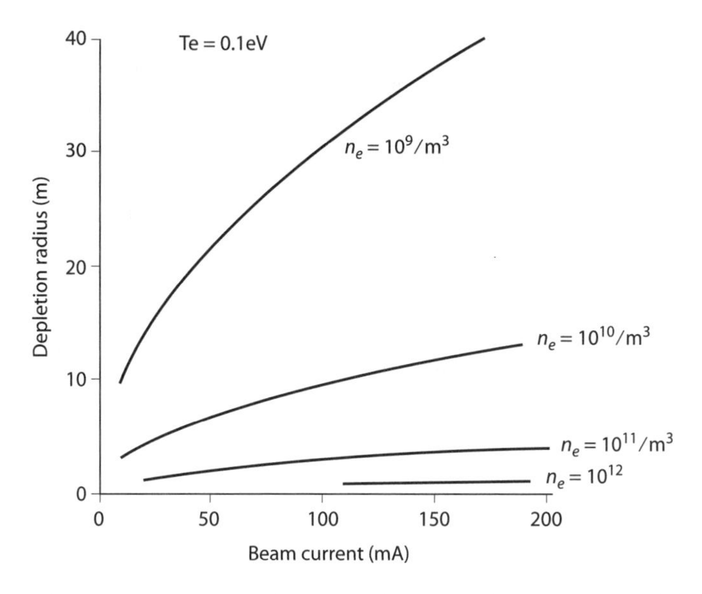

Figure 8.8 Examples of sheath radius r_o calculated by using equation (8.6) for various ambient plasma densities n_e.

8.6 Sheath Ionization

With spherical symmetry, the Poisson equation (8.11) becomes a radial equation:

$$\frac{1}{r^2}\frac{\partial}{\partial r}\left(r^2\frac{\partial\phi(r)}{\partial r}\right) = -\frac{\rho(r)}{\varepsilon_o} \tag{8.12}$$

where the gradient of the potential ϕ gives the electric field E:

$$\frac{\partial\phi}{\partial r} = -E(r) \tag{8.13}$$

Ionization by electron impact on neutrals creates electron and ion pairs in the sheath (figure 8.9):

$$e^- + O \rightarrow O^+ + 2e^- \tag{8.14}$$

The charge density ρ at any point r in the sheath is given by the sum of densities:

$$\rho(r) = e[n^+(r) - n^- - (r) - n_e(r)] \tag{8.15}$$

where n_e is the incoming ambient electron density, while n^+ and n^- are the ionization ion and electron densities, respectively. Current continuity gives the ambient electron density at r as follows:

$$4\pi r^2 n_e(r) \mathrm{v}_e(r) = 4\pi r_o^2 n_e \mathrm{v}_{th} \tag{8.16}$$

which gives

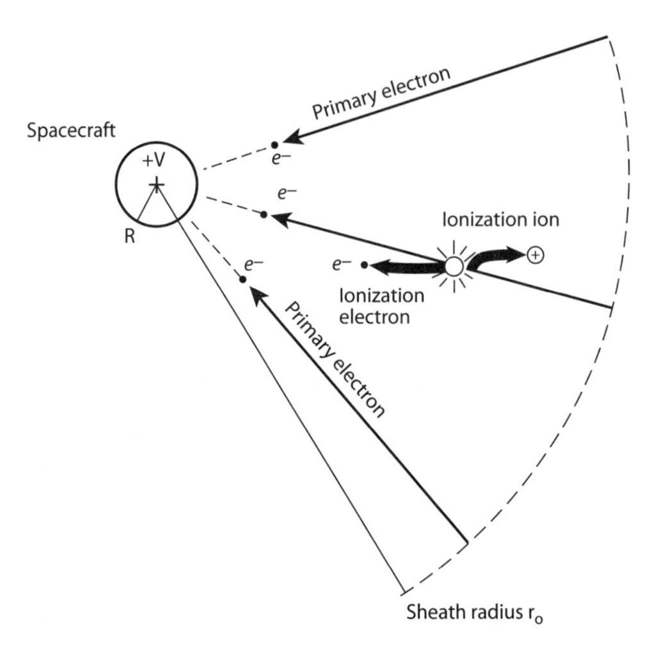

Figure 8.9 Electron impact ionization in the sheath. Electron and ion pairs are created. The electrons go toward the space body charged artificially to a positive potential, while the ions go outward.

$$n_e(r) = \frac{1}{r^2} \frac{n_e \mathrm{v}_{th} r_o^2}{[\mathrm{v}_{th}^2 + 2e\phi(r)/m_e]^{1/2}} \tag{8.17}$$

The ionization electron density $n^-(r)$ is due to all ionizations that occur outward of r, and the density $n^+(r)$ of ions at r is due to all ionizations that occur inward of r. Thus, for a spacecraft of radius R,

$$n^-(r) = \frac{1}{r^2}\int_r^{r_o} \frac{\left[\frac{dn}{dt}\right]_{r'} r'^2 dr'}{[2e|\phi(r) - \phi(r')|/m_e]^{1/2}} \tag{8.18}$$

and

$$n^+(r) = \frac{1}{r^2}\int_R^{r_o} \frac{\left[\frac{dn}{dt}\right]_{r'} r'^2 dr'}{[2e|\phi(r) - \phi(r')|/m_e]^{1/2}} \tag{8.19}$$

where the rate of ionization is a function of the mean free path λ, the incoming electron current, and the velocity-dependent probability P of ionization:

$$\left[\frac{dn}{dt}\right]_{r'} = \lambda^{-1} P[v_e(r')]\, n_e(r') v_e(r') \tag{8.20}$$

where the mean free path depends on the neutral density and cross section $\sigma(\mathrm{v}_e)$ of electron impact ionization:

$$\lambda^{-1} P[v_e(r')] = N(r')\sigma(v_e) \tag{8.21}$$

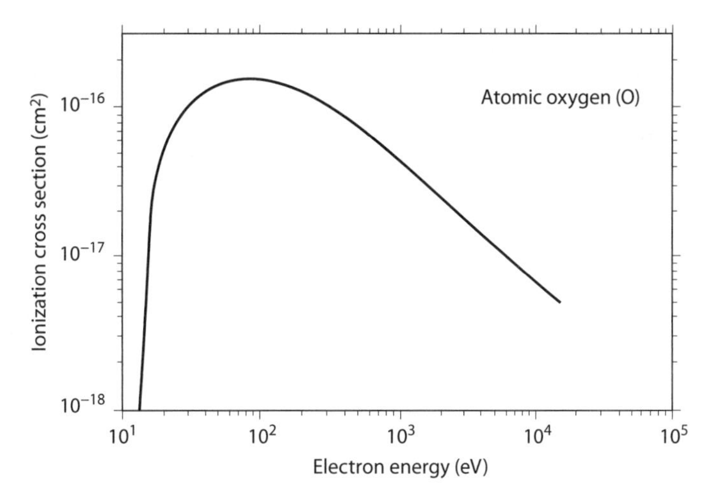

Figure 8.10 Cross section of electron impact ionization of atomic oxygen.

Figure 8.10 shows the ionization cross section $\sigma(E_e)$ of oxygen as a function of the electron energy E_e. The cross section has a threshold of about 14 eV and a maximum at about 90 eV for typical atmospheric neutrals. Since the newborn ion velocity is comparable to the spacecraft velocity v_s, we have added v_s to the ion velocity (equation 8.19) in the model.

8.7 Numerical Method for Sheath Ionization Model

To solve the system of equations [equations (8.12) to (8.21)], one divides the sheath into N concentric shells of equal thickness and sets up N coupled equations for the N unknowns ϕ_i (figure 8.11). In view of the complexity, it is impossible solve the N coupled equations exactly. One can solve the equations approximately.

A Newton-Ralphson method[14] for solving the set of equations (8.12) to (8.21) is as follows. From the Poisson equation (8), one defines a function f_i:

$$f_i(E_1, \ldots, E_N) = (r^2 E)_{i+1} - (r^2 E)_i - \frac{1}{\varepsilon_o}\left[r^2 \rho(E_1, \ldots, E_N)\right]_i \Delta r \tag{8.22}$$

where the electric field E is constructed in a finite difference scheme:

$$\phi_i - \phi_{i+1} = \Delta r(E_i + 2E_{i+1} + E_{i+2})/4 \tag{8.23}$$

The numerical method used to solve equations (8.22) and (8.23) is the Newton-Ralphson method of iteration:

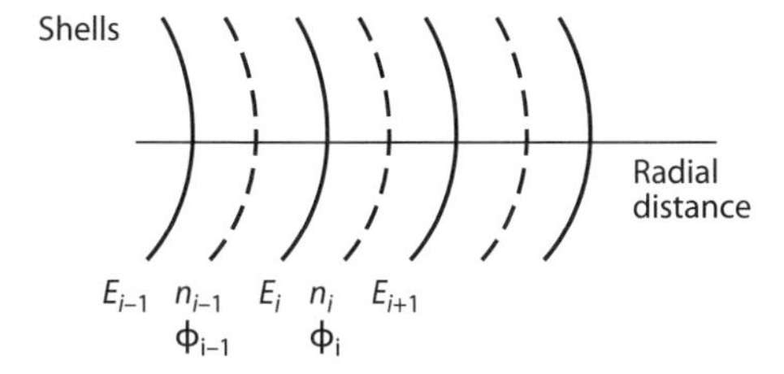

Figure 8.11 Decomposition of the sheath into shells.

$$E_i^{(j+1)} = E_i^{(j)} - \frac{f_i(E_1^{(j)}, \ldots, E_N^{(j)})}{\partial f_i(E_1^{(j)}, \ldots, E_N^{(j)})/\partial E_i} \tag{8.24}$$

To start, a set of trial solutions is used in the iteration process and a convergent set of solutions is sought.

8.8 Results of Sheath Ionization

Without ionization, the sheath is governed by the classical Langmuir probe equation in the space charge regime. The electron and ion pairs created by electron impact in the sheath alter the space charge flow. As mentioned earlier, the same model can be applied similarly to beam emission from spacecraft in the ionosphere or current conduction from the plasma contactor at the end of a tether. Figure 8.12 shows the numerical results obtained for typical spacecraft or rockets at ionospheric altitudes with various beam current. The beam is assumed to be of high energy so that negligible beam ionization occurs. The ionization is solely due to the electrons falling toward the spacecraft through the sheath.

An important result shows up in figure 8.12. It is the nonmonotonic behavior of the spacecraft potential as a function of electron beam current. At low current, the spacecraft potential increases with the beam current, as one expects by solving the Poisson equation equation (8.11) with negligible or no ionization. There exists a critical current beyond which the spacecraft potential decreases. This is a nonmonotonic current-voltage behavior. However, at high beam current, the calculated spacecraft potential rises again. At higher beam currents, the iteration fails to converge. The ions produced are too slow compared with the electrons and tend to build up local positive potential structures (virtual anodes) in the sheath, resulting in two-way flows of electrons locally. We will not pursue this physical regime, which is beyond the assumptions in the model described earlier.

We now return to study the important result: the nonmonotonic current-voltage behavior. Figure 8.13 shows the computed sheath potential profiles for various beam currents. For beam current I of 1 mA, the potential ϕ_S at the spacecraft surface is about 48 V. For $I = 2$ mA,

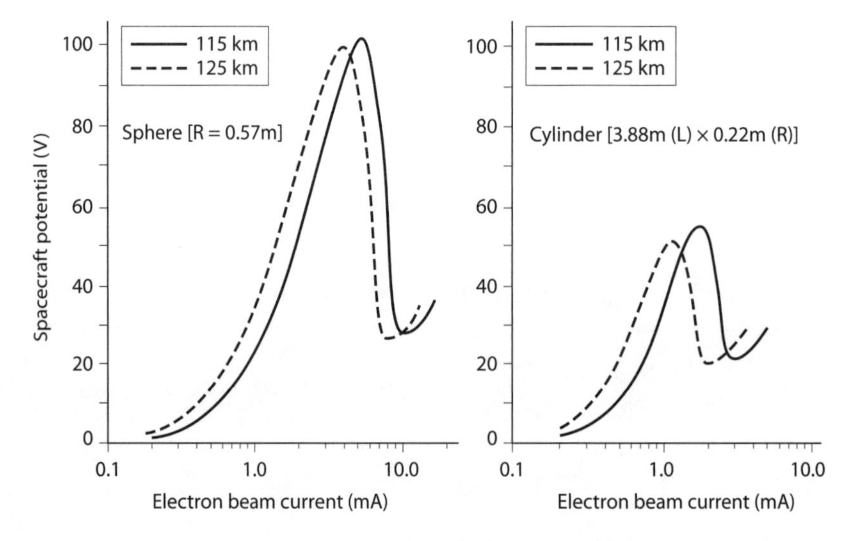

Figure 8.12 Nonmonotonic current-voltage behavior. As the beam current increases, the spacecraft potential increases. When the beam current is high, the potential decreases as a result of ionization of the neutrals in the atmosphere. (Adapted from reference 15.)

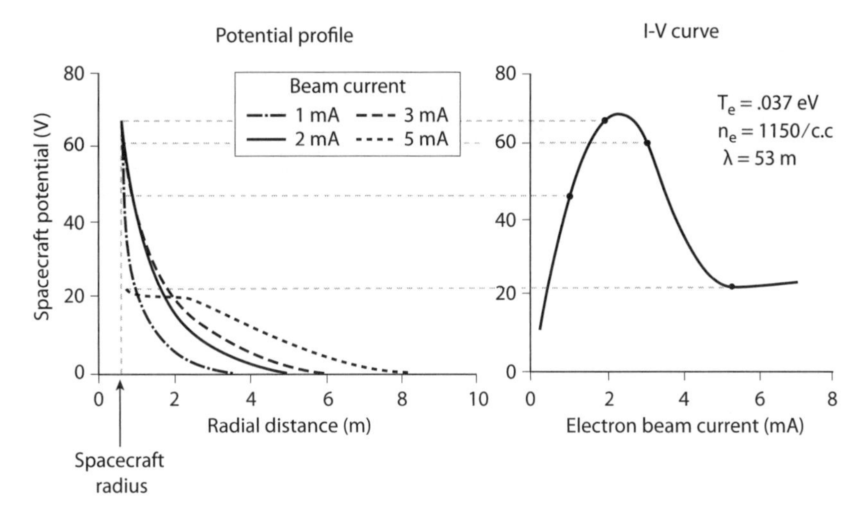

Figure 8.13 Relation between potential profile and current-voltage behavior. (Adapted from reference 15.)

the sheath size increases, and the potential ϕ_S climbs to about 68 V. For I = 3 mA, the sheath expands further, ionization becomes important, and the potential ϕ_S decreases to about 60 V. At I = 5 mA, the sheath expands, ionization reaches saturation, and the sheath potential profile shows a flat region. Further increase in current would lengthen the flat region and tend to build up local positive potential fluctuations.

8.9 Comparison of Theory with Space Experiment

The Beam Emission Rocket Test 1 (BERT-1) rocket experiment was conducted at about 130-km altitude in June 1985 to test the theory by means of electron beam emission. The spacecraft potential was measured by means of extended booms. The beam energy was 1.9 keV. Figure 8.14 shows the data obtained.[15] Figure 8.15 shows the experimental data plotted as an I–V

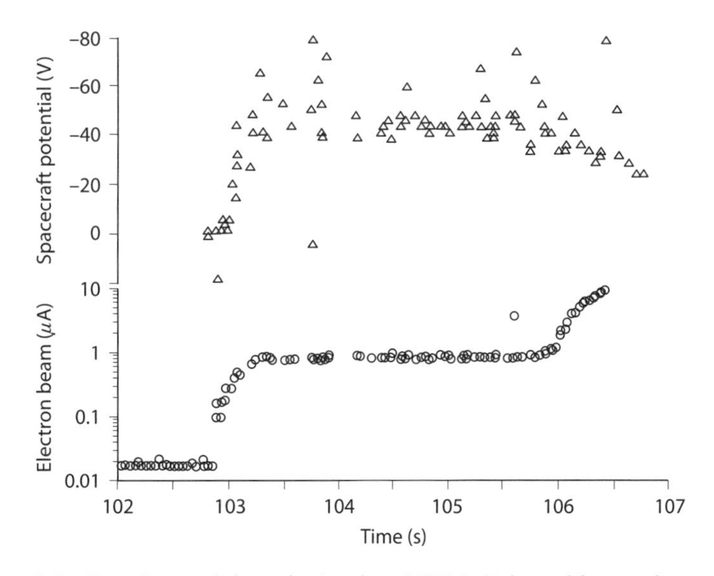

Figure 8.14 Experimental data obtained on BERT-1. (Adapted from reference 15.)

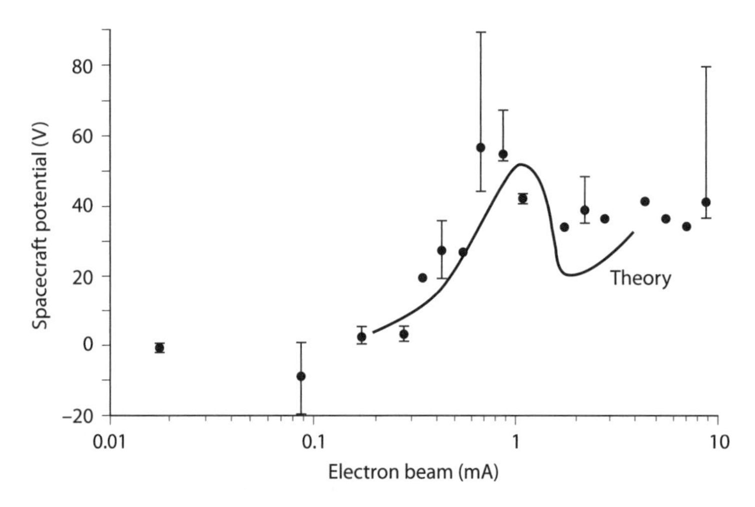

Figure 8.15 BERT-1 spacecraft potential plotted as a function of electron beam current. The solid line is from theory. (Adapted from reference 15.)

curve, together with the theoretical curve for comparison. In obtaining the I–V curve, one lets the beam current data fall into equal size cells in the x axis. With each nonempty cell, the mean values of the potential data and the upper and lower root mean deviation bars are shown. Despite the simplicity of the model, it is surprising that the comparison is so good.[15]

8.10 Exercises

1. Describe in your own words how a tether can be used to decelerate or propel a spacecraft.
2. Formulate the ionization model for a cylindrical object by using cylindrical coordinates. One may neglect the end effects for simplicity.
3. Solve the cylindrical sheath ionization model by using numerical methods.

8.11 References

1. Landau and Lifshitz, "The classical theory of fields," *Lectures in Theoretical Physics*, vol. 2, p. 62, Pergamon Press, NY (1959).
2. Griffiths, D. J., *Introduction to Electrodynamics*, Prentice Hall, Englewood Cliffs, NJ (1981).
3. Ahedo, E., and J. R. Sanmartin, "Analysis of bare-tether systems for deorbiting low-Earth-orbit satellites," *J. Spacecraft and Rockets* 39, no. 2: 198–205 (2002).
4. Mott-Smith, H. M., and I. Langmuir, "The theory of collectors in gaseous discharges," *Phys. Rev.* 28: 727–763 (1926).
5. Sanmartin, J. R. and R. D. Estes, "The orbit-motion-limited regime of cylindrical Langmuir probes," *Phys. Plasmas* 6, no. 1, 395–405 (1999).
6. Alpert, Y. L., A. V. Gurevich, and L. P. Pitaevskii, "Disturbance of the plasma and the electric field in the vicinity of a charged body at rest," in *Space Physics with Artificial Satellites*, p. 186, Consultants Bureau, New York (1965).
7. Parker, L. W., and B. L. Murphy, "Potential buildup on an electron emitting ionospheric satellite," *J. Geophys. Res.* 72, no. 5, 1631–1636 (1967).
8. Cooke, D. L., and I. Katz, "TSS-1R electron currents: magnetic limited current collection in a heated presheath," *Geophys. Res. Lett.* 25, no. 5: 753–756 (1998).

9. Roy, S. R., and D. E. Hastings, "A brief review of electrodynamic tethers," in *The Behavior of Systems in Space Environments*, eds. R. N. DeWitt, D. Dwight, and A. K. Hyder, pp. 825–836, Kluwer Academic, Norwell, MA (1993).
10. Sanmartin, J. R., M. Martinez-Sanchez, and E. Ahedo, "Bare-wire anodes for electrodynamic tethers," *J. Propulsion and Power* 9, no. 3, 353–360 (1993).
11. Sanmartin, J. R., E. C. Lorenzini, and M. Martinez-Sanchez, "Electrodynamic tether applications and constraints," *J. Spacecraft Rockets* 47, no. 3, 442–456 (2010).
12. Sanmartin, J. R., "A review of electrodynamic tether applications in science and applications," *Plasma Sources Sci Technol.* 19, 034022 (7 pp.), doi:10.1088/0963-0252/19/3/034022 (2010).
13. "Special issue on TSS-1R: Electrodynamic tether-ionospheric interactions," *Geophys. Res. Lett.* 25, nos. 4–5 (1998).
14. Lai, S. T., H. B. Cohen, and W. J. McNeil, "Spacecraft sheath modification during electron beam ejection," AFGL-TR-85-0215, pp. 1–15, Air Force Geophysics Laboratory, Hanscom AFB, MA (1985); presented at the International School for Space Simulations, Kauai Beach Boy Hotel, Kauai, Hawaii, 3–16 February 1985 (DTIC# ADA-166604).
15. Lai, S. T., "Sheath ionization during electron beam emission from spacecraft," *Phys. Space Plasmas* 11: 411–419 (1991).
16. Hastings, D. E., and H. Garrett, *Spacecraft-Environmental Interactions*, Cambridge University Press, Cambridge, UK (1996).

9

Surface Charging Induced by Electron Beam Impact

The beam impact energy together with the secondary and backscattered electron coefficients are important for determining the spacecraft potential induced by incoming mono-energetic electron beams.

9.1 Impact Energy of an Electron Beam

Consider an electron beam impacting a surface. For simplicity, the beam is assumed well collimated and has an initial energy E_{beam} with negligible energy spread. All other currents, such as ambient electron, ambient ion, and photoelectron currents, are assumed small compared with the beam current and are therefore ignored. If the surface is initially charged, the potential repels or attracts the beam electrons depending on the sign of the potential. Therefore, the beam energy changes as the beam electrons approach the surface. The beam impact energy E (not the energy E_{beam} of the beam before it approaches the spacecraft) is an important factor in charging the surface:

$$E = E_{beam} - q_e \phi \tag{9.1}$$

where the electron charge is $q_e = -e$ and e is the elementary charge ($e > 0$). The beam impact energy is also called the primary electron energy E.

Secondary electrons[1-4] and backscattered electrons,[5] especially the former, also play an important role in beam impact charging. If the primary electron energy E lies in the range ($E_1 < E < E_2$), the total outgoing secondary and backscattered electron flux normalized by the incident electron flux exceeds unity (figure 9.1):

$$\delta(E) + \eta(E) > 1 \qquad \text{for } E_1 < E < E_2 \tag{9.2}$$

Here $\delta(E)$ is the secondary electron coefficient, and $\eta(E)$ the backscattered electron coefficient. The energies E_1 and E_2 are the crossover points of $\delta(E) + \eta(E)$ (for more details, see chapter 3).*

In the rest of this chapter, we will consider, as an exercise, an electron beam impacting a surface of a given initial potential. In what way would the surface potential respond? For simplicity, we assume that the electron beam current is much larger than the ambient electron and ion currents. In other words, the electron beam, secondary electrons, backscattered

*Perhaps it should be mentioned that there are various $\delta(E)$ measurements and formulae published in the literature. The measurement results of E_1 and E_2 by various authors are fairly consistent. However, the formulae by various authors are different in the high-energy regime above about 10 keV. Since E_1 and E_2 of $\delta(E)$ are sufficient for discussing the direction of change of the surface potential by electron beam impacts, there is no need to choose a preferred $\delta(E)$ formula.

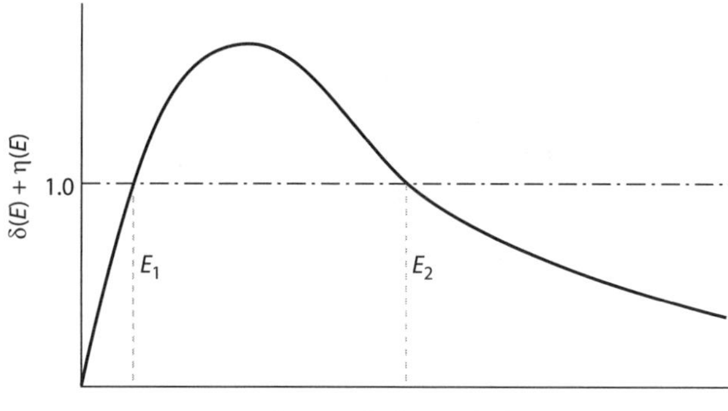

Figure 9.1 Secondary electron emission coefficient. The energies E_1 and E_2 are the points crossing unity. With the addition of backscattered electrons, the locations of E_1 and E_2 are practically unchanged.

electrons, and surface potential are the sole components in this model. We then ask ourselves what the response would be.

9.2 Electron Beam Impact on an Initially Uncharged Surface

Suppose that the initial surface potential is $\phi_i = 0$ V. Let us consider what would happen. As the electron beam impacts on the surface, secondary and backscattered electrons go out. Electrons impacting a surface would normally charge the surface to negative volts. However, since the beam energy E lies in the range (E_1, E_2), the outgoing electron flux exceeds the incoming electron flux. Therefore, the resultant potential, ϕ_f, of the surface increases and remains positive—that is, $\phi_f > 0$ V. However, secondary electrons have typically energies of a few eV only. Backscattered electrons are much less abundant. With positive surface charging, most of the secondary and backscattered electrons cannot leave because they are attracted by the positive surface potential. As a result, the preceding process has to stop when the surface potential ϕ_f reaches a few volts positive:

$$\phi_f = \text{a few volts (positive)} \tag{9.3}$$

The beam energy increases by a few eV only:

$$E_f = E + \text{a few eV} \tag{9.4}$$

Surface charging is sometimes up to thousands of volts at geosynchronous altitudes and is therefore of concern. For most practical purposes, a surface potential ϕ_f of a few volts is insignificant and can be regarded as nearly zero.

9.3 Electron Impact on an Initially Negatively Charged Surface

9.3.1 Case 1

If the surface is initially at a negative potential $\phi_I (< 0)$, and the beam impact energy E lies in the range (E_1, E_2), the process described in section 9.2 occurs. The final potential ϕ_f is again

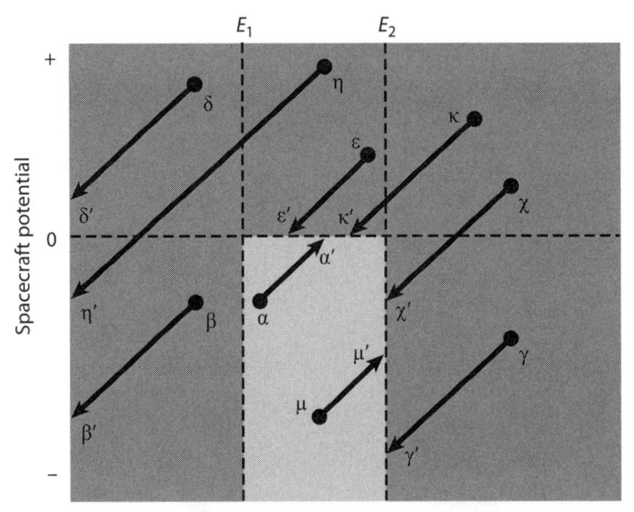

Figure 9.2 Surface charging behavior by mono-energetic electron beam impact. The initial and final conditions are indicated by the beginning and the end of an arrow, respectively. All arrows are pointing in the southwest direction except those in the region ($\phi < 0$; $E_1 < E < E_2$), in which arrows are in the opposite direction.

a few volts positive. The change in surface potential is $\Delta\phi = \phi_f - \phi_i$. Therefore, the change in beam energy is $\Delta E = e\Delta\phi$. For most purposes, a few volts can be ignored. Thus, one can write

$$\phi_f \approx 0 \tag{9.5}$$

and

$$E_f = E + e\Delta\phi \approx E + e|\phi_i| \tag{9.6}$$

In this figure, each state point is represented by (ϕ, E). The arrow (α, α') in figure 9.2 illustrates the initial and final states discussed in this case. Since the slope is unity in equation (9.6), the arrow is at an angle of $\tan^{-1}(1) = 45°$.

9.3.2 Case 2

If the beam impact energy E is less than E_1, the incoming electron flux exceeds the outgoing electron flux. As a result, the surface accumulates more electrons, charging to a higher negative potential ($\phi_f < 0$). Because of the higher negative potential, the electron beam impacting on the surface is repelled and becomes less energetic. This process continues until the negative potential increases (in the negative direction) to a value that the final beam energy E_f becomes zero. Then, the process stops. This is the final condition:

$$e\phi_f = e\phi_i - E \tag{9.7}$$

$$E_f = E + e(\phi_f - \phi_i) = 0 \tag{9.8}$$

Note that e is the elementary charge ($e > 0$). In equations (9.7) and (9.8), ϕ_f, ϕ_i, and $e(\phi_f - \phi_i)$ are negative. The arrow (β, β') in figure 9.2 illustrates the initial and final states discussed in this case. The arrow is at 45°.

9.3.3 Case 3

If the beam energy E is greater than E_2, the incoming electron flux exceeds the outgoing electron flux. Like case 2 (section 9.3.2), the surface accumulates more and more electrons as the electron beam continues to impact on the surface. As a result, the surface potential increases in the negative direction and the beam impact energy decreases. The process continues until the final beam energy E_f decreases to E_2, the second crossing point (figure 9.1). Therefore, the final condition is given by

$$e(\phi_f - \phi_i) = E_2 - E \tag{9.9}$$

$$E_f = E + e(\phi_f - \phi_i) = E_2 \tag{9.10}$$

where ϕ_f, ϕ_i, and $e(\phi_f - \phi_i)$ are negative. The arrow (γ,γ') in figure 9.2 illustrates the initial and final states discussed in this case.

9.4 Electron Impact on an Initially Positively Charged Surface

Consider a beam impacting an initially positively charged surface. Let E be less than E_1, the first crossing point of $\delta(E) + \eta(E)$. In this energy range ($0 < E < E_1$), the incoming electron flux exceeds the outgoing electron flux. Therefore, the surface receives more electrons and therefore charges more negatively. Accordingly, the primary electron energy is reduced upon impact. This process would continue until the primary electron energy is zero—i.e., there are no beam electrons coming in. In this case, the final beam energy is $E_f = 0$, and the spacecraft final potential is given by $e\phi_f = -E$. The arrow (δ,δ') in figure 9.2 illustrates the initial and final states discussed in this case.

Suppose that the beam impact energy lies in the range, $E_1 < E$, and the initial positive surface potential ϕ_i is well above a few volts. The beam impact generates abundant secondary and backscattered electrons. However, most of the generated electrons cannot leave because they are attracted back to the surface (figure 9.3). Therefore, the surface receives more and more electrons from the beam, reducing the positive surface potential.

When the positive potential eventually reaches nearly zero, more low-energy (secondary and backscattered) electrons can escape. The electron beam energy is decreasing because the spacecraft surface is accumulating more electrons. The process reaches equilibrium when the outgoing electron flux equals the incoming electron flux. There can be four possibilities, as follows.

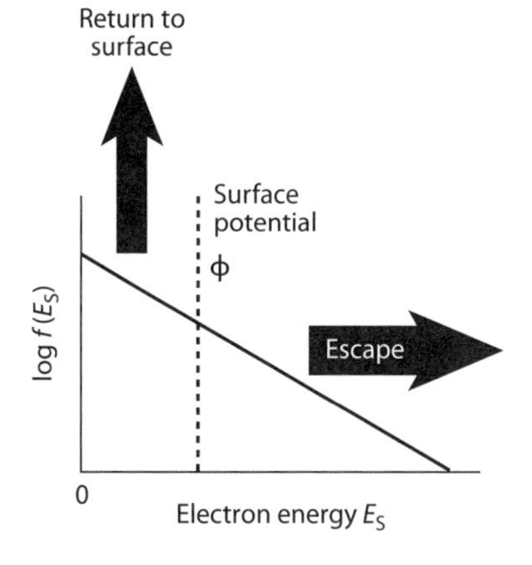

Figure 9.3 Escape of low-energy electrons from a positively charged surface. A Maxwellian electron energy distribution is shown. The low-energy electrons with energy E_S below the $q_e\phi$, where q_e is the electron charge (negative), are attracted by the surface potential (dashed line) and therefore return to the surface.

9.4.1 Case 1

If the primary beam electron energy E is in the range, $E_1 < E < E_2$, there are more secondary and backscattered electrons outgoing out than the beam electrons coming in. A positive surface potential beyond a few volts is able to attract most of the secondary electrons back to the surface. As a result, the surface potential decreases because the beam electrons are coming in while most of the secondary electrons are not leaving. At equilibrium, the final potential is a few volts positive, below which most of the secondary electrons can leave. For practical purposes, a surface potential ϕ of a few volts can be regarded as zero. Thus, the change $\Delta\phi$ in potential is given by

$$\Delta\phi = \phi_f - \phi_i = 0 - \phi_i = -\phi_i \tag{9.11}$$

The final beam impact energy is given by

$$E_f = E - e\phi_i \tag{9.12}$$

The arrow $(\varepsilon, \varepsilon')$ in figure 9.2 illustrates the initial and final states discussed in this case.

9.4.2 Case 2

If the primary electron energy of the preceding case continues to decrease to below E_1, the outgoing (secondary and backscattered) electron flux cannot compensate the incoming (beam) electron flux. As a result, the surface potential continues to fall toward the negative direction while the beam impact energy continues to decrease until the final beam impact energy E_f is 0:

$$E_f = E - q_e\Delta\phi = 0 \tag{9.13}$$

The final potential is given by

$$e\phi_f = e\phi_i - E \tag{9.14}$$

If the initial beam energy E exceeds the initial surface potential energy $e\phi_i$ (in eV), the final surface potential ϕ_f is negative. The arrow (η, η') in figure 9.2 illustrates the initial and final states discussed in this case.

9.4.3 Case 3

If the primary electron energy is above E_2, the outgoing (secondary and backscattered) electron flux cannot compensate the incoming (beam) electron flux. As a result, the surface potential continues to fall and the beam impact energy continues to decrease until the final beam energy is E_2. At E_2, the outgoing flux equals the incoming flux. Therefore, the final condition is given by

$$E_f = E - e\Delta\phi = E_2 \tag{9.15}$$

$$e\phi_f = e\phi_i - E \tag{9.16}$$

The arrow (χ, χ') in figure 9.2 illustrates the initial and final states discussed in this case.

9.4.4 Case 4

If the initial beam energy E is above E_2 and the surface potential is well above a few positive volts, there are more incoming electrons than outgoing electrons because the secondary electrons are attracted back to the surface. Both the surface potential and the beam energy decrease. If the decreasing beam energy reaches E_2 but the surface potential is still above a few positive volts, the process continues until the surface potential is below a few positive volts—i.e., nearly zero. At that point, the secondary electrons can leave. The arrow (κ,κ') in figure 9.2 illustrates the initial and final states discussed in this case.

9.5 Summary

When an electron beam of energy E_{beam} bombards a spacecraft surface, the beam impact energy E is different from E_{beam} if the surface is charged. The impact energy E is less (more) than E_{beam}, depending on the repulsion (attraction) energy of the surface potential. For simplicity, we have assumed that the beam energy has negligible spread. Upon impact, the beam generates secondary and backscattered electrons. The beam impact energy is also called the *primary electron energy*. Depending on the various initial conditions of surface potential and primary electron energy, the final condition can be deduced. As the saying goes, "A picture is worth a thousand words"; the results are summarized in figure 9.2.

There is a remarkable behavior. Although the secondary electron coefficient $\delta(E)$, or the combined $\delta(E) + \eta(E)$ coefficient, has two crossings of zero (figure 9.1), the crossings have different properties. The first crossing (at E_1) has a positive slope and the second crossing (at E_2) has a negative slope (figure 9.1). The beam impact energy never ends at the first crossing point E_1, no matter what the initial beam impact energy and initial surface potential are (figure 9.2).

9.6 Limitation

For clarity in illustrating the main physical mechanism involved in the phenomenon of electron beam impact, we have limited the number of game pieces to two only: the incoming (or primary) electron beam and the outgoing secondary (and backscattered) electrons. All other currents or factors have been neglected. This approximation would be valid if the beam current is much larger than all other currents, including the electron and ion currents from the ambient space plasma. If the incoming electron beam current is high and the spacecraft potential is negative, the ambient electrons from the space plasma are repelled, and as a result, the ambient electron current is reduced by a Boltzmann factor. The ambient ion current is usually much smaller than the ambient electron current at the geosynchronous environment.[6,7] However, when the spacecraft potential ϕ is high and negative, the ambient ions are attracted toward the spacecraft. In the Mott-Smith and Langmuir model, the ambient ion current increases linearly with the potential ϕ. The ion current may compete with the incoming electron beam current if the magnitude of the negative spacecraft potential is sufficiently high.

There are other factors that may come into play, depending on the situation. High-current beams may interact with the space plasma, resulting in plasma instability waves. Beam interaction with atmospheric neutral species and spacecraft exhausts may generate ionization. The newly created electron and ion pairs from ionization may be abundant, thus affecting the spacecraft potential. In sunlight, photoemission is abundant and capable of competing with the beam currents, unless the photoelectrons are trapped by potential wells. Also, the effect of deep dielectric charging by prolonged beam bombardments would add a major factor. Spacecraft charging is more than a two-dimensional board game; it is rich in physics, applications, and systems engineering.

9.7 Exercises

1. An electron beam of initial energy 1 keV is impacting continuously on a spacecraft surface, which is initially at -400 V. Therefore, the beam impact energy is 600 eV. In response to the beam impact, the spacecraft potential changes and so does the beam impact energy.

Suppose that the secondary electron emission coefficient crossover points ($\delta = 1$) of the surface material are at 60 eV and 1600 eV. Suppose also that the backscattered electrons can be neglected. Suppose further that the electron beam current is so large that the ambient ion current can also be neglected. What is the final spacecraft potential? What is the final beam impact energy?

2. Repeat problem 1, but with initial beam energy 1.7 keV. All other assumptions remain the same. What is the final spacecraft potential? What is the final beam impact energy?

9.8 References

1. Sternglass, E. J., "Theory of secondary electron emission," Paper 1772, Westinghouse Research Laboratories, Pittsburgh, PA (1954).
2. Sternglass, E. J., "Backscattering of kilovolt electrons from solids," *Phys. Rev.* 95, no. 2: 345 (1954).
3. Sanders, N. L., and G. T. Inouye, "Secondary emission effects on spacecraft charging: energy distribution consideration," in *Spacecraft Charging Technology 1978*, eds. R. C. Finke and C. P. Pike, NASA-2071,ADA-084626, pp. 747–755, U.S. Air Force Geophysics Laboratory, Hanscom AFB, MA (1978).
4. Scholtz, J. J., D. Dijkkamp, and R.W.A. Schmitz, "Secondary electron emission properties," *Philips J. Res.* 50, nos. 3–4: 375–389 (1996).
5. Reagan, J. B., R. E. Meyerott, R. W. Nightingale, P. C. Filbert, and W. L. Imhoff, "Spacecraft charging currents and their effects on space systems," *IEEE Trans. Electr. Insulations* 18: 354–365 (1983).
6. Lai, S. T., and D. Della-Rose, "Spacecraft charging at geosynchronous altitudes: new evidence of existence of critical temperature," *J. Spacecraft and Rockets* 38: 922–928 (2001).
7. Lin, Y., and D. G. Joy, "A new examination of secondary electron yield data," *Surf. Interface Anal.* 37: 895–900 (2005).

10

Spacecraft Charging Induced by Electron Beam Emission

At equilibrium, the potential of a spacecraft is determined by the balance of currents. When charged particle beams are emitted from a spacecraft, the beam currents participate in the current balance. Depending on the currents, electron beam emission from spacecraft can induce charging to positive potentials, while positive ion beam emission can induce negative potential charging. If the beam current is much higher than all other currents, the beam controls the spacecraft potential.

10.1 Current Balance without Beam Emission

First, let us review the theory of spacecraft charging without photoemission or beam emission. Then, we will introduce a small beam current and study how the potential changes as the beam current increases. Consider a geosynchronous spacecraft at a negative potential ϕ at equilibrium. The potential ϕ is determined by current balance. Ambient electrons are coming in; so are ambient ions. With a negative potential, the spacecraft repels the ambient electrons. Those ambient electrons that are energetic enough to overcome the repulsion can still come in. When these incoming electrons impact on the surface, they generate secondary electrons and backscattered electrons, both of which are outgoing. The negative potential also attracts ambient ions. In the Mott-Langmuir formulation,[1,2] which is often valid at geosynchronous altitudes, the current-balance equation is given by

$$J_T(\phi) = I_e(0)\exp\left(-\frac{q_e\phi}{kT_e}\right) - I_S(\phi) - I_b(\phi) - I_i(0)\left(1 - \frac{q_i\phi}{kT_i}\right) = 0 \tag{10.1}$$

where the total current $J_T(\phi)$ denotes the total current. The notations in equation (10.1) are as in chapters 1 and 2. In equation (10.1), since $q_e\phi$ is positive, the exponential factor is less than unity, implying that the ambient electron current collected by the charged spacecraft is less than that intercepted by the uncharged spacecraft. Since $q_i\phi$ is negative, the ion current is greater than that intercepted by the uncharged spacecraft. The secondary[3,4] and backscattered[5,6] electron currents are proportional to the ambient electron current and therefore can be written as fractions of the ambient (primary) electron current. Equation (10.1) can be written as follows:

$$J_T(\phi) = I_e(0)[1 - \langle\delta + \eta\rangle]\exp\left(-\frac{q_e\phi}{kT_e}\right) - I_i(0)\left(1 - \frac{q_i\phi}{kT_i}\right)^{\alpha} = 0 \tag{10.2}$$

where

$$\langle\delta + \eta\rangle = \frac{\int_0^\infty dEEf(E)[\delta(E) + \eta(E)]}{\int_0^\infty dEEf(E)} \tag{10.3}$$

In equation (10.2), the orbit-limited ion attraction factor has a exponent $\alpha = 1/2$; for a sphere, $\alpha = 1$. For SCATHA, which is a short cylinder, α equals 0.7 approximately.[7] In equations (10.2) and (10.3), δ is the secondary electron coefficient and η is the backscattered electron coefficient. The rest of the notations in equations (10.1) to (10.3) are as in chapters 1 and 2. In equation (10.2), the net electron current collected is given by

$$I_e(\phi) = I_e(0)\,[1 - \langle \delta + \eta \rangle]\exp\left(-\frac{q_e \phi}{kT_e}\right) \tag{10.4}$$

and the net ion current collected is given by

$$I_i(\phi) = I_i(0)\left(1 - \frac{q_i \phi}{kT_i}\right)^{\alpha} \tag{10.5}$$

The solution (root) of the current-balance equation, equation (10.1) or equation (10.2), gives the spacecraft potential ϕ.

It is instructive to see the roots graphically. Figure 10.1 shows schematically the total current, $I_I(\phi) - I_e(\phi)$, plotted as a function of the potential ϕ. As the positive potential ϕ increases, more and more electrons are coming in, rendering the magnitude of the total current increasing in the downward (negative) direction (figure 10.1). The root is at the zero crossing (where the total current equals 0) of the curve.

As the magnitude of negative potential ϕ increases, more and more positive ions are coming in, rendering the total current $J_T(\phi)$ increasing in the upward (positive) direction. Note that, for ambient plasmas of multiple energy distributions, there can be multiple roots for a current-voltage curve (see chapter 5). For simplicity, we assume in this chapter that there is one root only.

10.2 Electron Beam Emission

With electron beam emission from a spacecraft, the sum of all currents includes the electron beam current I_{beam}. For electron beam emission, the beam electrons are going out and therefore I_{beam} has the same sign (positive) as a current of ions coming in. The current balance equation becomes[8]

$$I_e(\phi) - I_i(\phi) - I_{beam}(\phi) = 0 \tag{10.6}$$

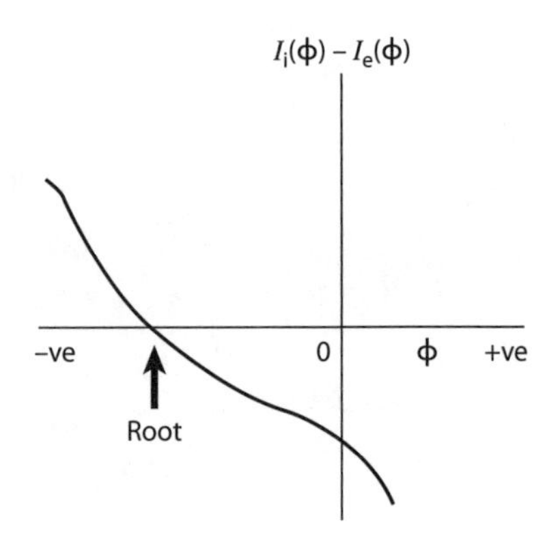

Figure 10.1 Current-voltage curve. The root is located where the total current equals zero.

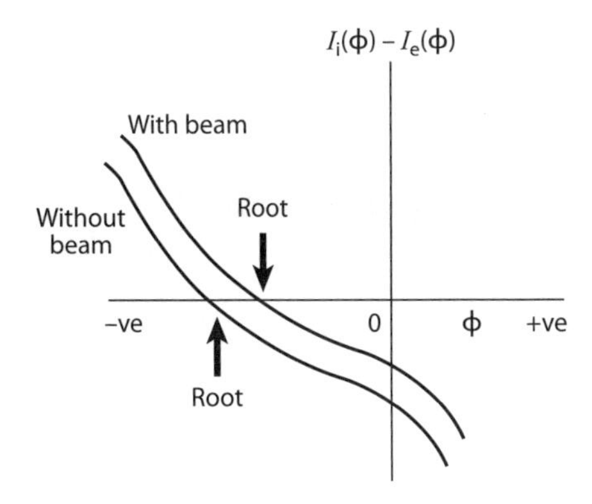

Figure 10.2 Current-voltage curves with and without electron beam emission. The root moves to a new location given by equation (10.6).

With the addition of I_{beam}, the total current curve shifts upward (figure 10.2). As a result, the root moves to the right-hand side—i.e., toward the positive side. If the beam current is small, the new root is still be at a negative voltage value. Likewise, if an ion beam is emitted, the curve would move downward—i.e., the new root ϕ (< 0) would shift toward the left, that is, the negative root increases in magnitude.

To estimate the magnitude of the shift $\Delta\phi$, let us assume that the incoming ion current is negligible compared with the incoming electron current (see chapters 4 and 11). In this approximation, the current balance equation (10.6) for an electron beam emitted from a spacecraft becomes

$$I_e(0)[1 - \langle\delta + \eta\rangle]\exp\left(-\frac{q_e\phi}{kT}\right) = I_{beam} \tag{10.7}$$

For Maxwellian distributions, the outgoing electron current $\langle\delta + \eta\rangle$ of secondary and backscattered electrons is independent of the charging potential. Differentiating both sides, one obtains

$$-\frac{q_e}{kT} I_e(0)[1 - \langle\delta + \eta\rangle]\exp\left(-\frac{q_e\phi}{kT}\right)\Delta\phi = \Delta I_{beam} \tag{10.8}$$

Therefore, the amount $\Delta\phi$ of shift in the spacecraft potential due to an increase in electron beam emission

$$\Delta\phi = -\left(\frac{kT}{q_e}\right)\frac{\Delta I_{beam}}{I_{beam}} \tag{10.9}$$

where the charge q_e of an electron is negative, rendering the shift $\Delta\phi$ positive. If the emitted electron beam current is so large that the root shifts to a positive value, the formulation in equation (10.2) becomes invalid. One needs a different formulation for the positive potential regime, in which the spacecraft repels ions and attracts electrons. We discuss this regime in the next section.

10.3 Charging to Positive Potentials

Consider a spacecraft emitting a large electron beam current I_{beam}. Suppose that the electron beam current is so large that the spacecraft potential ϕ becomes positive. The Mott-Smith and Langmuir equation of current balance is of the form

$$I_e[1 - \langle \delta + \eta \rangle]\left(1 - \frac{q_e\phi}{kT_e}\right)^{\alpha} - I_i \exp\left(-\frac{q_i\phi}{kT_i}\right) - I_{beam}(\phi) = 0 \tag{10.10}$$

where the ϕ is the potential (> 0) and the other notations are as in previous chapters. Because the potential is positive, ambient electrons are attracted and ambient ions are repelled. The repulsion is described by the exponential factor in which both the ion charge q_i and the spacecraft potential ϕ are positive, rendering the exponential less than unity. Since secondary electrons have typically a few eV in energy, they cannot leave the spacecraft if the spacecraft potential is higher than a few volts positive. Therefore, if the potential is higher than about 20 V, for example, the secondary electron current can be neglected. Since the backscattered electron current is much smaller than the ambient electron current, it is often neglected for most surface materials. Therefore, it is often a good approximation to neglect both secondary and backscattered electrons if the spacecraft potential exceeds a few positive volts. In this approximation, equation (10.7) becomes

$$I_e(0)\left(1 - \frac{q_e\phi}{kT_e}\right)^{\alpha} - I_i(0) \exp\left(-\frac{q_i\phi}{kT_i}\right) - I_{beam}(\phi) = 0 \tag{10.11}$$

In equation (10.11), $q_e = -e$ and $q_i = e$, where e is the elementary charge. The ambient ion current $I_i(0)$ is smaller than that of the electrons, because of the electron-ion mass difference.[9,10] Furthermore, the ions are repelled by the spacecraft's positive potential as a result of the large electron beam emission. Therefore, neglecting the ion current $I_i(\phi)$ in this large electron beam approximation, equation (10.11) becomes

$$I_e(0)\left(1 + \frac{q_e\phi}{kT_e}\right)^{\alpha} - I_{beam} = 0 \tag{10.12}$$

For a spherical spacecraft, $\alpha = 1$. Substituting $\alpha = 1$ in equation (10.9), one obtains a simple equation for the spacecraft potential during electron beam emission:

$$q_e\phi = kT_e\left(\frac{I_{beam}}{I_e(0)} - 1\right) \tag{10.13}$$

where $q_e\phi$ is positive and in units of electron volt. In equation (10.13), the potential ϕ is governed by the competition between I_{beam} and $I_e(0)$. Equation (10.13) requires that $I_{beam} > I_e(0)$ in order for $e\phi > 0$ (i.e., positive charging). In this approximation, when the electron beam current exceeds the ambient electron current, the spacecraft potential becomes positive and is controlled by the electron beam current. As the beam current increases, the spacecraft potential ϕ, which is positive, increases linearly (figure 10.3):

$$\Delta\phi = \left(\frac{kT_e}{q_e}\right)\frac{\Delta I_{beam}}{I_e(0)} \tag{10.14}$$

There is a maximum limit of the spacecraft potential ϕ induced by beam emission. This will be the topic of the next chapter.

10.4 Remarks

We have considered a simple system of electron beam emission from a spacecraft. It should be noted that we have assumed that the beam would leave the spacecraft completely.

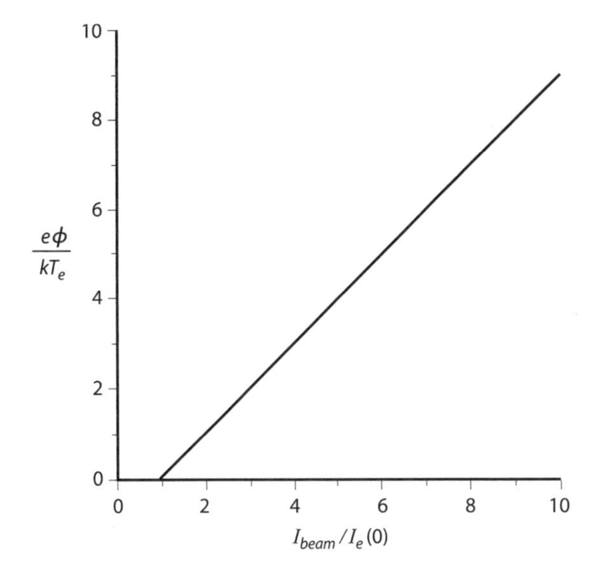

Figure 10.3 Positive spacecraft potential as a function of electron beam current [equation (10.10)].

Photoelectrons emitted from surfaces in sunlight have very low initial energy. The photoelectrons can leave the spacecraft if they are repelled and accelerated by negative spacecraft potentials. However, the photoelectrons can also be trapped by local potential wells, even if the overall charging voltage of the spacecraft is negative. For positive spacecraft charging, the photoelectrons would be totally or partially trapped, depending on the charging level. See chapter 7 for more details on spacecraft charging in sunlight.

The following remarks describe two mechanisms that can make beam emission from a spacecraft not too simple.

Suppose that a spacecraft has different surfaces electrically unconnected and that a large electron beam current is emitted from one of the surfaces. The surface from which the beam is emitted would charge to a high positive potential. If the potential is so high that it engulfs part of the neighboring surfaces, the secondary electrons and photoelectrons from the engulfed surfaces may be attracted by the high positive potential. This phenomenon is called *sheath engulfment*, which is described in appendix 6.

An electron beam can return partially, depending on the beam energy and the level of the positive potential of the spacecraft. For example, if one emits a 2 keV electron beam from a spacecraft that is charged to +50 V relative to the ambient plasma, the beam would leave completely, enabling equations (10.10) to (10.14) to be applicable. Depending on how large the current is, the increasing spacecraft potential may eventually approach the beam energy. When this phenomenon happens, one needs to consider partial beam return. With a beam returning to the spacecraft, the system becomes more complicated. More on this topic will be discussed in later chapters.

10.5 Exercises

1. With electron beam emission, the current balance equations in the positive and negative potential regimes are different. Are their current-voltage graphs continuously joint at zero potential?

2. Write down a current balance equation for ion beam emission in the (a) negative potential regime or (b) positive potential regime.

3. Write down a current-balance equation for simultaneous electron and ion beam emission in the (a) positive potential regime and (b) negative potential regime.

10.6 References

1. Mott-Smith, H. M., and I. Langmuir, "The theory of collectors in gaseous discharges," *Phys. Rev.* 28: 727–763 (1926).
2. Langmuir, I., *Collected Works of Irving Langmuir*, ed. C. G. Suits, Pergamon Press, New York (1960).
3. Sanders, N. L., and G. T.Inouye, "Secondary emission effects on spacecraft charging: energy distribution consideration," in *Spacecraft Charging Technology 1978*, eds. R. C. Finke and C. P. Pike, NASA-2071, ADA ADA-084626, pp. 747–755, U.S. Air Force Geophysics Laboratory, Hanscom AFB, MA (1978).
4. Sternglass, E. J., "Theory of secondary electron emission," Scientific Paper 1772, Westinghouse Research Laboratories, Pittsburgh, PA (1954).
5. Sternglass, E. J., "Backscattering of kilovolt electrons from solids," *Phys. Rev.* 95: 345–358 (1954).
6. Darlington, E. H., and V. E. Cosslett, "Backscattering of 0.5–10 keV electrons from solid targets," *J. Phys. D, Appl. Phys.* 5: 1961–1981 (1972).
7. Lai, S. T., "An improved Langmuir probe formula for modeling satellite interactions with near geostationary environment," *J. Geophys.Res.* 99: 459–468 (1994).
8. Lai, S. T., "An overview of electron and ion beam effects in charging and discharging of spacecraft," *IEEE Trans. Nuclear Sci.* 36, no. 6: 2027–2032 (1989).
9. Lai, S. T., and D. Della-Rose, "Spacecraft charging at geosynchronous altitudes: new evidence of existence of critical temperature," *J. Spacecraft and Rockets* 38, 922–928 (2001).
10. Reagan, J. B., R. E. Meyerott, R. W. Nightingale, P. C. Filbert, and W. L. Imhoff, "Spacecraft charging currents and their effects on space systems," *IEEE Trans. Electr. Insulations* 18: 354–365 (1983).

11

Supercharging

Beam emission from a spacecraft changes the charging level (voltage) of the spacecraft. As the beam current increases, the charging level increases accordingly. There exists a maximum charging level induced by beam emissions. The maximum level is given by the beam energy. No matter how large the beam current emitted from a spacecraft is, the spacecraft potential energy cannot exceed the beam energy. When the maximum spacecraft potential is reached, attempts to emit more current from the emitter will result in partial beam return to the spacecraft, while both the maximum spacecraft potential and the net current leaving a spacecraft remain unchanged. If the induced spacecraft potential exceeds that given by the beam energy, the phenomenon is called *supercharging*. Supercharging cannot exist unless there is an extra energy source unaccounted for. Although supercharging cannot exist theoretically according to physical laws, there are occasional reports on observations of supercharging.[1–3] One must be careful in providing physical interpretations to experimental results.

11.1 Charging Induced by Large Beam Current Emission

In the previous chapter, equation (10.13) states that the spacecraft potential increases with the beam current. The physics content of this statement is easy to understand, as follows. Beam emission throws out electrons (or ions) from a spacecraft. As a result, the spacecraft becomes deficient in electrons (or ions). Therefore, it develops a potential relative to the ambient plasma. Throwing out more and more electrons (or ions) increases the positive (or negative) potential. The situation is analogous to digging a hole in the ground. If one throws out more and more soil, the hole becomes deeper and deeper (figure 11.1).

From common experience, we have the rule: "You can dig as deep as you can throw." When the depth of the hole reaches its maximum, any attempt to throw more soil upward would result in soil return because of gravitational attraction. The maximum depth of the hole is limited by the strength of one's arms, because physics requires the sum of the kinetic and potential energies of a system to be constant. With this analogy, we have the rule: "You can charge a spacecraft to a potential as high as the beam energy only but not higher." In other words, a spacecraft cannot be charged by beam emission to a potential ϕ higher than that given by the beam energy E_{beam}. For example, the maximum potential $\phi_{\max}$ induced by electron emission is given by

$$q_e \phi_{\max} = E_{beam} \tag{11.1}$$

With this rule, equation (10.13) of the previous chapter should be modified as follows:

$$I_e(0)\left(1 + \frac{q_e \phi}{kT_e}\right)^{\alpha} = I_{beam} - I_r \Theta(q_e \phi - E_{beam}) \tag{11.2}$$

where I_r is the returning beam current, and $\Theta(x)$ is a step function that equals 1 if $x > 0$, and $\Theta(x) = 0$ if $x \leq 0$. When a spacecraft is charged to the maximum potential $\phi_{\max}$, any attempt to emit more beam current would result in beam return (figure 11.2). The net current I_{net} given

Figure 11.1 Digging a hole. You can dig as deep as you can throw.

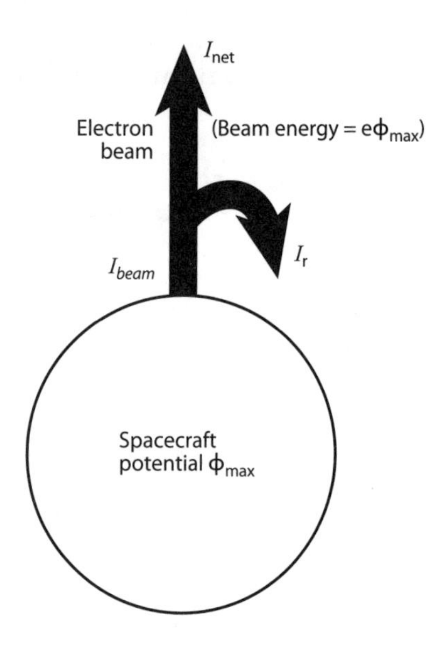

Figure 11.2 Emission of excessively large current from spacecraft. Part of the current returns.

by the right-hand side of equation (11.2) is the maximum current I_{max} leaving the spacecraft. Any attempt to emit more current beyond I_{max} would change neither the spacecraft potential ϕ nor the maximum current I_{max} leaving the spacecraft (figure 11.3). This is because the excess current beyond I_{max} returns to the spacecraft. That is, the return current I_r equals the excess current:

$$I_r = I_{beam} - I_{\max} \tag{11.3}$$

The net current I_{net} leaving the spacecraft is also unchanged:

$$I_{net} = I_{\max} \tag{11.4}$$

At the maximum potential ϕ_{max}, the current balance equation becomes

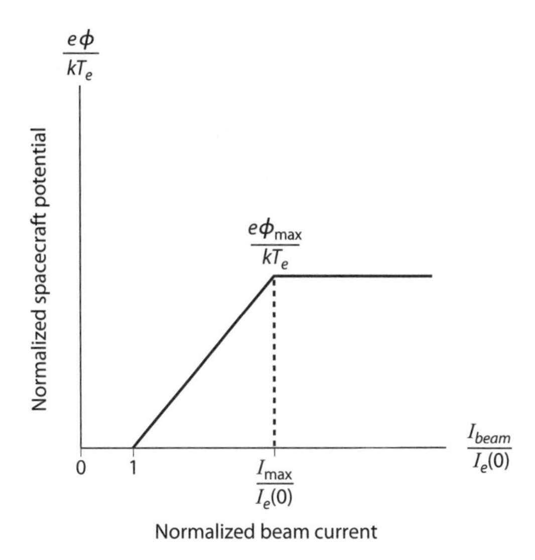

Figure 11.3 Current-voltage characteristics of large current beam emission from spacecraft. The measurement is given by equation (11.5). The maximum spacecraft potential ϕ_{max} is controlled by the beam energy.

$$I_e(0)\left(1 + \frac{e\phi_{max}}{kT_e}\right)^{\alpha} = I_{max} \tag{11.5}$$

A typical question might be: Why have the secondary and backscattered electron currents been neglected in the preceding equations? The preceding equations concern electrons returning to the spacecraft. If the spacecraft potential is negative, the electron beam emitted from the spacecraft would not return. This is because the beam electrons are repelled by the negative spacecraft potential. On the other hand, if the spacecraft potential is positive and beyond a few volts, the secondary electrons cannot leave because they, with their low energies, are attracted back to the spacecraft. Some backscattered electrons may be energetic enough to leave, but their current, depending on the backscattered electron coefficient, is usually much smaller than those of the secondary electrons and the ambient electrons. Therefore, it is reasonably justified to neglect both secondary and backscattered electrons in the preceding approximation.

11.2 Supercharging

If the induced spacecraft potential exceeds that given by the beam energy, the phenomenon is called *supercharging*. Physics requires energy to be conserved. Energy can be converted from one form to another, but the total energy must be constant. (A common example of energy conversion is colliding two masses together to get heat and sound, but the total energy is constant.) Supercharging cannot exist by physical law. However, one can observe supercharging if one misinterprets the observation. Einstein was quoted as saying: "A theory is something nobody believes, except the person who made it. An experiment is something everybody believes, except the person who made it."

11.3 Physical Interpretation of Experimental Results

In seeking nature's truth, a researcher needs to be bold to think out of the box, so to speak. Future experiments can uphold, refute, or enable improvement in the theory. In the spirit of research, let us consider possible sources of error in physical interpretation of the results of this type of experiment, supercharging. One possible source is energy. Another is surface charging, which will be studied in the next section.

In considering the total energy, one must know how to count all the sources. It is difficult to isolate a system so that the total energy considered is constant. For an example, figure 11.1 illustrates the principle "you can dig only as deep as you can throw." The soil that goes up must stop at some height and then falls down. What would happen if a strong wind blows the soil away? What if a second person catches the soil when it comes up? Or, what if the soil explodes? These are some possible scenarios of extra energy sources that one must remember to count. There are more scenarios. We will confine our scope here to electrostatics only and do not consider further complications.

11.4. Surface Charging of Booms

Long booms were often used for measuring spacecraft potentials (figure 11.4). The booms are electrically isolated from the spacecraft body. The potential difference $\Delta\phi$ between the tip of a boom and the spacecraft body gives an approximate value of the spacecraft potential ϕ_s:

$$\Delta\phi = \phi_s - \phi_{boom} \tag{11.6}$$

where ϕ_{boom} is the potential of the tip of a boom. Since the boom is long, one assumes that the tip of the boom is at the ambient plasma potential, which is zero. With this assumption, equation (11.6) is approximated as

$$\Delta\phi \approx \phi_s \tag{11.7}$$

This approximation breaks down if the tip of the boom charges to high voltage (figure 11.5).

Suppose that a large beam current is emitted, charging a spacecraft to the beam energy. Suppose that more current is emitted; the excess current must return. The beam expands

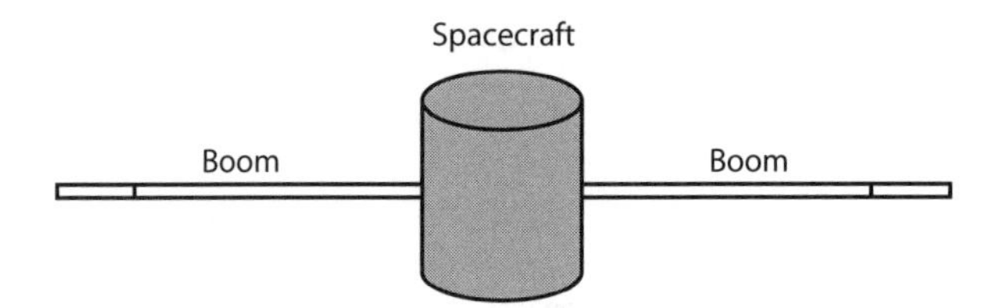

Figure 11.4 Satellite with long booms for measuring potential difference between the tip of a boom and the satellite body.

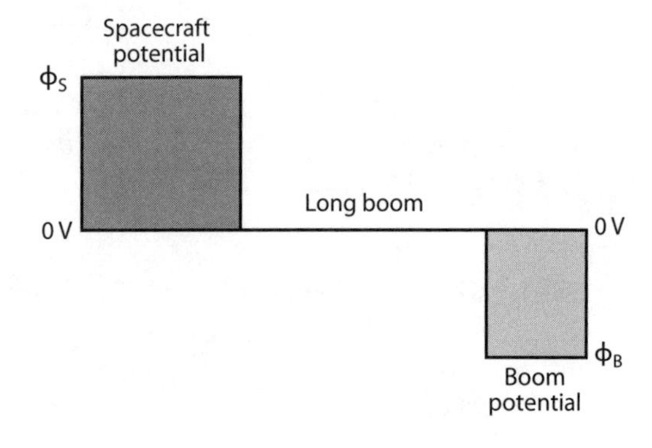

Figure 11.5 Potential difference between the tip of a boom and the satellite body. If the satellite charges to a positive potential while the boom tip charges to a negative potential, the difference exceeds the magnitude of the satellite potential. Measurement $\Delta\phi = \phi_B - \phi_S$.

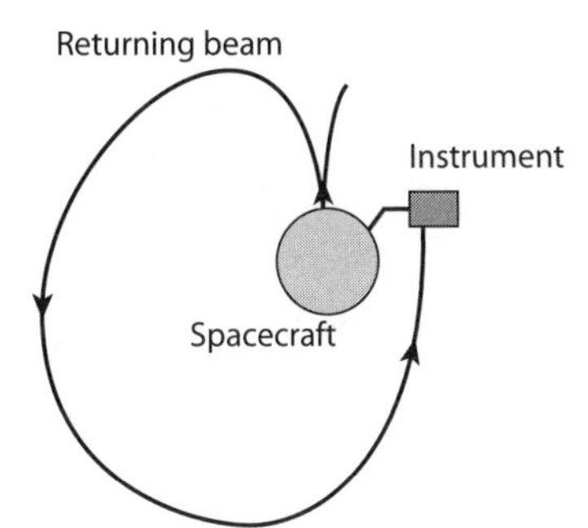

Figure 11.6 Returning electrons hitting instruments and booms.

because of the beam space charge. The returning beam electrons acquire large angular momentum and may go around the spacecraft a few times before returning to the spacecraft surface. In this situation, the spacecraft has acquired its own plasma environment in which the beam electrons are going around. These electrons may hit the tip of the boom and charge it to high potentials. As a corollary, any instrument in the vicinity of the spacecraft may be hit by such electrons going around the spacecraft and become charged by the electrons (figure 11.6).

As an illustration, suppose that a 6 keV electron beam is emitted, charging the spacecraft body to +4 kV and the booms to −4 kV. Equation (11.6) gives

$$\Delta\phi = 4 - (-4) = 8 \text{ kV} \tag{11.8}$$

If one uses equation (11.7) for the determination of the spacecraft potential, the difference $\Delta\phi$ (= 8 kV) in equation (11.8) would be interpreted as the approximate spacecraft potential ϕ_S. Such an interpretation is erroneous, because it has not taken the charging of the boom into account.

11.5 Summary

In summary, a spacecraft can charge only as high as the beam energy. In an isolated system, the total energy is constant. The beam energy is the initial kinetic energy plus the potential energy due to the beam space charge. Beam expansion converts the space charge potential energy to kinetic energy. Emitting excessive beam current beyond the point of maximum charging will not change the maximum charging potential but the excessive beam current will return. The returning electrons impacting on the instruments on or outside the spacecraft surface can charge the instruments. One must be careful[4] in providing physical interpretations to experimental results.

11.6 Exercises

1. Consider a cylindrical electron beam of 1 mm diameter, 1 keV energy, and 100 mA current. Use the Gauss law to calculate the transverse electric field.

2. An electron beam of energy E can charge a spacecraft to the maximum potential given by $e\phi_{max} = E$. Write down an expression of the beam current I_{max} required to charge the spacecraft to its maximum potential. You may ignore the ambient currents.

3. In problem 2, if the beam current I_{beam} exceeds $I_{max,}$ part of the beam has to return while the rest escapes. What is the amount of the return current? What is the amount of the escape current?

11.7 References

1. Maehlum, B. N., J. Troim, N. C. Maynard, W. F. Denig, M. Frederich, and K. M. Torkar, "Studies of the electrical charging of the tethered electron accelerator mother–daughter rocket Maimak," *Geophys. Res. Lett.* 15: 725–728 (1988).
2. Managadze, G. G., V. M. Balebanov, A. A. Burchudladze, T. I. Gagua, N. A. Leonov, S. B. Lyakhov, A. A. Martinson, A. D. Mayorov, W. K. Riedler, M. F. Frederich, K. M. Torkar, A. N. Laliashvili, Z. Klos, and Z. Zbyszynski, "Potential observations of an electron beam emitting rocket payload and other related plasma measurements," *Planet. Space Sci.* 36: 399–410 (1988).
3. Denig, W. F., N. C. Maynard, W. J. Burke, and B. N. Maehlum, "Electric field measurements during supercharging events on the Maimak rocket experiment," *J. Geophys. Res.* 96: 3601–3610 (1991).
4. Lai, S. T., "On supercharging: electrostatic aspects," *J. Geophys. Res.* 107, no. A4: doi: 101029/2000JA000333 (2002).

12

Ion Beam Emission from Spacecraft

Emitting positive ion beams from a spacecraft swings the spacecraft potential toward negative voltages. At equilibrium, the spacecraft potential is, of course, determined by the current balance of all currents. If the beam current exceeds all other currents, the beam current dominates. As in the case of electron beam emissions, if the beam ions are less energetic than the potential energy of the charged spacecraft, the beam ions return to the spacecraft. It is possible to emit a steady ion beam with a substantial current for controlling the spacecraft potential at a steady voltage. Since ions are slower than electrons, space charge effects show up more readily in ion beams than in electron beams. Chemical effects, such as charge exchange, can occur in ion beams and affect the spacecraft potential.

12.1 Active Control of Spacecraft Potential

Ions are considered positive in this chapter. Suppose that a spacecraft charges to a negative potential $\phi(<0)$ in a Maxwellian space plasma at geosynchronous altitudes in eclipse. The current balance between the net electron current coming in and the ambient ions attracted toward the spacecraft is given by the equation

$$I_i(0)\left(1-\frac{q_i\phi}{kT_i}\right)+I_e(0)(1-\langle\delta+\eta\rangle)\exp\left(-\frac{q_e\phi}{kT_e}\right)=0 \tag{12.1}$$

The preceding equation and its symbols are as explained in chapters 3 and 4. We remind ourselves that $q_i\phi$ is negative, $q_e\phi$ is positive, $I_i(\phi=0)\ (>0)$ is the ambient ion current without charging, $I_e(\phi=0)\ (<0)$ is the ambient electron current without charging, and $\langle\delta+\eta\rangle$ is the normalized Maxwellian average of the secondary and backscattered electrons (see chapter 4). For a Maxwellian distribution, $\langle\delta+\eta\rangle$ is independent of the spacecraft potential ϕ and depends on the electron temperature only (see chapter 4).

Now, suppose that an ion beam with a current I_b is emitted from the spacecraft. We assume that the ion beam is energetic enough to leave completely. We also assume that the beam current density is so small that space charge effects in the beam are negligible. The current balance equation in the Mott-Smith and Langmuir model (chapter 4) is of the form:

$$I_i(0)\left(1-\frac{q_i\phi}{kT_i}\right)-I_b+I_e(0)(1-\langle\delta+\eta\rangle)\exp\left(-\frac{q_e\phi}{kT_e}\right)=0 \tag{12.2}$$

The sum of the first two terms in equation (12.2) describes the net incoming ion current. The ambient ions are attracted by the negative spacecraft potential, while the ion beam is going out. The net incoming ion current is reduced by the ion beam current. The net incoming electron current is reduced by the negative potential induced by the beam emission. Mathematically, the exponential factor is reduced because the magnitude of $q_e\phi\ (>0)$ is enhanced.

For an increase ΔI_b in ion beam current, the change $\Delta\phi$ in spacecraft potential ϕ is obtained by differentiating equation (12.2):

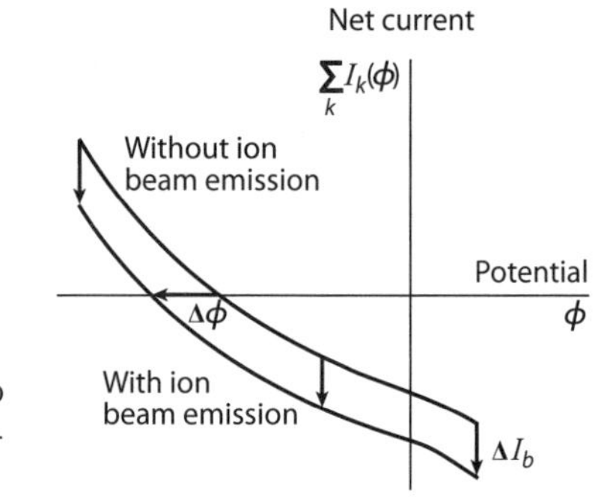

Figure 12.1 Shift in the IV curve as a result of ion beam emission. The curve shifts downward uniformly by the amount of the ion beam. The spacecraft potential (the zero crossing) shifts to the left by an amount $\Delta\phi$ given by equation (12.3).

$$\Delta\phi = \frac{\Delta I_b}{\left[\frac{q_e I_e(0)(1-\langle\delta+\eta\rangle)}{kT_e} - \frac{q_i I_i(0)}{kT_i}\right]} \tag{12.3}$$

Since $q_e < 0$, $q_i > 0$, $I_e(0) > 0$, $I_i(0) > 0$, and $1 - \langle\delta + \eta\rangle > 0$, the denominator of equation (12.3) is negative. The spacecraft potential change $\Delta\phi$ due to the increase ΔI_b in ion beam current is, of course, negative (figure 12.1).

With high ion beam current emission, taking away the ions from the spacecraft to the far ambient environment, the spacecraft potential (negative) must be high accordingly. When the spacecraft potential is very high and negative at a very high ion beam current, the ambient electrons are mostly repelled by the spacecraft potential. In the limiting case that the electron current approaches zero, the current balance is between the incoming ions attracted by the negative spacecraft potential and the high outgoing beam ions only. At equilibrium, we have the balance of ion currents:

$$I_i(0)\left(1 - \frac{q_i\phi}{kT_i}\right) = I_b \tag{12.4}$$

The resultant equilibrium potential ϕ (negative volts) of the spacecraft is of the form

$$-q_i\phi = kT_i\left(\frac{I_b - I_i(0)}{I_i(0)}\right) \tag{12.5}$$

We remind ourselves that in equations (14.4), $q_i > 0$, $\phi < 0$ and the term in parentheses is greater than unity. In order to satisfy equation (12.4), a necessary condition is that the beam current I_b must exceed the ambient ion current $I_i(0)$:

$$I_b > I_i(0) \tag{12.6}$$

If the condition (12.6) is not satisfied, neither is equation (12.4).

It is possible to emit a steady current of ions in order to clamp the spacecraft potential at a steady value. If the spacecraft potential is initially positive (due to photoemission in sunlight), it is possible to clamp the potential at near zero volts by using ion beam emission.[1] If the ion beam current is very large, the spacecraft potential can be clamped at a finite negative value.[2] Space experiments[1, 2] have successfully demonstrated these two ideas in the low plasma density (low ambient electron and ion currents) space plasma environment.

12.2 Return of Ion Beam

If the beam current increases indefinitely, can the spacecraft potential (negative) increase indefinitely? Apparently, equation (12.4) says yes. However, we have assumed in deriving equations (12.1) to (12.5) that the beam is energetic enough to leave completely. In reality, the beam energy E_b is finite. When the spacecraft potential energy $|q_i\phi|$ reaches the beam energy E_b, the beam can no longer leave. This poses an upper limit on the magnitude of the spacecraft potential induced by beam emission. This situation is similar to that of electron beam return, as discussed in chapter 11. Refer to figure 11.1 for an illustration of the rule: "You can dig only as deep as you can throw."*

Here is an aside: If an external hand catches the soil being thrown upward and takes it away, you can certainly dig deeper. The hand is an external energy source. However, we are considering only electrostatic aspects of beam emission from a spacecraft into the ambient space plasma without external systems. In the language of classical or quantum mechanics, the Hamiltonian is constant in our system.

It is difficult to manufacture a mono-energetic ion beam device. In practice, the beam ions are not mono-energetic; they have an energy distribution $F(E)$. Those ions more energetic than the spacecraft potential energy will leave, while the less energetic ones will return to the spacecraft. In this manner, the energy distribution is partitioned at the spacecraft charging energy and therefore the current balance equation (12.2) can be written as follows:

$$I_i(0)\left(1-\frac{q_i\phi}{kT_i}\right)-fI_b+(1-f)I_b+I_e(0)(1-\langle\delta+\eta\rangle)\exp\left(-\frac{q_e\phi}{kT_e}\right)=0 \qquad (12.7)$$

where f is the fraction of the ion beam current going out. That is, fI_b is the net ion beam current going out and $(1-f)I_b$ is the ion current returning to the spacecraft:

$$f=\frac{\int_{|q_i\phi|}^{\infty}dEF(E)}{\int_0^{\infty}dEF(E)} \qquad (12.8)$$

In equation (12.8), we have used the normalized energy distribution $F(E)$:

$$I_b=\int_0^{\infty}dEF(E) \qquad (12.9)$$

Suppose that the ion beam emission drives the spacecraft potential to so high a value (negative) that practically all ambient electrons cannot come in—that is, we have a case of ion current balance. For an ion beam with an energy distribution, the equation of ion current balance [equation (12.4)] is modified to the form

$$I_i(0)\left(1-\frac{q_i\phi}{kT_i}\right)=fI_b-(1-f)I_b \qquad (12.10)$$

where f is given by equation (12.8).

*If one emits an ion beam current that is so small that it is not dominating over with the incoming ambient ion current, the magnitude of the potential is below the upper limit. In this situation, the limiting depth has not been reached.

12.3 Lower Limit of the Reduced Potential

Last, we remark that if a spacecraft is initially charged to a negative potential ϕ, emission of high-energy ion beams of energy E_b ($> q_i\phi$) will not reduce the magnitude of the potential ϕ. This is because the high-energy ion beam leaves, driving the potential to even higher negative values. However, emission of low-energy ion beams ($E_b < q_i\phi$) can reduce the magnitude of the potential, because the ions return, impact the surfaces, and generate secondary electrons, which will leave (see chapter 6). Although this "beam return" mechanism[3] can reduce the magnitude of the spacecraft potential (negative volts), it cannot reduce the magnitude below E_b. This restriction can be relaxed to some extent if (1) charge exchange, electron impact ionization, or some other chemical reactions, also occur, or (2) plasmas (electrons and ions) are emitted instead of purely ions (see chapter 6).

12.4 Space Charge Effect

If a beam has high charge density of one sign predominantly, the mutual repulsion between the charges slows down the beam velocity and limits the current flow. This is analogous to a traffic jam on a highway.

Child[4] derived the classic law of current density limitation of a space charge flow in one dimension. The governing equations are as follows:

$$\frac{d^2\phi(x)}{dx^2} = -\frac{\rho(x)}{\varepsilon_0} \tag{12.11}$$

$$J = q\text{v}(x)\rho(x) \tag{12.12}$$

$$\frac{1}{2}M\text{v}_0^2 = \frac{1}{2}M\text{v}^2(x) + q\phi(x) \tag{12.13}$$

where the notations are standard, viz, $\phi(x)$ is the electrostatic potential at location x, ρ the charge density, J the current density, v the charge velocity, q the elementary charge, and M the mass of the charged particle. Equations (12.11) to (12.13) are the Gauss law, current continuity, and conservation of energy. The boundary conditions used are (1) zero charge velocity initially, and (2) known potential ϕ_d (or energy) at the target electrode located at a given distance $x = d$. The results (chapter 2) are a space charging limiting current density J and a monotonic behavior of the potential $\phi(x)$ as a function of the distance x. The Child result is applicable to, for example, a diode in which electrons are generated with zero initial velocity at a cold cathode while the anode is at a given applied potential.

The classic paper[5] of Fay, Samuel, and Shockley considered the same equations (12.11) to (12.13) for a beamlike charge flow in a diode, in which the charged particles come out from the exit point at a finite velocity governed by a given potential $\phi(0)$. They found the existence of a potential extremum $\phi(x_m)$ located somewhere between the cathode and the anode (figure 12.2).

To solve the set of equations (12.11) to (12.13), one uses the boundary conditions as follows:

$$\phi(0) = \phi_0 \tag{12.14}$$

$$\phi(d) = \phi_d \tag{12.15}$$

$$\phi_{\text{I}}(x_m) = \phi_{\text{II}}(x_m) \tag{12.16}$$

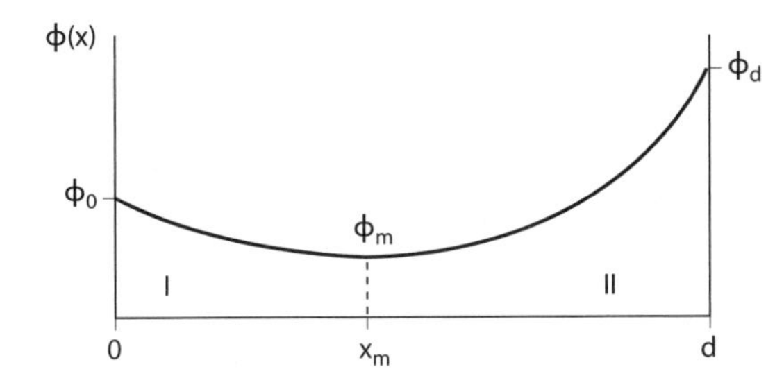

Figure 12.2 Potential profile in a space charge flow. The initial velocity of the charged particles is finite. The potentials at the end points are given. A potential extremum ϕ_m develops at $0 < x < d$.

$$\frac{d}{dx}\phi_{\mathrm{I}}(x_m) = \frac{d}{dx}\phi_{\mathrm{II}}(x_m) \tag{12.17}$$

The solving of equations (12.11) to (12.13) using the boundary conditions equations (12.14) to (12.17) will be left as an exercise for those graduate students who are interested. One needs to integrate equation (12.11) twice and determine the integration constants satisfying the boundary conditions. Some algebraic details can be found in references 4–8. An important feature of the results is discussed as follows.

As the applied current density J increases from low values, a potential extremum $\phi(x_m)$ shows up gradually. At a critical current density J^*, the value of $\phi(x_m)$ equals ϕ_0. If the beam is mono-energetic, the charged particle velocity slows down to a halt at x_m. Some will decide to go forward and some backward. If the beam has an energy distribution, those particles energetic enough will go forward and those less energetic will return. In this way, the space charge flow develops into a two-way flow. Only a fraction of the beam current will be transmitted through the virtual electrode at x_m.

To summarize, there is a nonmonotonic behavior in the net beam current emitted as a function of current density. At low values of current density, the net current emitted increases with the current density. At the critical current density, a two-way flow commences while the net current transmitted drops suddenly to a lower value (figure 12.3).

For space applications, it has been proposed that virtual electrodes[9] can occur in high current emissions of electrons or ions. Since the net current emitted is related to the spacecraft potential, one observes that the spacecraft potential behaving in a nonmonotonic manner as the applied beam current increases (figure 12.3). Indeed, this nonmonotonic behavior was observed[3] in the ion and electron beam emission experiments on the SCATHA satellite. Numerical simulations[8] on large computers have also demonstrated the formation of virtual

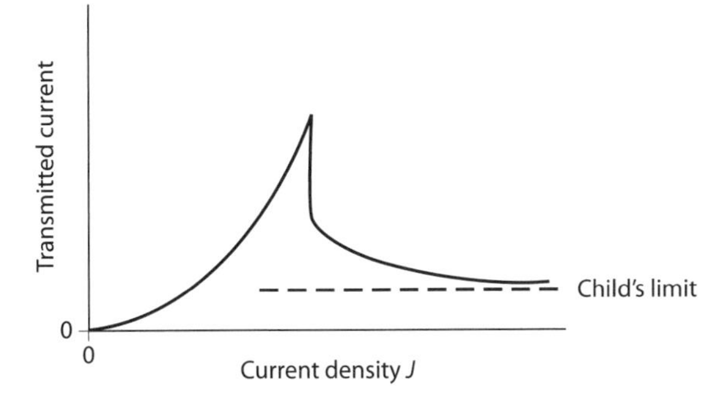

Figure 12.3 Nonmonotonic behavior of the potential profile as a function of the applied current density J. A critical current density J^* exists at which the potential extremum $\phi_m = \phi_o$. Above J^*, the space charge flow is limited.

electrodes in charged particle beams. The location x_m of the virtual electrode is very near the exit point of the beam.

Last, we remark that beam divergence reduces the space charge build up in the beam. For a narrow beam, the space charge divergence of the beam dilutes the space charge density and therefore tends to eliminate the virtual electrode formation. With beam divergence, the cross section of the beam is not uniform. For a beam broader than a critical diameter b^*, the region in the interior part of the beam near the exit point would behave like a one-dimensional beam. There, the formation of virtual electrode can occur. To our knowledge, no analytical or computational result on the functional dependence of the potential and location of the virtual electrode on the critical diameter b^*, beam energy, beam particle mass, ambient plasma temperature and density, and spacecraft potential is available at the time of writing. We conjecture that the critical diameter is of the order of $2x_m$, depending on various beam and ambient parameters. We suggest this problem as a thesis topic for graduate students.

12.5 Charge Exchange in Charged Particle Beams

In today's technology, ion beam devices are unable to generate beams of 100% ions. The ratio of the number of ions and neutral atoms coming out of the exit point of an ion beam device is less than 10% typically. For the ion beam[10] on SCATHA, the ratio was about 6%.

In an ion beam device, the ions are extracted from a plasma generated in a discharge chamber (figure 12.4). An opening with a negatively charged grid extracts positive ions through the grid. The ions extracted are further accelerated through an acceleration grid and a controlling grid. The accelerated ions coming out form an ion beam. The electrons in the plasma are guarded by the grids against any escape. However, the plasma in the discharge chamber is not fully ionized. There are neutral atoms or molecules mixed with the plasma. More importantly, the neutrals can wander out through the grids without any notice by the guards. The neutrals are at thermal velocities and mixed with the ion beam in the vicinity of the exit point.

Since the beam consists of fast ions and slow neutrals, charge exchange[11, 12] between the ions and the neutrals can occur. To our knowledge, reference 13 was probably the first ever to realize that charge exchange can occur in ion beams emitted from spacecraft and that this has important consequence to spacecraft charging. Let us take a xenon ion beam, for example, because xenon has been commonly used in space experiments.[3, 10, 14] The chemical equation of the charge exchange is as follows:

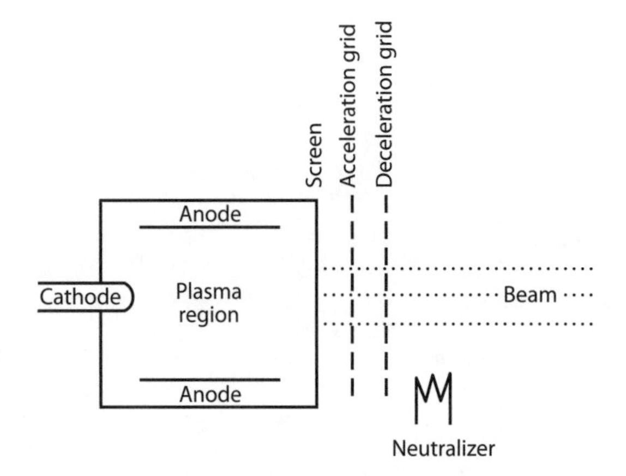

Figure 12.4 An ion beam device. Positive ions are extracted from a plasma chamber through negatively charged grids. An acceleration grid controls the fine-tuning of the beam energy. A neutralizer can be turned on for controlling the level of beam neutralization. The neutrals in the plasma chamber can wander out at will.

$$\mathrm{Xe}_f^+ + \mathrm{Xe}_s \rightarrow \mathrm{Xe}_f + \mathrm{Xe}_s^+ \tag{12.18}$$

where the subscripts f and s denote fast and slow, respectively. As a result of charge exchange, the fast beam ions have become fast neutrals, while the slow neutrals have become slow ions. Typically, the fast beam ions are keV in energy, while the slow ions are well below 1 eV in energy because the slow ions are generated from the neutral atoms/molecules at thermal energies. The latest measurement of the xenon charge exchange (equation 12.18) cross section is given in reference 15.

If a spacecraft charges to a negative potential ϕ, those slow ions (equation 12.18) that are generated by charge exchange and less energetic than $q_e\phi$ will return to the spacecraft (figure 12.5). As a result, the net ion current emitted is reduced. Thus, taking the charge exchange effect into account, the current balance equation equation (12.2) becomes

$$I_i(0)\left(1 - \frac{q_i\phi}{kT_i}\right) - I_{b,f} + I_{b,s} = I_e(0)(1 - \langle\delta + \eta\rangle)\exp\left(-\frac{q_e\phi}{kT_e}\right) \tag{12.19}$$

where $I_{b,f}$ is the current of fast ions going out, while $I_{b,s}$ is the current of the returning slow ions generated by charge exchange. The neutrals, of course, contribute nothing directly to spacecraft charge.

Last, we remark that in computer simulations for uncovering further details of charge exchange effects in spacecraft charging, one should take into account the dynamics of the returning ions. The returning ions tend to interact with the fast beam ions in a two-stream instability, changing the beam energy distribution. Also, the returning ions will increase the space charge density, in analogy to a traffic jam. Slowing down the beam ions favors more change exchange, depending on of the energy dependence of the charge exchange cross section. The newly born slow ions will carry on a similar cycle and so on. This cyclic process will lead to more and more charge exchange, which can be halted eventually by some loss mechanisms. In this model, the resultant spacecraft potential can be calculated. We suggest this topic as a thesis for graduate students.

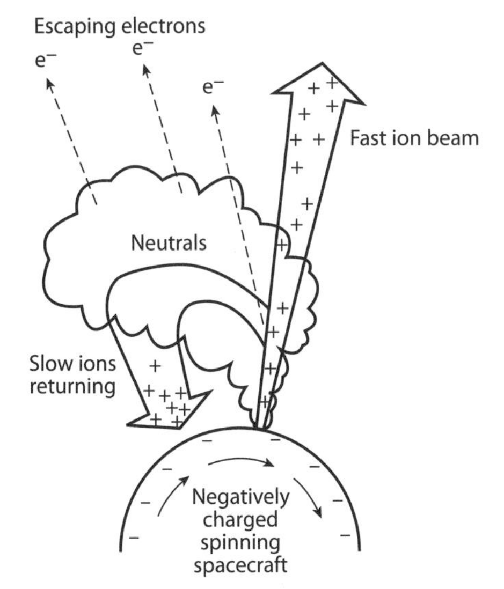

Figure 12.5 Return of energy ions. Charge exchange generates low-energy ions, which return to the spacecraft and reduce the spacecraft (negative) voltage.

12.6 Chemical Reactions in Ion Beams

There are other chemical reactions[11,16,17] in addition to charge exchange. For example, ion-neutral reaction between the fast xenon beam ions and the xenon beam neutrals is as follows:

$$Xe_s + Xe_f^+ \rightarrow Xe_s^+ + Xe_f^+ + e^- \tag{12.20}$$

which generates slow ions and electrons. The latter are expelled by the negatively charged spacecraft. Reaction (12.20) includes ionization, whereas reaction (12.18) does not. The cross section of reaction (12.20) is given in reference 17.

The three-body reaction [equation (12.21)] has a much lower cross section than reaction (12.20). Thus, it has very low probability to occur, unless the beam is very dense:

$$Xe^+ + 2Xe \rightarrow Xe_2^+ + Xe \tag{12.21}$$

The ion molecule reaction of equation (12.21) can be followed by molecular ion recombination:

$$Xe_2^+ + e^- \rightarrow 2Xe \tag{12.22}$$

which is fast.[11] For spacecraft charging consideration, reactions (12.21) and (12.22) tend to reduce the ion beam current.

Interactions between beam ions and beam neutrals with atmospheric species are insignificant at geosynchronous altitudes because the atmosphere is thin there. In the lower ionosphere,[15] however, ion beam interactions with the atmospheric species are more significant. More work needs to be done in this area. The level of natural charging of spacecraft is very low in the ionosphere. However, the effects of beam-atmosphere interactions during artificial charging of spacecraft in the ionosphere need systematic and careful experimental studies for better understanding of the physics and chemistry involved.

12.7 Ion Beam in Sunlight

Two effects of ion beam emission in sunlight will be discussed in this section. The first concerns electron impact ionization. The second concerns unblocking of the photoelectrons generated from the sunlit surface of a spacecraft. Both effects can influence the spacecraft potential.[3,13]

12.7.1 Electron Impact Ionization

The preceding sections pointed out that ion beams have an extra player that should not be overlooked. That player is the beam neutrals. Electron impact ionization of the neutrals generates a new ion and adds an electron.

$$Xe + e^- \rightarrow Xe^+ + 2e^- \tag{12.23}$$

Cross sections of the electron impact ionization of xenon and other inert gases are given in the literature.[18,19] Typically, the cross section of electron impact ionization of most elements has a threshold at about ten to few tens of eV and a peak at about one hundred to a few hundreds of eV.

If a spacecraft charges to a few hundreds of negative volts, the photoelectrons generated on the sunlit surface of a spacecraft are accelerated outward to a few hundred eV. So are the secondary and backscattered electrons generated from the surface, but they are less abundant than photoelectrons. The ambient electron current at geosynchronous altitudes is also much smaller than that of the photoelectrons. If the photoelectrons accelerated to hundreds of eV in energy travel through the neutral gas cloud of the ion beam, electron impact ionization can occur. A newborn ion has nearly the same thermal energy (less than 1 eV) as the neutral particle from which it was born. As a result, the newborn ions constitute an ion current returning to the spacecraft, affecting the current balance, and altering the spacecraft potential.

12.7.2 Unblocking the Photoelectrons

A spacecraft covered with conducting surfaces charges to typically a few volts positive in sunlight, because the outgoing photoelectron current exceeds the incoming ambient electron current. However, a spacecraft covered with nonconducting surfaces charges to a dipole-monopole[20] potential distribution in sunlight. This is because the sunlit surface, emitting photoelectrons, charges to a few positive volts while the surfaces on the dark side charge to high negative volts due to the impact of the energetic ambient electrons. If the spacecraft is spinning in sunlight, monopole-multipole potential configurations[21–25] can also form.

In a monopole-dipole, or monopole-multipole, potential distribution, the potential contours of the highly charged dark side can wrap around to the sunlit side, thus forming a potential barrier on the sunlit side. Although photoelectrons are very abundant, they are of low energy (electron temperature $\approx$ 1.5 eV). A barrier of even a few negative volts can block the photoelectrons. Since most of the photoelectrons cannot leave, the magnitude of the average (negative) potential of the spacecraft becomes higher in magnitude than that without blocking.

When an ion beam is facing the sun, the beam passes through a photoelectron cloud. There, the positively charged ions can pull some of the electrons along and carry them through the barrier. This is a general physics description. More advanced students may want to formulate the problem as traveling ion waves propagating through the electron cloud, thus creating two-stream plasma oscillations that carry the electrons up the potential ladder in an escalator manner. In the presence of an ambient magnetic field, a modified two-stream instability may occur. At any rate, the escape of the photoelectrons will affect the current balance and therefore lower the charging level (negative) of the spacecraft during the ion beam emission (figure 12.6). In this model, the final spacecraft potential can be calculated.

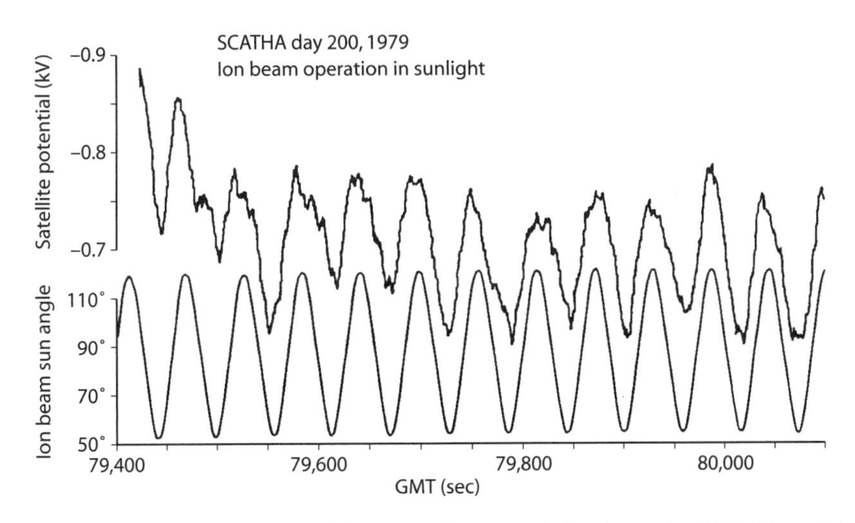

Figure 12.6 Spacecraft potential during ion beam emission from the SCATHA satellite. The potential and the ion beam Sun angle are in phase. (Adapted from reference 13.)

12.8 Exercises

1. Plot a current-voltage curve of spacecraft charging schematically. You may assume that the charging voltage is negative.
2. Modify the curve if an electron beam current I_e is emitted and if an ion beam current I_b is emitted.
3. Modify the curve further if the ion beam returns to the spacecraft at a certain spacecraft potential.
4. Using equation (12.10), express the equilibrium spacecraft potential ϕ in terms of all other variables, including the fraction f of the ion beam returned. Modify the result if the emitted ion beam current I_b is much larger than the ambient ion beam current $I_i(0)$. Modify the result further if the ion beam returns completely.
5. Conduct a computer simulation of ion beam emission from a spacecraft charged a priori to a negative voltage. What would happen if the returning ions enhance the beam space charge?

12.9 References

1. R. Schmidt, H. Arends, W. Riedler, K. Torkar, F. Rüdenauer, M. Fehringer, B. Maehlum, and B. Narheim, "Energetic ion emission for active spacecraft control," *Adv. Space Res.* 12, no. 12: 61–64 (1992).
2. Torkar, K., A. Fazakerley, and W. Steiger, "Active spacecraft potential control: results from the Double Star project," *IEEE Trans. Plasma Sci.* 34, no. 5: 2046–2052 (2006).
3. Lai, S. T., "An overview of electron and ion beam effects in charging and discharging of spacecraft," *IEEE Trans. Nucl. Sci.* 36, no. 6: 2027–2032 (1989).
4. Child, C. D., "Discharge from hot CaO," *Phys. Rev.* 32: 492–511 (1911).
5. Fay, C. E., A. L. Samuel, and W. Shockley, "On the theory of space charge between parallel electrodes," *Bell Sys. Tech. J.* 17, no. 49: 49–79 (1938).
6. Salzberg, B., and A. Haeff, "Effects of space charge in the grid-anode region of vacuum tubes," *RCA Rev.* 2, no. 4: 336–374 (1938).
7. Kirstein, P. T., G. S. Kino, and W. E. Waters, *Space-Charge Flow*, McGraw-Hill, New York (1967).
8. Wang, J., and S. T. Lai, "Numerical simulations on virtual anodes in ion beam emissions in space," *J. Spacecraft and Rockets* 34, no. 6: 829–836 (1997).
9. Lai, S. T., W. J. Burke, and H. A. Cohen, "A theoretical investigation of virtual electrode formation near the SCATHA satellite," *EOS*, 60, no. 46: 923, abstract (1979).
10. Masek, T., *Rocket Model Satellite Positive Ion Beam System*, AFGL-TR-78-0179, ADA063253, U.S. Air Force Geophysics Laboratory, Hanscom AFB, MA (1978).
11. Hasted, J. B., "Charge transfer and collisional detachment," in *Atomic and Molecular Processes*, ed. D. R. Bates, pp. 696–720 Academic Press, New York (1962).
12. Bond, J. W., K. M. Watson, and J. A. Welch, *Atomic Theory of Gas Dynamics*, Addison Wesley, Reading, MA (1965).
13. Lai, S. T., W. J. McNeil, and T. L. Aggson, "Spacecraft charging during ion beam emission in sunlight," Paper AIAA-90-0636, presented at AIAA Aerospace Science Meeting, Reno, NV (1990).
14. Lai, S. T., E. Murad, C. P. Pike, W. J. McNeil, and A. Setayesh, "A feasibility study on the xenon and carbon dioxide gas release experiments on the ARGOS satellite," *Adv. Space Res.* 13, no. 10: 81–89 (1993).

15. Miller, J. S., S. H. Pullins, D. J. Levandier, Y. H. Chiu, and R. A. Dressler, "Xenon charge exchange cross sections for electrostatic thruster models, *J. A Phys.* 91, no. 3: 984–991 (2002).
16. Kikiani, B. I., G. M. Mirianashvili, Z. E. Salia, and I. G. Bagdasarova, "Ionization in collisions between ions and atoms of noble gases in the energy range 200–4000 eV," in *Electronic and Atomic Collisions*, eds. J. S. Risley and R. Geballe, p. 613, University of Washington Press, Seattle (1975).
17. Kikiani, B. I., G. M. Mirianashvili, Z. E. Saliia, and I. G. Bagdasarova, "Ionization in collisions of ions and atoms of inert gases in the 200–4000 eV energy range," *Z. Tekhnich. Fiziki* 46: 1077–1081 (1976).
18. Cohen, H. A., C. Sherman, and E. G. Mullen, "Spacecraft charging due to positive ion emission: an experimental study," *Geophys. Res. Lett.* 6, no. 6: 515–518 (1979).
19. Ton-That, D., and M. R. Flannery, "Cross sections for ionization of rare gas atoms," *Phys. Rev.* A15: 517–526 (1977).
20. Stephan, K., and T. D. Mark, "Absolute and partial electron impact ionization cross sections of Xe from threshold up to 180 eV," *J. Chem. Phys.* 81: 3116–3117 (1984).
21. Besse, A. L., and A. G. Rubin, "A simple analysis of spacecraft charging involving blacked photoelectron currents," *J. Geophys. Res.* 85, no. A5: 2324–2328 (1980).
22. Tautz, M., and S. T. Lai, "Analytic models for a rapidly spinning spherical satellite charging in sunlight," *J. Geophys. Res.* 110, no. A07: 220–229 (2005).
23. Lai, S. T., and M. Tautz, "Aspects of spacecraft charging in sunlight," *IEEE Trans. Plasma Sci.* 34, no. 5, 2053–2061 (2006).
24. Tautz, M., and S. T. Lai, "Analytic models for a spherical satellite charging in sunlight at any spin," *Ann. Geophys.* 24: 2599–2610 (2006).
25. Tautz, M., and S. T. Lai, "Charging of fast spinning spheroidal satellites in sunlight," *J. Appl. Phys.* 102: 024905-1-10 (2007).

13

Discharges on Spacecraft

13.1 Introduction

The word *discharge* is used in the spacecraft charging literature to describe two different but related phenomena: (1) Discharging a spacecraft can mean mitigating a charged spacecraft. For example, one says "It is effective to use plasma emission for discharging a spacecraft that is charged a priori to a high voltage." (2) Discharge can mean a sudden flow of charges between two surfaces which are charged a priori to different potentials. This chapter concerns the latter. Electrostatic discharge is commonly abbreviated ESD. If the discharge current flow is both small and of short duration, there is generally no important consequence. If the current can sustain or even increase for some time, the consequence can be significant or harmful. If a discharge is visible, it is commonly called an *arc*. If an arc sustains, it is a *sustained arc*. Mitigation has been discussed in an earlier chapter. This chapter discusses harmful discharges on spacecraft.

13.2 Location of Discharges on Spacecraft

If discharges occur, they likely begin at locations of high electric fields. Between two surfaces of different potentials, the electric field $E = -\Delta\phi/d$, where $\Delta\phi$ is the voltage difference and d the distance between the surfaces. Therefore, large potential difference $\Delta\phi$ and short distance d give high electric fields E.

The electric field E of a sphere is proportional to r^{-2}, where r is the radius. Electric fields are high at sharp points, where the radius of curvature is small. As an example, lightning rods on buildings tend to catch lightning strikes.

In short, spacecraft discharges are most likely to occur at sharp points, sharp corners, sharp edges, short distances between neighboring solar cells, the connecting point between a long solar panel and a spacecraft body, the junctions of cables, the hinges of long tethers, tin whiskers protruding from solder sites, and short separations between differentially charged surfaces with large potential differences. The separation between neighboring solar cells is usually about 1 mm only (figure 13.1).

Indeed, spacecraft discharge data obtained in space and in the laboratory indicate that discharges are likely to begin at the locations listed earlier. In addition, there are some configurations whose importance has received much attention in recent years.[1,2] We discuss them as follows:

Triple junction. This is located between a metal surface, an insulating layer covering the metal surface partially, and the ambient plasma (figure 13.2). If surface charging occurs in the ambient plasma, different surface materials charge to different potentials depending on the surface properties. If the metal and the insulator charge to different potentials, there is a possibility of discharge between them. The plasma also plays a role initiating the electron and ion flows for the discharge. One might ask whether both layers can be different metals in a triple junction. Yes, they can, but they need to be insulated from each other in order to have different voltages. However, the insulating material sandwiched between them would form a new triple junction. One might also

Figure 13.1 Damage by a discharge between adjacent solar cells of different voltages. (NASA photo. Courtesy of Dale Ferguson.)

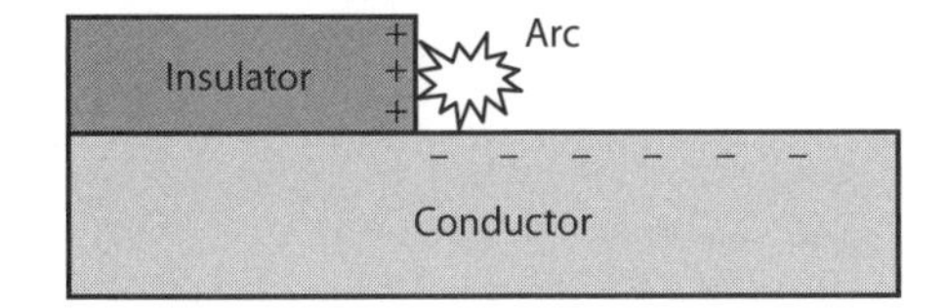

Figure 13.2 A triple junction. This is a favorite location for spacecraft discharges, or arcs, to occur. It is a junction between a metal surface, a dielectric layer partially covering the metal surface, and the ambient plasma.

ask whether both layers can be insulators of different properties. Yes, they can, and together with the ambient plasma, they form a triple junction.

Metal to dielectric. In the initiation of a discharge, it is more likely for electrons to travel from a metal surface to a dielectric one than the reverse. This is because it is difficult to extract electron current from dielectrics, which have low conductivity. Also, it is generally more difficult to take ions, compared with electrons, out of a surface.

Inverted gradient. (IVG, also called IVGD). It follows from the property discussed earlier that it is easier to initiate a discharge in an IVG (figure 13.2), where the metal is at a negative potential and the insulator positive than in the opposite gradient. This is because a metal surface with a negative potential implies an excess of electrons on the surface. It is easier to extract an electron current from the negatively charged metal to the positive charged dielectric than in the opposite direction. In the opposite gradient, which is called normal gradient (NVG), the metal is at a positive potential while the insulator is negative. It is unlikely for a discharge to initiate in an NVG.

13.3 Surface Discharge Scaling Law

The current I for discharging a capacitance C with a potential difference $\Delta\phi$ is given by

$$\tau I = C\Delta\phi \tag{13.1}$$

where τ is the duration of the discharge. From equation (13.1), one can deduce some simple scaling results.

Balmain[3] and Balmain and Dubois[4] reported observation of discharge pulses propagating at a speed of the order of 300 km/s for a number of surface materials. There is need for deeper understanding of the physical mechanism involved, how a discharge pulse propagates, and why its speed is so slow. With the observed speed, one can estimate the duration τ for a discharge pulse sweeping over a given surface area. For a given duration τ, the discharge current I scales linearly with the capacitance C or with the potential difference $\Delta\phi$ [equation (13.1)]. For a flat surface, the capacitance C is proportional to the area A divided by the thickness d. For various surface areas of the same thickness, the total charge Q of a discharge scales as the surface area A (figure 13.3):

$$Q = \int_0^{\tau} I(t)\,dt \propto A \tag{13.2}$$

The scaling law is useful for discharge estimates. It is not known how generally the scaling law can be applied. For example, surface inhomogeneity affects its applicability. In practice, the exact area swept by a discharge, the material depth involved in the discharge, and the plasma interactions may also be uncertain.

Surface capacitance is usually small compared with that of a thin dielectric layer. The latter is proportional to d^{-1}, where d is the thickness of the layer. Since thin layers have large capacitance C, it is not surprising that spacecraft discharges occurring on thin thermal blankets are often harmful.[5] As another example of large capacitances, when two large spacecrafts of very different potentials are approaching each other and if an astronaut is touching both, the discharge current through the astronaut may be large.

13.4 Differential Charging

As mentioned earlier in this chapter, differential charging is an important cause of spacecraft discharging. Surface materials with different properties of secondary emission, backscattered emission, and photoemission charge to different potentials in space. For example,

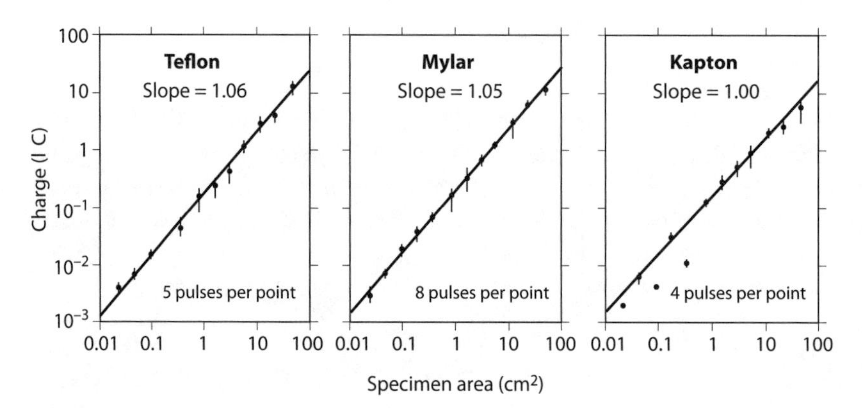

Figure 13.3 The variation of the released charge $Q = \int_0^{\tau} I\,dt$ with specimen area A. (From reference 4.)

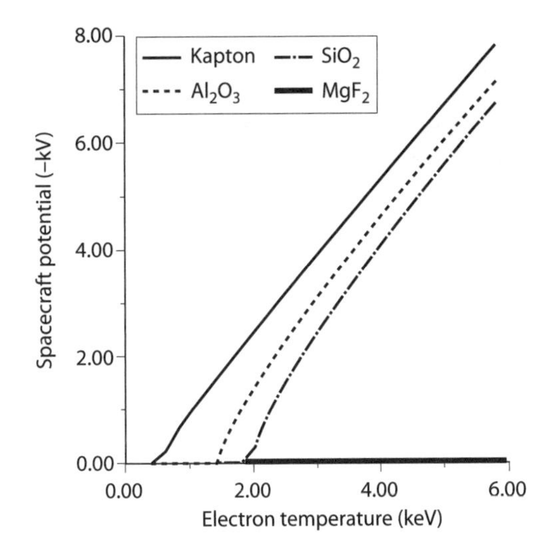

Figure 13.4 Potential of surface materials at geosynchronous altitudes as calculated by using a simple orbit-limited model.

magnesium fluoride (MgF_2), because of its large secondary emission coefficient $\delta(E)$, is often used on top of solar cells, whereas kapton, which has much lower $\delta(E)$ coefficient, is a common spacecraft surface material. As an example, a simple calculation, using Mott-Smith and Langmuir's orbit limited model with $\delta(E)$ given in reference 6, shows (figure 13.4) that kapton and MgF_2 charge to −7 kV and zero V, respectively, in typical geosynchronous altitudes in eclipse. This example illustrates a worst-case scenario that differential charging of 7 kV is possible, although it does not occur every day. Note that this example is for illustrating differential charging and, for its purpose, needs not use the latest values of $\delta(E)$ measurements, exact geometry of spacecraft, and detailed measurements of space plasma.

In sunlight, spacecraft with all conducting surfaces charge to a few positive volts only and rarely to negative potentials, but spacecraft with nonconducting or mixed surfaces can charge differentially. A spacecraft in sunlight always has a shadowed side. The shadowed area is at least as large as the sunlit area. The shadowed surfaces can charge to high negative potentials if the space plasma temperature is high. High-voltage differential charging can occur between the shadowed areas and the sunlit areas.[7]

We remark that modern space mirrors are highly efficient in reflecting light. They are made of metals such as aluminum. Their reflectivity in the solar UV region is well over 80% even with protective coating. It has been conjectured[8] that such highly reflecting mirrors may emit little or no photoemission, because most of the photon energy is reflected and not transferred into the mirror material. Without photoemission, such mirrors would behave like in eclipse, charging to high negative potentials if the space plasma is hot. As a result, differential charging between mirrors and nearby surfaces may occur after eclipse exit in the morning sector.

Differential charging depends not only on the potential difference but also on the separation between the positions of the two different potentials. It is often difficult to pinpoint the cause of a spacecraft anomaly because there can be multiple possible causes.[9] For example, mirrors flanking solar cells may greatly enhance sunlight on the cells and, as a result, cause temperature effects.

13.5 "Brush Fire" Discharge

Figure 13.5 shows a scenario of "brush fire" discharge. In the figure, there is a potential gradient (in the x direction) on a surface, which can be a dielectric surface or pieces of disconnected metal and dielectric surfaces. A small number of electrons from a negative voltage (−V)

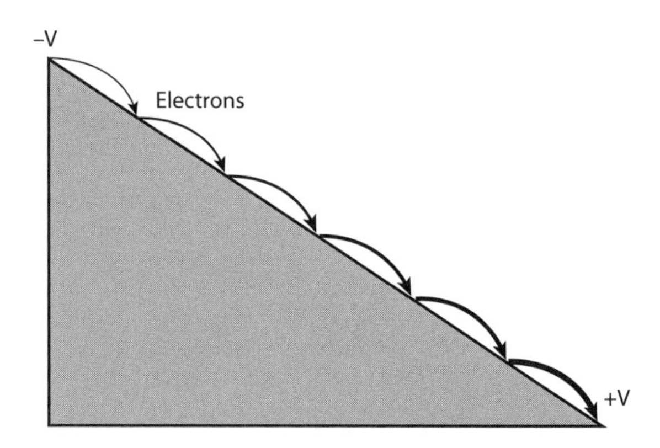

Figure 13.5 A scenario of the propagation of a discharge by hopping electrons generating more and more secondary electrons.

location start to move toward a positive voltage (+V) direction. As the electrons hop along the potential gradient, they gain energy. Depending on the secondary electron emission coefficient of the surface material, the electron energy may become sufficient to generate multiple secondary electrons. The primary electron is absorbed or backscattered by the surface. The newborn electrons continue to hop toward the +V direction, thereby gaining energy and continuing the process of generating secondary electrons. If the secondary electron coefficient $\delta(E)$ exceeds unity, the number of electrons generated continues to increase. As a result of the discharge, the voltage difference decreases. Depending on the capacitances involved, the voltage difference decreases and the decrease is usually fast. In summary, the brush fire discharge is one in which the electrons multiply as they hop along a potential gradient on a surface (figure 13.5).

This type of discharge can occur on a surface in vacuum and needs not be a gas discharge. In a plasma or a neutral gas environment, interactions with the plasma and collisions with the neutral gas modify the discharge process. With the presence of neutral gas, electron impact ionization can generate new electron and ion pairs. Loss mechanisms such as molecular recombination, excitation without ionization, and escapes of electrons by scattering need to be considered.

13.6 Paschen and Non-Paschen Discharges

Paschen discharge describes discharges in which the electrons and ions are energized by external electric fields applied to the discharge region. A Paschen discharge can be a gas discharge. It is not restricted to the scenarios of electrons hopping along surfaces only and is therefore more general than that shown in figure 13.5.

Not all discharges are driven by applied electric fields. As an example of non-Paschen discharge, we mention the critical ionization velocity (CIV) discharge. Alfvén[10] proposed that when the relative velocity component, perpendicular to the ambient magnetic field, between a neutral gas and a plasma exceeds a critical ionization velocity V^*, the gas is rapidly ionized. The critical ionization velocity V^* is given by

$$V^* = \sqrt{\frac{2e\Phi}{M}} \tag{13.3}$$

where Φ is the ionization potential of the neutral atomic species, and M the mass of the neutral atom. CIV is not a Paschen discharge because the energy in CIV comes from the relative velocity. The mechanical energy is converted to electrostatic energy which accelerates the electrons. The energy conversion mechanism involves plasma instabilities. For CIV reviews, see, for example, references 11 and 12.

13.7 The Townsend Criterion

Suppose that a few electrons travel from a negatively charged surface to another one of relatively positive voltage. If the process evolves no further, the result is a small voltage change without much significant consequence. Consider, however, a cyclic process in which an electron is lost, more than one electron is generated, and the newborn electrons carry on the process, multiplying further. The consequence can be significant. An example of electron multiplication is the brush fire discharge discussed earlier.

The Townsend criterion says that if the number of electrons generated equals or exceeds the number lost in a cycle, the discharge can sustain or increase, respectively. If the increase continues rapidly, the discharge becomes an avalanche discharge. The Townsend criterion applies to all discharges, and not just Paschen, non-Paschen, vacuum, or gas discharges.

Gas release at a discharge site is important for enhancing a discharge. We now review some fundamentals of discharge physics. Consider the initiation of a discharge in the neutral gas by an electron. Let μ ionization events be generated in the gas by the electron transit between a cathode and an anode. The cathode and anode can be two differentially charged surfaces. The μ electrons are accelerated toward the anode, and the μ ions are accelerated toward the cathode. Suppose that an ion impact on the cathode can produce γ electrons. For most metal cathodes and inert gas ion impact, the value of γ is of the order of 0.1 for the impact energy below 1 keV. μ ion impacts would produce $\mu\gamma$ electrons from the cathode. In an accounting balance, these newly generated electrons are counted against the loss of the original electron which is absorbed by the anode. A necessary condition to sustain the discharge is the inequality

$$\mu\gamma \geq 1 \tag{13.4}$$

The inequality, equation (13.4), is the basic equation for the Townsend criterion. The newly created electrons would start their journeys toward the anode, thereby creating at least μ new electron-ion pairs for every electron. If the electron multiplication continues and the inequality (equation 13.4) continues to hold, the discharge is sustained or even enhanced.

We now calculate the quantity μ. Let α ionizations be generated by an electron traveling per unit distance. For n electrons at x, the number of ionizations at $x + dx$ is given by

$$dn = n\alpha dx \tag{13.5}$$

Integrating equation (13.5) with x from the cathode ($x = 0$) to the anode (s), the total number of ionizations is given by

$$n(s) = n(0)\exp(\alpha s) \tag{13.6}$$

where $n(0)$ is the initial number of electrons starting the journey at $x = 0$.

Thus,

$$\mu = \exp(\alpha s) - 1 \tag{13.7}$$

Substituting equation (13.7) into equation (13.4), one obtains the Townsend criterion inequality:

$$[\exp(\alpha s) - 1]\gamma \geq 1 \tag{13.8}$$

One may prefer to write equation (13.8) in the form

$$\exp(\alpha s) \geq 1 + \frac{1}{\gamma} \tag{13.9}$$

If the system is inhomogeneous, α is not a constant along x, and equation (13.9) takes the form

$$\int_0^s dx\alpha(x) = \log\left(1 + \frac{1}{\gamma}\right) \tag{13.10}$$

For electron impact, the ionization rate of neutral gas is given by

$$\frac{dn_e}{dt} = N\int_0^\infty dEE^{1/2}f(E)\sigma(E) \tag{13.11}$$

where $f(E)$ is the plasma electron velocity distribution function with the electron velocity expressed in terms of the electron energy E, $\sigma(E)$ is the impact ionization cross section, and N is neutral density at the location x. The number α of ionizations per unit distance can be written as

$$\alpha = \frac{N}{v_d}\int_0^\infty dEE^{1/2}f(E)\sigma(E) \tag{13.12}$$

where v_d is the average electron velocity at every collision. Thus, equation (13.7) can be written as

$$\mu = \exp\left(\frac{Ns}{v_d}\int_0^\infty dEE^{1/2}f(E)\sigma(E)\right) - 1 \tag{13.13}$$

In equations (13.4) to (13.13), one notes that the neutral gas density N, the plasma electron distribution function $f(E)$, and the ionization cross section all play important roles in mediating a discharge.

For a spacecraft, the effect of a sustained low-level discharge can drain currents and degrade systems, whereas an avalanche discharge can short a circuit, drain the power supply, or burn out parts of the instruments.

The inequality [equation (13.4)] is necessary but not sufficient. To be sufficient, one needs to account for all loss mechanisms. Taking into account the losses, the Townsend criterion, equation (13.4), is still valid, but the number of electrons reaching the anode is reduced as a result of the loss mechanisms. As a result, the parameter μ of equation (13.4) is replaced by a net number, μ', after subtracting the losses. How the losses are accounted for is a case-by-case problem. Examples of loss mechanisms are (1) excitation of neutral atoms or molecules and (2) the escape of electrons from the discharge area. An excitation takes away an amount of energy less than the ionization energy. The excited neutral absorbs energy but does not produce an electron and ion pair. The excited energy is lost when de-excitation occurs later, emitting radiation. To explain mechanism 2, some electrons may scatter and escape sideways, leaving the theater and playing no role in the discharge. Mechanism 1 concerns energy loss, while mechanism 2 concerns particle loss.

13.8 Remark on Threshold Voltage

In space operations, one is often concerned about the threshold voltage for a current flow that drains current, a discharge that may cause undesirable damage, or an arc that burns out catastrophically.

There is no well-defined threshold for a small electron current leaking from a negatively charged surface through an ambient plasma to a positively charged one. A high-density, low-energy plasma, such as the ionospheric plasma, conducts better than a low-density, high-energy plasma, such as the ambient plasma in the geosynchronous environment. The keV electrons in the geosynchronous environment impact on satellite surfaces and generate secondary electrons.

If a satellite surface has sharp points, field emission[13] can emit electrons from the surface. For an extreme example, sharp points of tin whiskers at soldered joints have been blamed for the loss of Galaxy IV satellite[14,15] In addition to field emission, tin whiskers can also cause a direct shorting discharge between surfaces if the whiskers are long enough. There is no threshold voltage for shorting. Tin solder should be replaced by, or coated with, alloys that have less propensity of forming whiskers.

Starting from a small current flowing from one surface to another, the process does not necessarily lead to a sustained discharge. To sustain a discharge, one needs to satisfy Townsend's criterion. To satisfy Townsend's criterion, it is necessary to have a supply of new electrons. Ionization of the neutral gas between the electrodes gives new electron and ion pairs. In other words, one starts from the first phase (a small current discharge) and steps up to a second phase (a gas discharge). In a gas discharge, the voltage difference between the electrodes (or surfaces) must exceed the ionization potential of the gas species. Below the ionization potential, the cross section σ of ionization is zero, and therefore, no ionization is possible (equation 13.11). For some gas species, the ionization potential of typical elements is about 10 eV, and the maximum ionization cross section is at about 100 eV or more. Taking into account various losses, the threshold voltage must be above the ionization potential.

If the gas density is low, there is little probability of electron collision with the neutrals for ionization during an electron transit. If the gas density is too high, the electron collides too frequently with the neutrals, and as a result, the electron does not have sufficiently high energy for ionization. An optimal ionization depends on the neutral gas density, applied electric field, ionization cross section, etc. The energy reached by an electron per collision depends on the mean free path. Furthermore, the transit path depends on the geometry. In short, the threshold voltage and the optimal voltage for a gas discharge depend on the mean free path, the neutral species, the ionization cross section, the electron energy distribution, the geometry of the transit path, etc. The threshold voltage is not a physical constant.

13.9 Time Evolution of a Discharge

A discharge can evolve. The ionization rate of the neutral gas along the discharge current path is not a constant. The neutral density, the electron distribution, the geometry of the discharge path and the ionization rate evolve. When the neutral gas heats up, ionization can increase gradually by means of thermal ionization in addition to electron impact ionization. The Saha equation describes thermal ionization of gases (see, for example, reference 16). The discharge current prefers to follow the ionized path of least resistance albeit with random scattering.

An arc can evolve catastrophically. A primary arc can lead to secondary arcs. During a discharge, not only the neutral gas heats up but also the electrode surfaces. If the cathode surface melts, the molten surface can form small sharp points emitting electrons by field emission.[13] The sharp molten points may dance about, emitting electron currents. The high-current arc path dances accordingly. A high current can burn out a substantial area of the surface (see figure 13.1 as an example). The dancing arc may heat up and ionize neighboring surfaces, leading to secondary arcs and further evolutions. Not only can a large current short and burn the electrodes, it may even ignite the vapor generated from the electrode materials. This process of melting and burning of electrodes is often described as *pyrolysis*. The result can be catastrophic.

13.10 Laboratory Observations on Discharges

A schematic diagram of instrumentation used for measuring discharge pulses in the laboratory is shown in figure 13.6. The displacement current generated by a pulse is detected on the oscilloscope. We have mentioned repeatedly two key players in the discharges on spacecrafts: plasma and neutral gas. Indeed, laboratory experiment results[17,18] have shown that plasma and neutral gas from the surface material samples are always present at discharges. Laboratory experiments[18] confirmed that the spectral lines observed in the plasma and neutral gas generated in a discharge were from the elements of the sample material used.

For a discharge initiated deep inside a dielectric sample, neutral gas escaped from the plasma channels generated.[17] These laboratory results infer what would happen in space. Even if a discharge is initiated by a vacuum discharge, the neutral gas generated at the initial discharge site may enhance the probability of occurrence of a gas discharge. Once a gas discharge occurs, the discharge current may multiply more rapidly than in a vacuum discharge. The gas discharge takes over the discharge process. Frederickson (personal communication, 1996) repeatedly observed the presence of neutral gas released during arcing in spacecraft materials and believed that neutral gas was important for spacecraft discharges.

Figure 13.7 shows some measurements of surface voltage of a solar cell coverglass sample under electron irradiation to simulate charging in the geosynchronous environment. The figure shows that the surface potential dropped when discharges occurred and that discharges were more likely during high electron flux.[19]

The recent Spacecraft Charging Technology Conferences showed vigorous research efforts on spacecraft discharges.[20,21] Advances are being made in areas such as radiation-induced conductivity and cold-temperature effects in discharges. The results from the

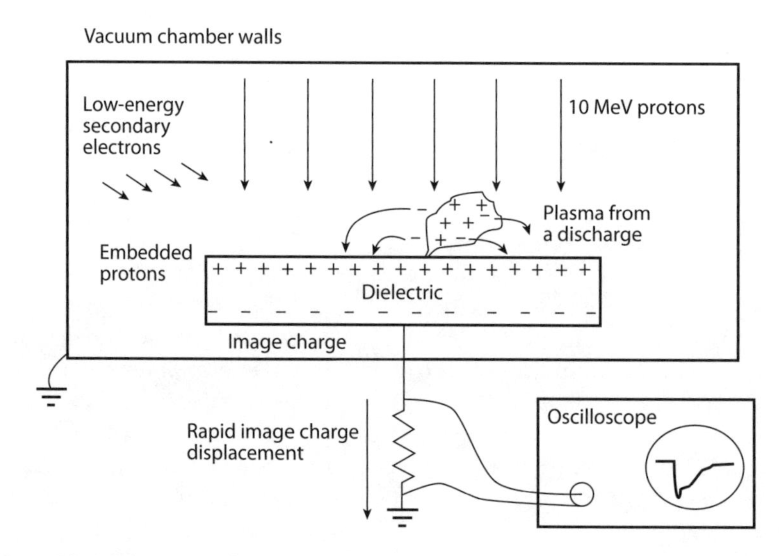

Figure 13.6 Schematic of instrumentation for measurements of dielectric discharge pulses during energetic proton bombardment experiments. (From reference 19.)

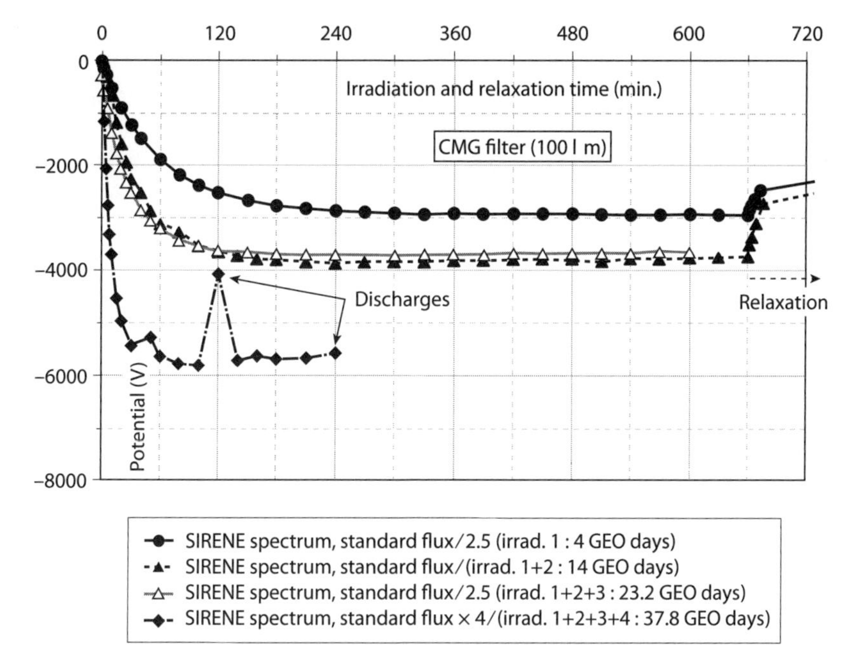

Figure 13.7 Surface voltage versus time on 100 μm CMG coverglass after different simulated days in orbit. (Adapted from reference 20.)

ongoing research efforts will hopefully enable better understanding of the underlying physics.

More discussion on spacecraft discharges can be found in chapter 15.

13.11 Discharges Initiated by Meteor or Debris Impacts

Meteoroids are fast. The average speed is about 15 to 18 km/s. Leonids meteoroids travel at about 72 km/s. At such speeds, a meteoroid hits a spacecraft like a bullet. It can penetrate and cause physical damage. It can generate neutral gas and plasmas, which, in turn, can cause discharges (figure 13.8). Measurement on plasma discharges on spacecraft is difficult because meteor hits are rare. In the laboratory, it is difficult to accelerate a neutral particulate to tens of km per sec. Debris are usually larger and slower than meteors. However, some debris are large and can cause worse damage on spacecraft. More details on meteors and meteor impacts are given in later chapters. For recent reviews on the subject of hypervelocity impacts on spacecraft, see, for example, reference 22.

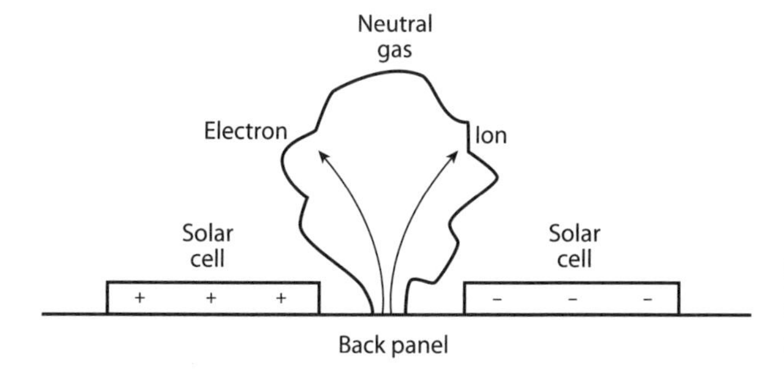

Figure 13.8 Electrons, ions, and neutral gas produced by a sudden meteor impact near solar cells.

13.12 Exercises

1. What is a triple junction? What is an inverted gradient?
2. What is a Paschen discharge? Give an example of a non-Paschen discharge.
3. Photoemission can mitigate the negative charging potential of a surface usually. Can sunlight mitigate the negative surface potential of a highly transparent lens?
4. What is the physical mechanism that may prevent photoemission from a highly reflective surface?
5. What are some loss mechanisms that may drain the electron energy or the electron population in a discharge?

13.13 References

1. Several papers in "Special issue on spacecraft charging technology," *IEEE Trans. Plasma Sci.*, 1946–2219 (2007).
2. Cho, M., chapter in book *A Guide to Spacecraft Charging and Mitigations*, AIAA, in progress.
3. Balmain, K. G., "Scaling laws and edge effects for polymer surface discharges," in *Spacecraft Charging Technology*, eds. R. C. Finke and C. P. Pike, NASA Conference Publication 2071, pp. 646–656, NASA and U.S. Air Force Geophysics Laboratory, Hanscom AFB, MA (1978).
4. Balmain, K. G., and G. R. Dubois, "Surface discharges on Teflon, Mylar, and Kapton," *IEEE Trans. Nuclear Sci.* NS-26: 5146–5151 (1979).
5. Anderson, P. C., and H. C. Koons, "Spacecraft charging anomaly on a low-altitude satellite in an aurora," *J. Spacecraft and Rockets* 33: 534–539 (1996).
6. Sanders, N. L., and G. T. Inouye, "Secondary emission effects on spacecraft charging: energy distribution considerations," eds. R. C. Finke and C. P. Pike, NASA-2071, ADA-084626, pp. 747–755, NASA and U.S. Air Force Geophysics Laboratory, Hanscom AFB, MA (1978).
7. Lai, S. T., and M. Tautz, "Aspects of spacecraft charging in sunlight," *IEEE Trans. Plasma Sci.* 34, no. 5: 2053–2061 (2006).
8. Lai, S. T., "Charging of mirror surfaces in space," *J. Geophys. Res.* 110: A01204 (2005).
9. Hopkins, J., (Ed.), "Boeing 702 satellites solar arrays possibly defective," *SpaceandTech.com* (2001), available at http://www.spaceandtech.com/digest/flash2001/flash2001-082.shtml.
10. Alfvén, H., "Collision between a nonionized gas and a magnetized plasma," *Rev. Mod. Phys.* 32: 710–713 (1960).
11. Brenning, N., "Review of the CIV phenomenon," *Space Sci. Rev.* 59: 209–314 (1992).
12. Lai, S. T., "A review of critical ionization velocity," *Rev. Geophys.* 39, no. 4: 471–506 (2001).
13. Fowler, R. H., and L. W. Nordheim, "Electron emission in intense electric fields," *Proc. Roy. Soc.* A119: 173–181 (1928).
14. Brusse, J., available at http://nepp.nasa.gov/whiskers/reference/tech_papers/Brusse2003_Zinc_Whisker_Awareness.pdf.
15. Available at http://www.wirelessweek.com/whiskers-caused-satellite-failure.aspx.
16. Chen, F. F., introduction to *Plasma Physics*, 2nd ed., Plenum, New York (1984).
17. Frederickson, A. R., C. E. Benson, E. M. Cooke, "Gaseous discharge plasmas produced by high-energy electron-irradiated insulators for spacecraft," *IEEE Trans. Plasma Sci.* 28, no. 6: 2037–2047 (2000).
18. Amorim E., L. Levy, and S. Vacquie, "Electrostatic discharges on solar arrays: common characteristics with vacuum arcs," *J. Phys. D: Appl. Phys.* 35, no. 7: L21–L23 (2002).

19. Green, N. W., and J. R. Dennison, "Deep dielectric charging of spacecraft polymers by energetic protons," presented at 10th Spacecraft Charging Technology Conference, Biarritz, France (2007).
20. "Special issue on spacecraft charging technology," eds. S. T. Lai et al., *IEEE Trans. Plasma Sci.* 34, no. 5: 1946 (2006).
21. "Special issue on spacecraft charging technology," eds. S. T. Lai et al., *IEEE Trans. Plasma Sci.* 36, no. 5: 2218 (2008).
22. Lai, S. T., E. Murad, and W. J. McNeil, "Hazards of hypervelocity impacts on spacecraft," *J. Spacecraft and Rockets* 39, no. 1: 106–114 (2002).

14

Energetic Particle Penetration into Matter

14.1 Introduction

In surface charging, excess charges of one sign accumulate on the surface. If the incoming charged particles are of low to medium energies (up to low tens of keVs), the charges cannot penetrate deeply into the material and therefore stay on or near the surface. In contrast, if the incoming charged particles are energetic (MeVs), they can penetrate deeply into the surface material. This property poses serious problems to spacecraft.

In a conductor, as the incoming energetic charged particles penetrate deeply into the conducting material, the excess charges will promptly move from deep inside the material to the surface, due to mutual repulsion according to Coulomb's law. As a result, despite the deep penetration of incoming charged particles, the excess charges stay on the surface. Unlike dielectrics, which are nonconducting, conductors can have surface charging only. Therefore, deep conductor charging never occurs.

In a dielectric, the incoming energetic (MeVs) charged particles that have penetrated inside will stay there because the conductivity of the material is very poor. At the energy range of up to tens of MeVs, electrons penetrate deeper than ions, forming a negative charge layer at greater depth than the ions (figure 14.1). For a spacecraft exposed to an energetic particle environment for days, months, or years, the accumulation of electrons inside dielectrics may build up high internal electric fields. This is the main idea describing how deep dielectric charging occurs. Some effects of high-energy charged particle penetration into dielectrics will be discussed in section 14.9 and the next chapters.

14.2 High-Energy Charged Particle Penetration into Solids

High-energy (MeV) electrons and ions impacting on a surface material penetrate to different depths depending on the particle energies, particle species, and material properties (figure 14.1). At keV energies, electrons and ions can penetrate only to shallow depths near the surface; there is no significant difference between their shallow depths. At MeV energy, electrons penetrate deeper than ions; the difference between their penetration depths becomes significant. At 100 MeV and above, ions penetrate to deeper depths than electrons; meson production and nuclear reactions may become possible. Some good references in English are Fermi,[1] Segre,[2] and Ziegler.[3]

In the Earth's space environment, we are not too concerned about electron and ion fluxes of 100 MeV or above, because such high-energy electron and ion fluxes are very small. Such very high energy electrons and ions can come from cosmic rays. If they score a hit on a sensitive part of the onboard electronics, they can cause significant damage. However, because of their very low flux, the chance of a direct hit is very small. That is not charging per se but accounts for some of the spacecraft anomalies due to high-energy charged particles penetrating into matter.

The difference between the penetration depths of high-energy (MeVs) electrons and ions in dielectrics enables an electron layer to form deep inside. (Note: Since the ambient electron

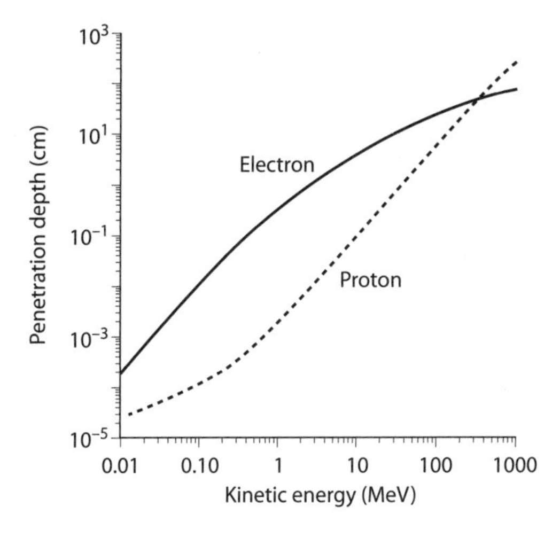

Figure 14.1 Penetration depths of electrons and protons in kapton polymide.

flux is two orders of magnitude higher than that of the ambient ions, as discussed in chapter 1, an ion layer has generally been ignored. See, however, the discussion of double layer formation inside dielectrics[5,6] in chapter 16.) The electrons deposited in the layer are immobile because dielectrics are poor conductors. As the charges accumulate, they build up electric fields. At sufficiently high electric fields, the dielectric material breaks down. The critical electric field is typically of the order of 10^6 V/m, depending on the material. Dielectric breakdown means a sudden jump in the conductivity locally, forming suddenly a local channel of conduction and resulting in an arc discharge.

14.3 Physics of High-Energy Charged Particle Penetration into Matter

When a high-energy charged particle travels in matter, it interacts via its Coulomb force with the charged particles in the matter. At first consideration, one might wonder why a charged particle can travel into a solid matter, which looks like a continuum. The matter inside a solid is not a continuum. From the perspective of a charged particle traveling in a solid, the atoms in the solid are far apart. If one imagines a nucleus as a billiard ball, its next neighbor (another billiard ball) is a few miles away. The chance of a direct collision between the projectile particle and the charged particles inside the solid is very rare.

Unlike collisions by direct contact between two billiard balls, Coulomb interaction is at a long distance. When a charged particle passes by the vicinity of another particle (electron or ion), momentum is transferred via Coulomb interaction. As a result, the charged particle slows down. However, the faster a particle is, the less time it would spend in the Coulomb field of interaction with its neighboring particle. The Coulomb field drops off as the distance between the particles increases. Therefore, a faster particle has less time of interaction with its neighboring particle.

14.4 The Bohr Model of Charged Particle Interaction

For pedagogical purposes, consider the Bohr model of a charged particle traveling along a straight trajectory and interacting electrostatically with a nearby electron at rest (figure 14.2). For a dielectric material, the electron is attached to an atom. The Bohr model is nonrelativistic.

The model assumes that the charged particle trajectory is unaffected by the electron and that the electron acquires an impulse without changing its position. The impulse acquired

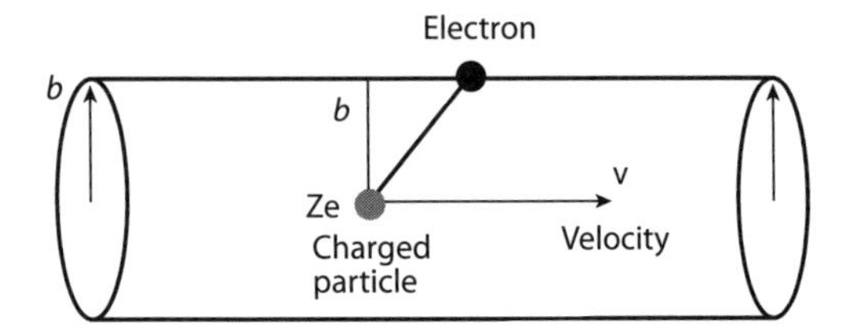

Figure 14.2 A charged particle passing by a stationary electron attached to an atom. The faster the particle is, the less energy is transferred.

by the electron is perpendicular to the trajectory. Consider a cylinder, with the axis along the trajectory and the electron on the surface of the cylinder (figure 14.2). The radius b of the cylinder is called the *impact parameter.* The impulse is given by the time integral of the force ($eE_\perp$) exerted by the charged particle to the electron in the material. The time integral of the force is the total momentum Δp transferred to the electron in the period of interaction.

$$\Delta p = \int_0^\infty eE_\perp dt = \int_{-\infty}^\infty eE_\perp \frac{dx}{\mathrm{v}} \tag{14.1}$$

where E is the electric field due to the charged particle, v the charged particle velocity, e the elementary charge, t the time, and x the distance along the projectile trajectory. Assuming the charged particle velocity v to be constant, the integral in equation (14.1) can be evaluated by using Gauss's law:

$$\int_{-\infty}^\infty dx E_\perp 2\pi b = \frac{Ze}{\varepsilon_o} \tag{14.2}$$

where Ze is the charge of the charged particle. Substituting equation (14.2) into equation (14.1), one obtains the total momentum Δp transferred to the electron as

$$\Delta p = \frac{e}{\mathrm{v}} \frac{Ze}{2\pi\varepsilon_0 b} = \frac{2Ze^2}{\mathrm{v}b(4\pi\varepsilon_0)} \tag{14.3}$$

The energy transferred ΔE to the electron of mass m is given by

$$\Delta E = \frac{(\Delta p)^2}{2m} = \frac{2Z^2}{m\mathrm{v}^2 b^2}\alpha^2 \tag{14.4}$$

where α is the fine structure constant ($\alpha = e^2/4\pi\varepsilon_0$).

The energy transferred to the electrons per unit distance along the projectile trajectory in the material is given by integrating ΔE over the volume of the cylinder of unit length as follows:

$$\int_{b\min}^{b\max} dA\, n_e \Delta E = 2\pi \int_{b\min}^{b\max} db\, b\, n_e \Delta E = \frac{4\pi Z^2 n_e \alpha^2}{m\mathrm{v}^2} \log \frac{b_{\max}}{b_{\min}} \tag{14.5}$$

where $A = \pi b^2$ is the cross-sectional area of the cylinder, n_e the electron density of the material, $b_{\min}$ the shortest impact parameter governed approximately by Rutherford's elastic scattering, and $b_{\max}$ the longest impact parameter governed approximately by electron oscillation time (see, for example, references 1–3).

In elastic scattering, the maximum momentum transfer is $2m\mathrm{v}$. As an analog, if a footballer's foot kicks with a velocity v_k toward a ball, the maximum resultant velocity with which the ball bounces away can not exceed $2\mathrm{v}_k$. Equating the electrostatic energy ΔE of equation (14.4) to $(2m\mathrm{v})^2/2m$, one obtains the lower cutoff $b_{\min}$ (> 0). The cutoff avoids the singularity in the Coulomb energy that would occur if the distance r is 0. To obtain the upper cutoff, $b_{\max}$, one assumes that at far distance greater than $b_{\max}$, the projectile's angular frequency is less than of the electron oscillation, the electron response is negligible.

For more elaborate formulations, one needs to include quantum mechanics, relativistic velocities, excitation and ionization of the molecules in the material, radiation loss, and if the energy is high enough, nuclear reactions. For gaining physical insight into deep dielectric charging, the Bohr model of charged particle interaction helps. The model equation (14.4) reveals nature's interesting behavior in high-energy charged particle passage in matter, even though the model is simple, nonrelativistic, and lacks elaborate details.

It is instructional to note that, in the Bohr model earlier, equation (14.4), the energy transferred ΔE is proportional to b^{-2}, implying that the energy transfer ΔE increases with the material density. Note also that the energy transferred is proportional to $1/\mathrm{v}^2$, implying that a faster projectile has less interaction with the material.

As a pedagogical analogy, think of a fireman walking by your door while shouting "Fire! Get out of the building!" You would be alarmed when you heard and interpreted the message. If, instead, the fireman runs by very fast while shouting his message at the same pitch and volume as he did when walking, you would be less likely to receive and react to the message.

14.5 Stopping Power

It is customary to refer the particle energy loss per unit distance as the *stopping power* $S(E)$:

$$S(E) = -\frac{dE}{dx} \tag{14.6}$$

where E is the kinetic energy of the particle, and x the distance. In spacecraft interaction physics and in biophysics, the term *linear energy transferred* (LET) is often used as an approximation of the stopping power of penetrating particles. There is a difference between stopping power and LET. Stopping power includes all energy loss mechanisms, including radiation. LET does not consider losses by radiation. At above 20 MeV, bremsstrahlung radiation becomes increasingly important. Above 100 MeV, nuclear reactions can occur. For deep dielectric, we are concerned with the energy range of about 20 keV to 20 MeV only, where bremsstrahlung is not important. In this energy range, the stopping power or LET is mainly due to momentum transfer, excitation, and ionization. In the Earth's space environment, the charged particle fluxes generally decrease rapidly as functions of particle energy. The fluxes above 20 MeV are usually negligible, as far as deep dielectric charging is concerned.

14.6 The Bethe-Bloch Equation

A more elaborate formulation of the stopping power of a charged particle with velocity v in a material is given by the Bethe-Bloch formula (see, for example, references 1–3). It is relativistic and is given as follows:

$$S(E) = -\frac{dE}{dx} = \frac{4\pi Z^2 n_e}{m\mathrm{v}^2}\alpha^2\left[\log\left(\frac{2m\mathrm{v}^2}{1-\beta^2}\right) - \beta^2 - \log I\right] \tag{14.7}$$

or

$$S(E) = \left(\frac{2\pi Z^2 M n_e \alpha^2}{m}\right)\frac{1}{E_p}\left[\log E_p + \log\left(\frac{4M}{m}\right) - \log(1-\beta^2) - \beta^2 - \log I\right] \quad (14.8)$$

where

$$E_p = \frac{1}{2}Mv^2 \quad (14.9)$$

In equations (14.7) to (14.9), m is the electron mass, e the elementary charge, z the charge of the projectile, v the projectile velocity, n_e the electron density of the material, α the fine structure constant, $\beta(= v/c)$ the relativistic factor, c the velocity of light, and E_p the projectile kinetic energy. The material properties are characterized by the average energy I of ion-impact excitation and ionization of the atoms in the material. The number density n_e of electrons per unit volume is related to the mass density ρ of the material by

$$n_e = \frac{Z\rho}{Am_n} \quad (14.10)$$

where Z is the number of protons in an atom, A the atomic number, ρ the mass density in g cm^{-3}, and m_n the neutron mass. There are additional terms for fine-tuning in various conditions. The stopping power for an electron projectile is similar to equation (14.7) except for some small difference in the logarithmic terms (see, for example, reference 2).

It is interesting that the stopping power $S(E)$ (equation 14.6) increases as the particle is slowing down. At nonrelativistic energies ($\beta \ll 1$), $S(E)$ varies as v^{-2} approximately, agreeing with the v^{-2} dependence in the Bohr model [equation (14.4)]. Qualitatively, the behavior is reminiscent of the fireman analogy (section 14.3): faster particles interact less with the neighboring particles. (The fireman analogy is still valid qualitatively in the relativistic regime, but the v^{-2} dependence is modified by the β terms. In the ultra-relativistic limit, $v \rightarrow c$, all particles are noninteracting.)

Note that the Bethe-Bloch equation applies to high-energy charged particle penetration into materials including dielectrics on spacecraft for primary particle energies up to about 20 MeVs. There are additional terms accounting for various corrections, such as shell corrections, scattering, and bremsstrahlung radiation, but they generally add less than about 6% only to the accuracy for energies up to 20 MeV approximately. Above 20 MeV, the correction terms become increasingly important, but we do not consider them here because the equation with corrections becomes very complicated and the flux of very high energy charged particles is very small in the Earth's space environment.

14.7 Range and Penetration Distance

Figure 14.3 shows a typical curve of the stopping power $S(E)$, or LET, of a MeV charged particle penetrating into matter. The stopping power increases as the particle penetrates and slows down. The curve is called a *Bragg curve*. There is typically a peak, which is called a *Bragg peak*. Beyond the peak, the particle slows down to a complete halt. The range of the particle in the material is defined as the distance from the surface to the midpoint between the Bragg peak and the point of stopping.

Note that since the interaction between a very fast particle (projectile) and the material is small at high velocities, the probability of ionization is small along the projectile path until

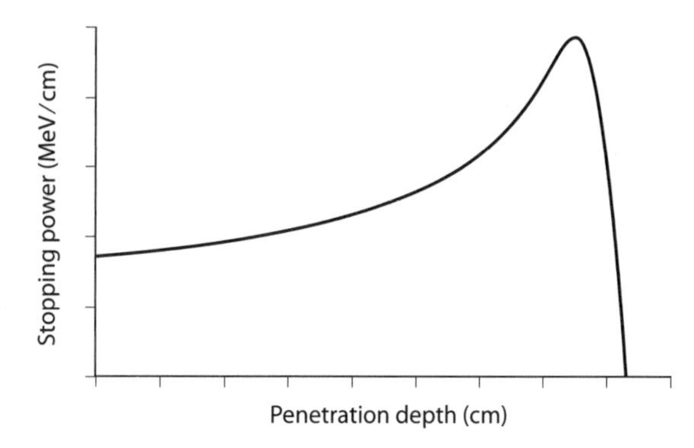

Figure 14.3 A typical curve of the stopping power as a function of penetration depth. The stopping power increases as the particle slows down, while the latter penetrates deeper and deeper into the material. Such a curve is called a Bragg curve. The peak is called the Bragg peak. Beyond it, the particle slows down rapidly to a halt. The range is defined conventionally as the depth at halfway between the Bragg peak and the zero of the stopping power.

the projectile slows down. Excitation requires less energy than ionization. Excitation occurs more prominently than ionization along the projectile path until the projectile is near the Bragg peak. The peak is located near the end of the particle journey. There, the particle velocity has slowed down from its initial energy, MeVs, to keVs or hundreds of eVs. There, interactions, momentum transfer and energy loss are at the maximum; so are the probabilities of atomic displacement in lattices, heat generation, excitation, and ionization of atoms in the material. Ionization may increase the number of ions and electrons by orders of magnitude in the narrow region around the Bragg peak.

The preceding property, together with the second property that the particle comes to a halt very shortly beyond the Bragg peak, enables medical uses of high-energy beams. For example, a proton beam can be aimed at a prostate cancer located at a depth where the beam interaction has been calculated to be maximum. As a further advantage, more unnecessary damage would not occur behind the cancer, because the beam particles will halt shortly beyond that location.

For a primary charged particle of mass m and initial energy E_p, the range $R(E_p)$ of penetration is related to the stopping power $S(E)$ by the following integral:

$$R(E_p) = \int_0^{E_p} dE \frac{1}{(dE/dx)} \tag{14.11}$$

Conventionally, the unit of range $R(E_p)$ is g.cm^{-2}, which means the material density ρ (g cm^{-3}) multiplied by distance (cm). To convert range (g cm^{-2}) to penetration depth p, one simply divides the range R by the density ρ.

$$p = \frac{R}{\rho} \tag{14.12}$$

Table 14.1 shows some typical ranges and penetration depths into materials by electrons and ions of 1 MeV in energy.

TABLE 14.1
Penetration Depth of 1-MeV Electrons and Ions into Materials

Material	*Density* ($g\,cm^{-3}$)	*e-Depth* p_e (*cm*)	*e-Range* R_e ($g\,cm^{-2}$)	*p-Depth* p_i (*cm*)	*p-Range* R_i ($g\,cm^{-2}$)
Aluminum	2.70	0.205	0.5546	0.00143	0.00387
Aluminum oxide	3.97	0.135	0.5367	0.00093	0.00370
Germanium	5.32	0.123	0.6560	0.00121	0.00643
Gold	19.32	0.040	0.7762	0.00065	0.01247
Graphite	1.70	0.292	0.4964	0.00162	0.00275
Iron	7.87	0.078	0.6159	0.00069	0.00544
Kapton polymide	1.42	0.337	0.4780	0.00191	0.00271
Lead	11.35	0.069	0.7843	0.00104	0.01183
Lucite	1.19	0.349	0.4150	0.00208	0.00247
Mylar	1.40	0.336	0.4702	0.00189	0.00265
Polyethylene	0.94	0.443	0.4160	0.00229	0.00215
Polypropylene	0.90	0.461	0.4150	0.00234	0.00211
PVC	1.30	0.380	0.4940	0.00225	0.00293
Pyrex glass	2.23	0.234	0.5219	0.00155	0.00345
Silicon	2.33	0.231	0.5386	0.00165	0.00384
Silver	10.50	0.066	0.6896	0.00073	0.00770
Teflon	2.20	0.238	0.5227	0.00149	0.00327
Titanium	4.54	0.133	0.6055	0.00105	0.00478

Calculated by using the interactive website http://physics.nist.gov/PhysRefData/Star/Text/contents.html.

14.8 Approximate Penetration Depth Formula

Electrons and protons penetrate to different depths. In table 14.1, the 1 MeV electron penetration depths are typically about a fraction of a centimeter, while the 1 MeV proton penetration depths are about two orders of magnitude smaller. Note also that the range R for electron penetration is approximately constant at 1 MeV. This property of approximately constant R holds over a wide range of energies from about 10 keV to 20 MeV. Therefore, if one knows the penetration depth of one material, one can deduce that of another material for the same primary particle energy by using a simple approximate formula. The reason for the small difference between the penetration distances for different materials is because ionization, which is a material property, occurs mainly in the short distance around the Bragg peak only. If this short distance is small compared with the total penetration distance, it would be a fair approximation to ignore any small variation in ionization within the short distance. One such formula is as follows:

$$S_1/S_2 = n_1/n_2 \tag{14.13}$$

where n_1 and n_2 are the charge number densities of the two materials, respectively. This approximation is rough, because the average excitation and ionization energy $\langle I \rangle$ varies from material to material.

To calculate energy-range curves in a given material, various algorithms are available on the Internet. They differ from one another according to the accuracies of various loss mechanisms (such as electronic excitation levels, ionization energies, angular scattering, displacement of lattice atoms, and radiations) included, the step sizes taken, relativistic effects, and shell corrections, etc. For example, some such interactive algorithms found on the Internet are (1) http://www.hpcalc.org/details.php?id=2207, (2) http://tvdg10.phy.bnl.gov/LETCalc.html, and (3) http://physics.nist.gov/PhysRefData/Star/Text/ASTAR.html. Table 14.1 is calculated by using interactive algorithm 3.

14.9 Effects of Charged Particle Penetration

A summary of some salient features of charged particle penetration into matter is as follows. The stopping power $S(E)$ of a particle penetrating into matter is defined by $S(E) = -dE/dx$. It is also called linear energy transfer (LET) if radiation loss is unimportant and neglected. The stopping power, or linear energy transfer, is proportional to v^{-2}, where v is the velocity of the penetrating charged particle. The faster the particle is, the less interaction it has with the particles in the matter. The Bethe-Bloch equation is the classical one for calculating the stopping power. However, the equation is invalid for low energies (below 10 keV) and needs corrections to account for radiation etc. at high energies (above several tens of MeVs). The range of the penetrating particle is given by the energy integral of the inverse of the stopping power. There is a Bragg peak in the curve of $S(E)$ as a function E. Since the particle energy slows down significantly at the end stage of its journey, the Bethe Bloch equation should not be used at that stage. There are interactive algorithms for calculating the range; some can be readily found on the Internet—for example, http://www-physics.lbl.gov/~spieler/physics_198_notes_1999/PDF/III-E-Deposition-1.pdf and http://science.nasa.gov/headlines/y2005/27jan_solarflares.htm?. One can calculate the penetration distance from the range if one knows the material density.

For spacecraft deep dielectric charging in the Earth's radiation belts, we are concerned mostly about electron and proton fluxes in the energy range of about 0.01 to 20 MeV. Fluxes of higher energy electrons and protons are much smaller. Depending on the energy, electrons and ions may penetrate to different depths. Below 0.01 MeV, the ranges of electrons and protons are approximately equal. Above about 100 MeV, protons penetrate deeper than electrons. Depending on the conductivity of the material, the electrons and ions penetrating to different depths inside may stay there for days or months, accumulating and escalating the internal electric charges and, by Gauss's law, their resultant electric fields. This property is the cause of the deep dielectric charging.

For spacecraft interactions concerns, charged particle penetration and deposition can cause five types of effects:

Electron-ion pairs. Since the penetrating particles generate excitation and ionization, the ionization level increases along and in the vicinity of the penetration path. This is especially so when the primary particle has slowed down to keVs and hundreds, or even tens, of eVs in energy. Depending on the conductivity of the material, the electron-ion pairs may recombine, albeit very slowly, in some dielectrics. The increase in the net charge, however, is independent of the amount of ionization.

Chemical effects. Since the excitation and ionization occur, especially when the particles have slowed down, there can be radiation-induced chemical effects such as coloring, luminescence, and chemical changes. As an exaggerated illustration, one might recall in some movies a luminescent green human after exposure to intense radioactivity.

Damage to lattices. The high-energy penetrating particle can displace atoms or even dislodge them from their lattice sites. As a result, lattice defects may occur, and therefore the impurity level of the material may increase.

Cascade effect. If an atom recoils with high momentum, it may cause cascade ionization or cascade damage to lattices. For example, a head-on collision of a fast proton with an atom in a material may give the atom a significant recoil momentum that, in turn, may cause cascade effects. Head-on collision is often the cause of single event upsets (SEUs) in spacecraft electronics.

High internal electric fields. When the electric fields are higher than some critical value, dielectric breakdown may occur. A sudden change in the material conductivity occurs locally, allowing an arc discharge to occur internally. This type of effect is one of the main reasons for studying deep dielectric charging.

14.10 Effects on Astronauts

This topic is, strictly speaking, not in the area of spacecraft interactions but is related to it. The discussion will be brief. For effects on astronauts, the unit adopted is rem (short for roentgen equivalent man). One rem is the radiation dose that causes the same injury to human tissue as 1 roentgen of x ray. The dose of a typical dental x ray is about 0.1 rem. During a severe solar storm, an astronaut outside a spacecraft in the Earth's space environment may receive up to several rems of radiation in a few days. Receiving a dose of 50 rems in a short time, perhaps minutes, would make one sick. Receiving 300 rems in a short time may be fatal, if no prompt medical treatment is rendered. Staying inside shielded spacecrafts or wearing protective gear would reduce the risk.

14.11 Research Questions in High-Energy Penetration of Charged Particles into Matter

As a first step to understand the effects of high-energy charged particle penetration into matter, we have discussed some fundamentals addressing the following question: How deeply do the charged particles penetrate into materials? There are many more questions: How long does it take to build up hazardous internal electric fields in the space environment? Where in space will spacecraft be exposed to high-energy (MeVs) charged particles? What is the typical flux level of high-energy charged particles in the Earth's space environment? What is the critical electric field that is hazardous? What have we observed in real measurements in space and in the laboratory? What concrete evidence do we have relating energetic charged particle fluxes to spacecraft anomalies? What physical interpretations can we provide to some of the features observed on spacecraft? What new physical ideas can we apply or explore in this area? What mitigation techniques can we use? This area is relatively new, its research begins to grow vigorously, and much more needs to be better understood. We will attempt to answer some of the preceding questions in later chapters.

14.12 Exercises

1. Derive equation (14.3) by writing down the electric field explicitly in equation (14.1) as follows:

$$E_{\perp}(r) = \frac{Ze}{4\pi\varepsilon_0 r^2}\cos\theta \tag{14.14}$$

where the distance r from the charged particle to the electron is given by

$$r = (b^2 + x^2)^{1/2} \tag{14.15}$$

and the angle θ subtended between r and impact parameter b is given by

$$\cos\theta = \frac{b}{(b^2 + x^2)^{1/2}} \tag{14.16}$$

One needs to evaluate the integral

$$\Delta p = e\int_{-\infty}^{\infty} dt E_\perp \tag{14.17}$$

where $x = \mathrm{v}t$, assuming that v is constant (figure 14.2).

Two possible approaches to this problem are as follows:

Method 1. Coulomb's Method

The perpendicular component of the electric field is given by

$$E_\perp(x) = \frac{Ze}{4\pi\varepsilon_o r^2}\cos\theta = \frac{Ze}{4\pi\varepsilon_o(b^2 + x^2)}\cos\theta \tag{14.18}$$

where Z is the number of charges of the charged particle, b is the impact parameter (i.e., the radius of the cylinder in the figure), x is the horizontal distance between the charged particle and the electron, and θ is the angle between b and r. The integral becomes

$$\Delta p = e\int_{-\infty}^{\infty}\frac{dx}{\mathrm{v}}E_\perp(x) = \int_{-\infty}^{\infty}\frac{dx}{\mathrm{v}}\frac{Ze^2}{4\pi\varepsilon_o(b^2 + x^2)}\frac{b}{(b^2 + x^2)^{1/2}} = \frac{Ze^2}{2\pi\varepsilon_o b\mathrm{v}} \tag{14.19}$$

Method 2. Gauss's Method

Using Gauss's law, we have the following integral:

$$\int_{-\infty}^{\infty} dx E_\perp 2\pi b = \frac{Ze}{\varepsilon_o} \tag{14.20}$$

Therefore, the integral [equation (14.7)] becomes

$$\Delta p = e\int_{-\infty}^{\infty} dt E_\perp = \frac{e}{\mathrm{v}}\int_{-\infty}^{\infty} dx E_\perp = \frac{Ze^2}{2\pi\varepsilon_o b\mathrm{v}} \tag{14.21}$$

We have recovered equation (14.3). Note that the preceding is in mks. In cgs, the $4\pi\varepsilon_o$ would be absent in the denominators of equation (14.18) and, accordingly, the results of Δp in equation (14.19) and equation (14.21) would be multiplied by $4\pi\varepsilon_o$ in the numerator.

2. Use a software available on the Internet to calculate the range and the penetration depth at a given energy of electrons and hydrogen ions into mylar, silicon, and a metal of your choice. For example, the National Institute of Standards software available at http://physics.nist.gov/PhysRefData/Star/Text/ASTAR.html will do.

14.13 References

1. Fermi, E., *Nuclear Physics,* University of Chicago Press, Chicago (1950).
2. Segre, E., *Nuclei and Particles*, 2nd ed., The Benjamin/Cummings Publishing Co., Reading, MA (1977).
3. J. F. Ziegler, "The stopping of energetic light ions in elemental matter," *J. Appl. Phys / Rev. Appl. Phys.* **85**: 1249–1272 (1999).

4. Goldstein, H., *Classical Mechanics*, Addison Wesley, Reading, MA. (1950).
5. Lai, S. T., "A mechanism of deep dielectric charging and discharging on spacecraft," in *Proc. IEEE International Conf. Plasma Sci.*, Raleigh, N.C. (1998). Conference Record of 1998 IEEE International Conference on Plasma Science, p. 154, doi: 10.1109/PLASMA.1998.677564, ISSN: 0730-9244 (1998).
6. Lai, S. T., E. Murad, and W. J. McNeil, "Hazards of hypervelocity impacts on spacecraft," *J. Spacecraft and Rockets*, 39, no. 1: 106–114. (2002).

15

Spacecraft Anomalies

15.1 Introduction

An anomaly is something that deviates from the normal or the expected. Spacecraft anomalies can be mechanical or electronic in nature. Here, we are concerned with the latter. Spacecraft anomalies may be harmful to the health of the onboard electronics, affect the functions of the instruments, and, in worst cases, may affect spacecraft navigation and spacecraft survivability. Spacecraft anomalies are often associated with high-level spacecraft charging. Not every event of high-level spacecraft charging is associated with anomalies. High voltages on spacecraft can occur without anomalies. Likewise, spacecraft anomalies can occur while little or no spacecraft charging occurs. Differential charging of spacecraft surfaces with respect to each other is often more hazardous than uniform surface charging with respect to the ambient plasma environment.

Although high-level spacecraft charging is a potential for space hazards, it does not necessarily produce hazards. As an analogy, a big pile of snow poised on a hilltop poses a potential hazard for avalanche, but the latter does not always occur. For the safety of the skiers on the hill slope, it is important to warn that an avalanche may be forthcoming if the snow pile continues to build up toward a dangerous level or when there is an imminent external force that may act as a trigger for avalanche. Likewise, some spacecraft may experience high potentials from time to time but do not necessarily suffer from any adverse effect, although it is more probable that adverse effects may occur at higher potentials. There are other causes, besides surface charging, that can cause adverse effects or damages on spacecraft.

15.2 Space Anomalies Due to Surface Charging

Figure 15.1 shows a compilation of cases of space missions terminated by various natural causes.[1] According to figure 15.1 (obtained from data in reference 1), the most likely cause of space mission termination is spacecraft surface charging. Later in the next chapter, we will discuss deep dielectric charging and the dangerous levels or external disturbances that may trigger avalanche discharges.

Figure 15.2 shows a case of good correlation between discharge signals detected on the SCATHA satellite and high surface potentials measured on the satellite.[2] The potential values in Figure 15.2 were in the range of hundreds of negative volts. Such a voltage range is common in surface charging.

Figure 15.3 shows a case in which the observed sudden deviations from the expected sequence of electronic states on SCATHA occurred at exactly the instants of discharges.[2] This case illustrates the adverse effects of the discharge signals, which, in turn, were caused by the high surface potentials.

As another case (figure 15.4), when the SC4-1 electron beam with a large current of 100 mA at 2 keV in energy was emitted from SCATHA, the SC2-2 instrument on the satellite surface promptly failed permanently.[3–5] In this case, the timing of the knock-out of the SC2-2 instrument immediately after the commencement of the high-current beam emission of SC4-1 suggests strongly a cause–effect relationship. The emission of the large-current beam induced a high potential on the spacecraft. To balance currents, the potential attracted large currents

Geosynchronous satellite

Vehicle	Date	Cause (*)
DSCS II	1973	ESD
GOES 4	1982	ESD
DSP Flight 7	1985	ESD
Feng Yun 1	1988	ESD
MARECS A	1991	ESD
MSTI	1993	SEU
Hipparcos	1993	MRD
Olympus	1993	Meteoroid
SEDS 2	1994	Meteoroid
MSTI 2	1994	Meteoroid
IRON 9906	1997	SEU
INSAT 2D	1997	ESD
TELSTAR401_	1998	MRD

(*) ESD = Electrostatic discharge
SEU = Single event upset
MRD = MeV radiation dosage

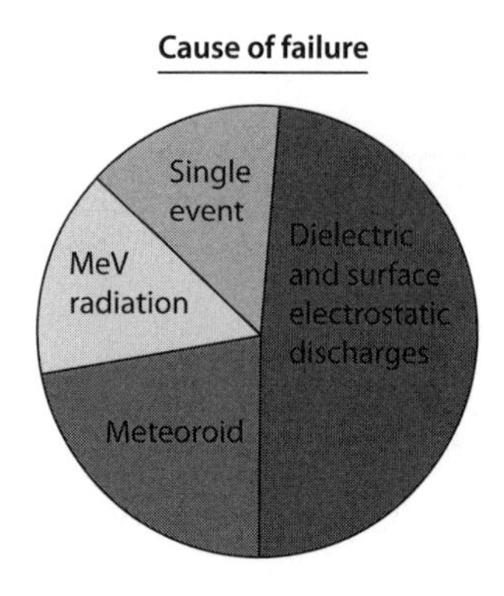

Figure 15.1 Space missions terminated due to the space environment. Most missions terminated are due to surface electrostatic discharges. (Adapted from reference 1.)

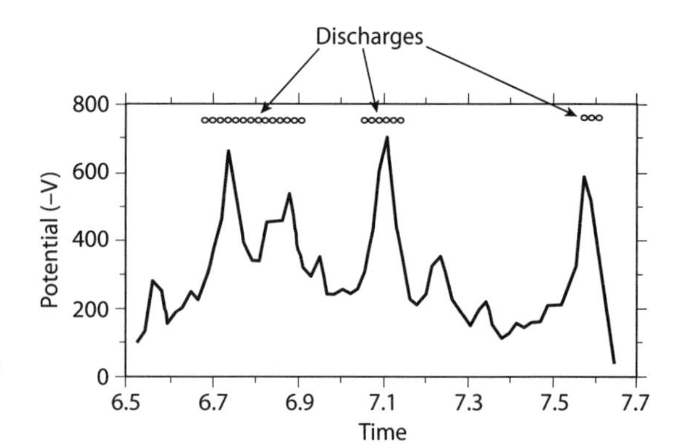

Figure 15.2 Correlation of discharge signals detected with high surface potentials measured on the SCATHA satellite. (Adapted from reference 2.)

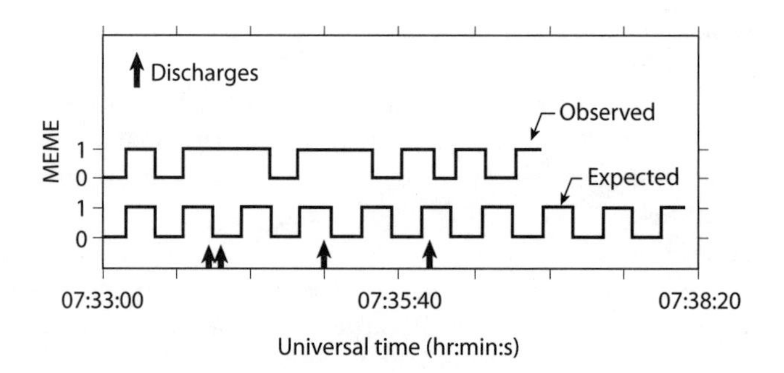

Figure 15.3 Correlation of discharges detected with the abrupt deviations from the expected sequence of electronic states. (Adapted from reference 2.)

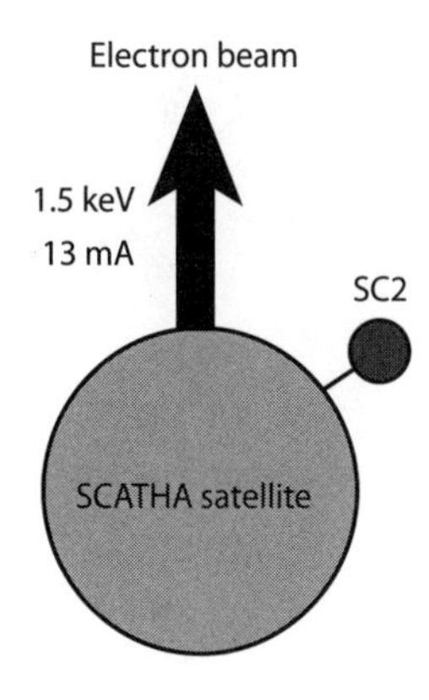

Figure 15.4 High current emitted from SCATHA. The SC2 instrument was knocked out promptly even though it is not in the path of the emitted beam. The large returning current hit and destroyed SC2.

to come in. The 2 keV voltage, being in the range for surface charging, rules out deep dielectric charging. The emitted beam current, 100 mA, exceeded the ambient current by over two orders of magnitude. The spacecraft potential reached that of the beam energy, 2 kV. The current emitted exceeded that for current balance required to maintain the 2 kV potential. Therefore, only a small fraction of the emitted beam escaped from the spacecraft; much of the emitted beam current had to return to the spacecraft surfaces. Evidently, the large incoming currents hit the SC2 instruments on the spacecraft surfaces and damaged it.

The three cases described here illustrate that surface charging may cause discharges. In turn, the discharges may disturb onboard electronic operations. High currents impinging the spacecraft may damage the instruments mounted on the surfaces. More evidence of discharges caused by surface charging has been obtained on other satellites.[6]

Deep dielectric charging and surface charging account for most of the spacecraft mission failures.[1] However, deep dielectric charging can be more damaging to spacecraft electronics than surface charging, even though the currents involved in deep dielectric charging are generally much smaller. This is because deep dielectric charging may cause adverse effects and damages deep beneath the surfaces or even in the electronics inside the spacecraft.

15.3. Energy of Surface Discharge

A capacitor can hold an amount of charge. When a discharge occurs, not only the charge stored in the surface capacitance, between the surface and the ambient plasma, is involved, but often the capacitance of the entire surface layer of material is also involved.[7] The capacitance C depends on the surface area A, the dielectric function ε, and the thickness d of the material between surfaces ($C = \varepsilon A/d$). For a given surface size and material, it is the thickness that controls the capacitance. Most of the undesired effects of electrostatic discharge include physical damage to sensitive electronics and electromagnetic wave or pulse generation. The magnitude of an electrostatic discharge depends on the energy β stored in a capacitance C:

$$\beta = CV^2/2 = \varepsilon A V^2/(2d) \tag{15.1}$$

The critical energy β^* above which damage may occur depends on the specifics of the system involved. Lacking any specifics, a value of β^* at 10 mJ is usually advised. For each value of β^*, one can plot the critical voltage V^* as a function of the thickness d. Charging to voltages above V^* is hazardous. Very thin (small d) dielectric layers sandwiched between conducting surfaces are often sources of high-energy ESD. Note that thin thermal blankets have been repeatedly reported as sources of ESD-induced anomalies on satellites.[1]

When a discharge initiates, the discharge current may begin with a small current only. Naturally, one asks "How can a small current cause any significant damage?" If the discharge current grows, a sustained arc may form. The requirement for a discharge to sustain and grow is governed by the Townsend criterion, which says that an electron has to generate at least an electron in order to maintain a chain reaction. In the chain reaction, an electron ionizes a neutral atom, or molecule, by impact. Thereby, an ionization pair is generated. The newborn electron must be accelerated by electric fields so that it may ionize further. Particle loss occurs when, for example, an electron is absorbed by an electrode or escapes from the interaction region. Energy loss occurs when, for example, an electron transfers its kinetic energy to excite an atom which, in turn, loses its energy by de-excitation with radiation emission.

If a sustained arc expands in area or spreads to wider locations, the situation becomes worse, similar to cancer metastasis. The level of damage to instruments depends mostly on the discharge current and its duration. The total charge Q is given by $Q = \int dt I(t)$, where $I(t)$ is the current. Fortunately, arc discharges on spacecraft often burn themselves out after a short while because the energy source (due to the voltage difference) decreases as a discharge progresses. As the voltage decreases, the kinetic energy of an electron in the discharge region decreases, so that the probability of ionization decreases accordingly. As a result, the Townsend criterion is no longer satisfied, and therefore the arc discharge extinguishes.

15.4 Correlation with Space Environment

It is often difficult to link spacecraft anomalies with the causes. The reasons are obvious. The spacecraft and their instruments are seldom returned to the ground for further investigation. Most spacecraft have no appropriate sensors for the space environmental parameters. Examples of common spacecraft anomalies are telemetry glitch, phantom commands, logic upsets, and status indicator errors. These anomalies are recoverable—the instrument returns to normal. Severe spacecraft anomalies involve sustained arc discharges, burnouts, sudden or stepwise loss of power, and damage to sensitive instruments. These anomalies are not recoverable. The symptoms of these anomalies are sometimes similar to each other.

The recoverable type of spacecraft anomalies are less harmful, often self-recoverable, and involve smaller currents, but they are more frequent. The nonrecoverable anomalies, which are more severe, usually involve large discharge currents, and the resultant damages can be permanent, although rare. Surface charging is more likely to involve larger capacitances and, therefore, larger discharge currents, than deep dielectric charging.

15.5 Evidence of Deep Dielectric Charging on CRRES

The CRRES satellite provided a good opportunity for experiencing deep dielectric charging. A good part of the elliptical orbit of the satellite traversed the radiation belts (also called the Van Allen radiation belts). There are high-energy (MeVs and higher) electrons and ions in the radiation belts. Indeed, CRRES experienced both surface charging and deep dielectric charging.

CRRES had many surface charging events (>30 V) during the first period (first 500 orbits) but with almost no discharge signal detected by the internal monitor.[8] Very few CRRES anomalies occurred during the first period, in which the high-energy (>300 keV) electron fluence was much lower than in the next 500 orbits. However, during the next 500 orbits, the (>300 keV) electron fluence rose to a new level (see the top panel in figure 15.5). The ion fluence also rose in about the same period. There were many internal-discharge signals and many CRRES anomalies detected (second and third panels). Interestingly, the three major peaks of discharges (at about orbits 500, 790, and 850) occurred at the same time as the three major peaks of high-energy (>30 keV) electron fluence (top panel). Also interestingly, the

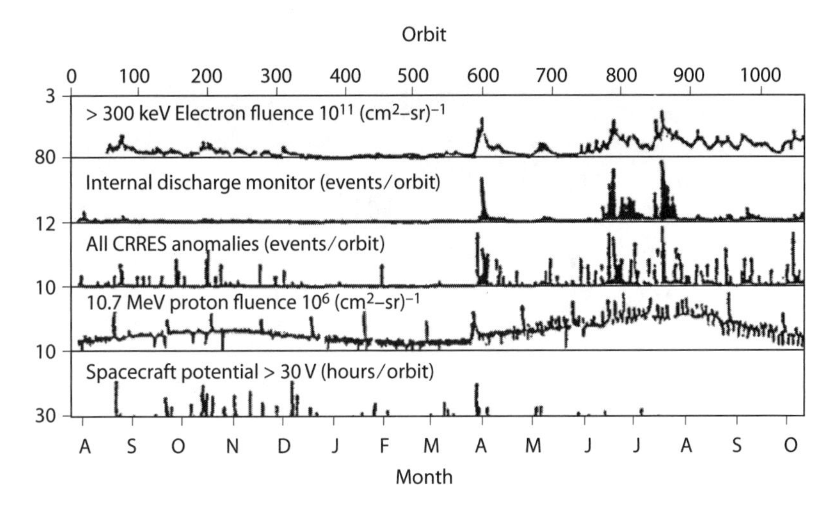

Figure 15.5 CRRES data. Many CRRES anomalies occurred during periods of high fluence of energetic electrons and ions. In these periods, the spacecraft surface potential was low. This implies that the anomalies were due to deep dielectric charging and not surface charging. (Adapted from reference 8.)

surface potential remained low throughout the discharges and anomalies in the high-fluence period (>500 orbits).

The CRRES events and correlations are important findings in the radiation belts. In summary: (1) The surface charging events had *insignificant* correlation with the anomalies and discharges. (2) The high-energy (>30 keV) electron fluence was *highly correlated* with the anomalies and discharges. (3) The correlations of the observed anomalies or discharges with the high-energy electrons infer that the cause of the anomalies and discharges is probably deep dielectric charging or other mechanisms related to high-energy electrons or ions. (4) The surface charging level of a spacecraft can be low when deep dielectric charging occurs.

Any comprehensive theory of spacecraft charging (including surface charging, deep dielectric charging, and anomalies caused by electrons and ions at low or high energies) to be developed now or in the future must be capable of explaining, at least, the preceding four findings.

15.6 Conclusive Evidence of Deep Dielectric Charging

In addition to the CRRES anomalies, the anomalies observed on the ANIK E1 and ANIK E2 satellites[9,10] were recognized to correlate well with high-energy electron fluence and attributed the cause to deep dielectric charging. Wrenn[11] analyzed the anomalies observed on the DRAS satellite and reported that the evidence of deep dielectric charging was conclusive. Figure 15.6 shows the high-energy (>2 MeV) electron two-day fluence during a year, 1994. The anomaly events of an altitude measurement equipment (AME) circuit are plotted as solid triangles. They are located almost perfectly at the fluence peaks, inferring that "high fluence of high-energy electrons caused the anomalies." Since the data used span a full year and the correlation is so nearly perfect, one has to conclude that the statement inferred is true. Since high fluence of high-energy electrons can cause deep dielectric charging, Wrenn[11] concluded that deep dielectric charging, without any doubt, caused the DRAS satellite anomalies.

Wrenn and Smith[12] suggested two thresholds for space hazards:

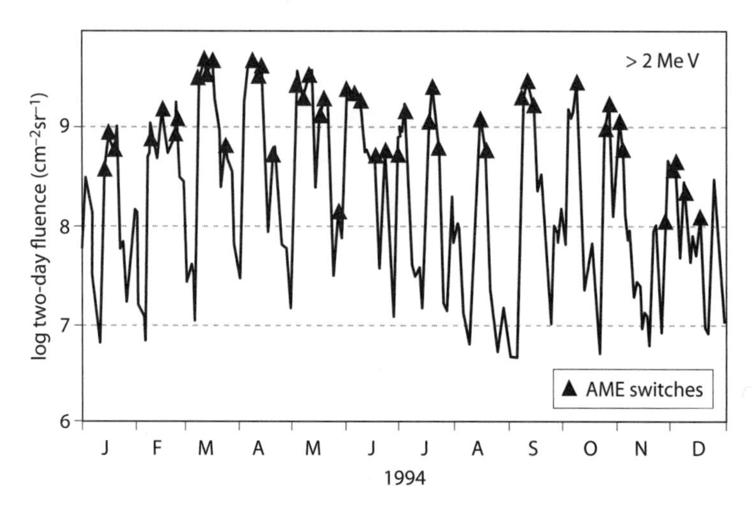

Figure 15.6 Near-perfect correlation between high-energy (>2 MeV) electron two-day fluence with the AME switch anomalies on the DRA satellite. (Adapted from reference 11.)

Threshold I for significant probability of hazard: "Daily fluence $\chi_1{}^* = 3 \times 10^8$ e/cm^2 sr for ambient electrons of energy greater than 2 MeV." It amounts to an average ambient flux F of about 3.5×10^3 e/cm^2 sr per sec for approximately constant flux.

Threshold II for extremely significant probability of hazard: "Daily fluence $\chi_2{}^* = 3 \times 10^9$ e/cm^2 sr for ambient electrons of energy greater than 2 MeV. It amounts to an average ambient flux F of about 3.5×10^4 e/cm^2 sr per sec, for approximately constant flux.

Note that the value of threshold fluence is useful as an order of magnitude estimate for guidance. The value is not a physical constant. The threshold fluence on each satellite depends on its many parameters, such as the material properties, geometry, thickness, and types of circuits, etc.

15.7 Anomalies Observed on Twin Satellites in the Radiation Belts

The Double Star[13] satellites, TC1 and TC2, were launched into near equatorial and near polar elliptical orbits, respectively (figure 15.7). They both experienced anomalies in their fluxgate magnetometers and remote terminals (figure 15.8). Because of the near polar orbit, TC2 did not cross the inner belt region. TC1 spent more time in the radiation belts and therefore had more anomalies than TC2. Four of the anomalies occurred nearly simultaneously on both satellites. TC2 had more anomalies during the outbound trips than the inbound ones, perhaps because it takes time for accumulating ambient high-energy electron fluxes inside the dielectrics to reach thresholds for hazards.

Interestingly, figure 15.8 shows that the electron flux in the period July 23 to August 7 satisfies threshold I of Wrenn and Smith,[12] and that in July 27 to August 6 satisfies threshold II. The anomalies occurred during periods of high electron flux.

To calculate the net fluence $F(t)$ deposited in time t, the equation used is as follows[13]:

$$F(t + dt) = F(t) \exp(-dt/\tau) + j(t)\,dt \qquad (15.2)$$

where $F(t)$ is the time integrated flux deposited inside the dielectric, τ is the leakage time, $j(t)$ is the ambient electron flux at time t. Reference 13 puts $\tau = 1$ h, following the CRRES results.

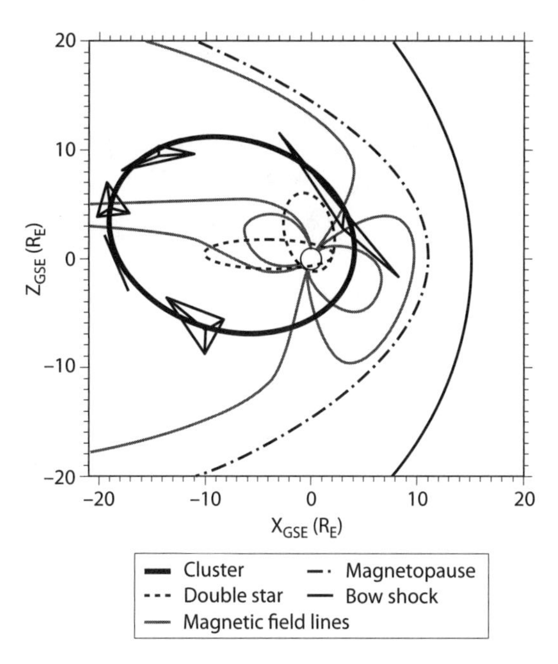

Figure 15.7 TC1 and TC2 orbits relative to the radiation belts and the Cluster orbit. (Adapted from "Orbits of the Double Star and Cluster satellites during the magnetotail," http://sci.esa.int/science-e/www/object/index.cfm?fobjectid=27711.)

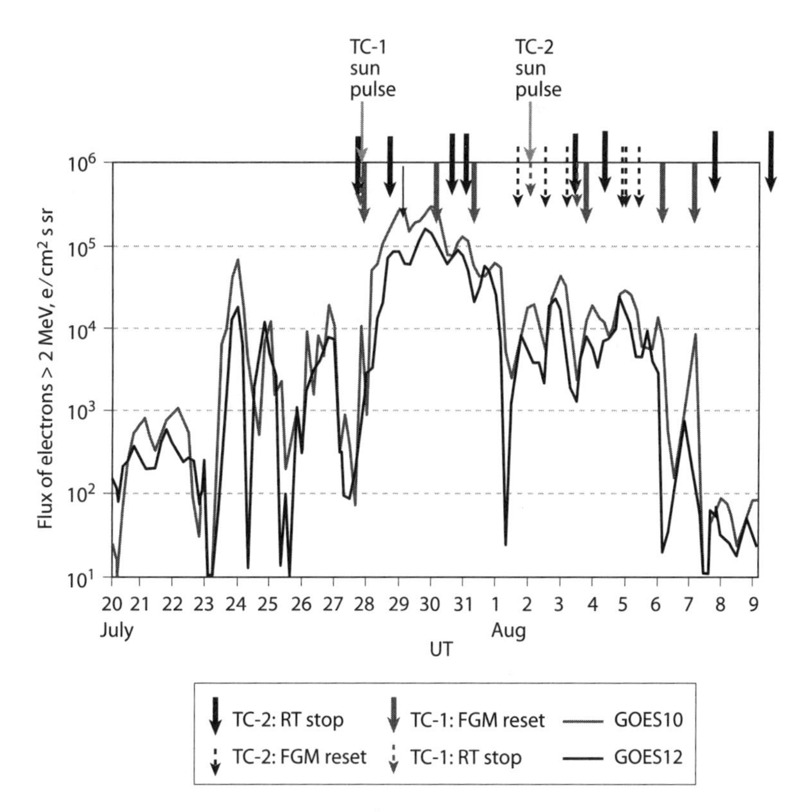

Figure 15.8 Hourly averaged flux (e/cm^2 sr s) of electrons measured by GOES10 and GOES12 satellites at geosynchronous altitudes. (Adapted from reference 13.)

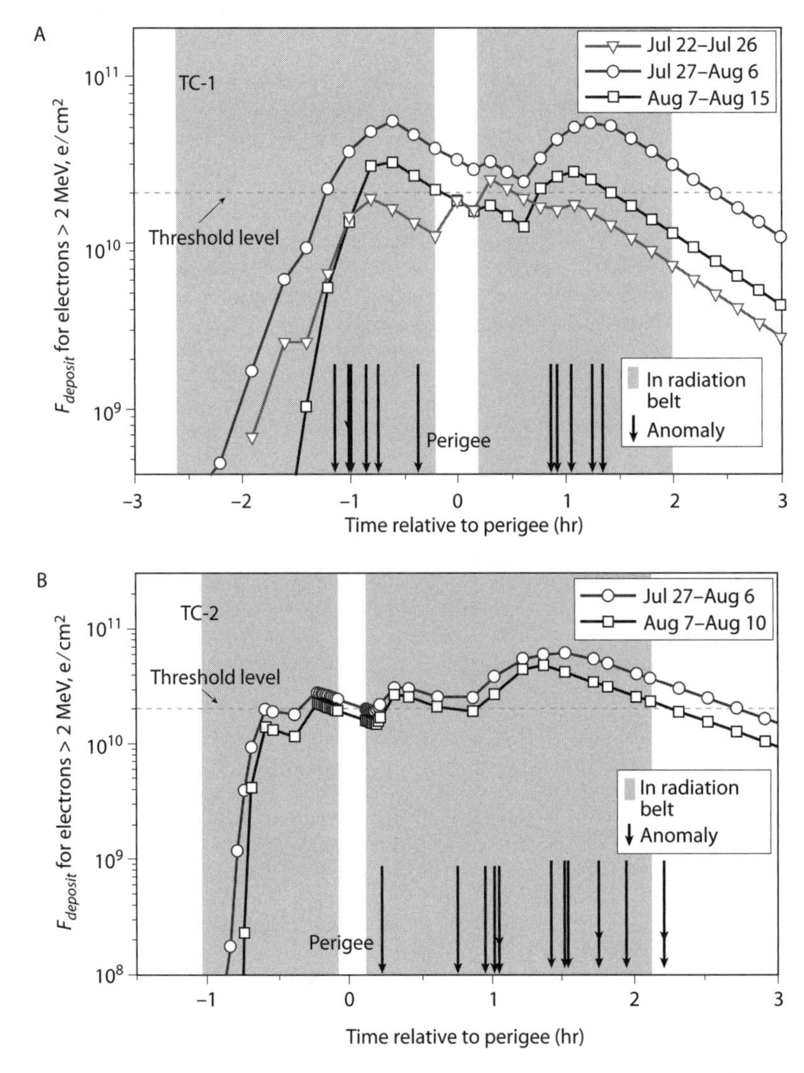

Figure 15.9 Net fluence *F(t)* of ambient electrons (over 2 MeV in energy) deposited inside dielectrics. (Inbound time is negative.) TC-2, because of its polar orbit, spends too little time in the first radiation belt, and accordingly, the integrated flux in the first radiation belt is insufficient to build up high electric fields for causing anomalies. (Adapted from reference 13.)

The results[13] obtained by using equation (15.2) are shown in figure 15.9. Note that τ is not a physical constant but depends on many factors such as material properties, geometry, and radiation induced conductivity.

From figure 15.9, the threshold daily flux of ambient electrons over 2 MeV in energy is about 2×10^{10} e/cm^2. Therefore, the anomaly environmental correlation and threshold values are consistent with the observations on CRRES[8] and DRAS.[11]

15.8 Exercises

1. The magnitude of an electrostatic discharge depends on the energy β stored in a capacitance C. Suppose that the critical charge for causing damage to a system is known. Show

that the critical voltage V of the capacitance varies with the capacitance thickness d as follows:

$$\log V = (1/2)\log d + \text{constant}$$

2. It is sometimes difficult to identify whether a spacecraft anomaly is due to surface charging or deep dielectric charging. Some indicators may be helpful:

   a. Suppose that the anomalies correlate well with high spacecraft surface potentials. What is the likely cause of the anomalies?

   b. Suppose that the anomalies correlate well with electron fluxes below about 30 keV. What is the likely cause of the anomalies?

   c. Suppose that the anomalies occur at or shortly after peaks of high-energy (MeV) electron fluencies, what is the likely cause of the anomalies?

15.9 References

1. Koons, H. C., J. E. Mazur, R. S. Selesnick, J. B. Blake, J. F. Fennell, J. L. Roeder, and P. C. Anderson, "The impact of space environment on space systems," Aerospace Report No. TR-99(1670)-1, Aerospace Corporation, El Segundo, Los Angeles, CA (1999).
2. Koons, H. C., P. F. Mizera, J. L. Roeder, and J. F. Fennell, "Several spacecraft-charging events on SCATHA in September 1982," *J. Spacecraft and Rockets* 25, no. 3: 239–243 (1988).
3. Cohen, H. A., and S. T. Lai, "Discharging the P78-2 satellite using ions and electrons," AIAA Paper 82-0266, presented at AIAA 20th Aerospace Science Meeting, Orlando, FL (1982).
4. Cohen, H. A., et al.,"P78-2 satellite and payload responses to electron beam operations on March 30, 1979," in *Spacecraft Charging Technology—1980*, eds. N. J. Stevens and C. P. Pike, NASA CP-2182, pp. 509–559, NASA Lewis Research Center, Cleveland, OH (1980).
5. Lai, S. T., "Spacecraft plasma environment induced by high current beam emission," AIAA 2003-1227, presented at 41st AIAA Aerospace Science Meeting, Reno, NV (2003).
6. Capart, J. J., and J. J. Dumesnil, "The electrostatic-discharge phenomena on Marecs-A," *ESA Bull.* 34: 22–27 (1983).
7. Ferguson, D. C., B. V. Vagner, J. T. Galofaro, and G. B. Hillard, "Arcing in LEO—does the whole array discharge?" AIAA-2005-481, presented at AIAA Aerospace Science Meeting, Reno, NV (2005).
8. Violet, M. D., and A. R. Frederickson, "Spacecraft anomalies on the CRRES satellite correlated with the environment and insulator samples," *IEEE Trans. Nucl. Sci.* 40, no. 6: 512–1520 (1993).
9. Vampola, A. L., "Thick dielectric charging on high altitude spacecraft," *J. Electrostatics* 20: 21–30 (1987).
10. Robinson, P. A., Jr., and P. Coakley, "Spacecraft charging—progress in the study of dielectrics and plasmas," *IEEE Trans. Electrical Insulation* 27, no. 5: 944–960 (1992).
11. Wrenn, G. L., "Conclusive evidence for internal dielectric charging anomalies on geosynchronous communications spacecraft," *J. Spacecraft and Rockets* 32: 514–520 (1995).
12. Wrenn G. L., and R.J.K. Smith, "Probability factors governing ESD effects in geosynchronous orbit," *IEEE Trans. Nucl. Sci.* 43, no. 6: 2783–2789 (1996).
13. Han, J., et al., "Correlation of double star anomalies with space environment," *J. Spacecraft and Rockets* 42, no. 6: 1061–1065 (2005).

16

Deep Dielectric Charging

16.1 Introduction

The previous chapter cited conclusive evidence of geosynchronous satellite anomalies due to deep dielectric charging. We need to better understand deep dielectric charging.

High-energy (MeV) electron and ion fluxes in the magnetosphere are orders of magnitude lower than those of space plasmas (electrons and ions of several keV and lower). Both high-energy electrons/ions and space plasmas exist in magnetosphere. They can coexist in the same region at the same time. Unlike space plasmas, high-energy electrons and ions do not readily charge spacecraft materials. Depending on the surface capacitances, spacecraft surface potentials respond readily to the varying space plasma condition. In contrast, measurements on or outside spacecraft surfaces do not show significant charging by high-energy electrons and ions that have penetrated deeply into the dielectrics on the satellite. Two reasons are (1) the high-energy charge fluxes are much lower than those of space plasmas, and (2) the electric fields of the electrons and ions inside compensate for each other grossly, although not necessarily locally.

Discharging may occur when the electric field in a dielectric material exceeds a critical value. We will attempt to understand the physical mechanisms and obtain estimates of the order of magnitude of the critical electric field.

16.2 The Importance of Deep Dielectric Charging

Deep dielectric charging results from the buildup of electrical charges inside dielectrics due to the penetration of high energy charged particles. Polymers, such as polyvinyl chloride (PVC), polyethylene, and plexiglass, are examples of dielectric materials. Dielectrics are insulators, having very low conductivity. High-energy electrons and ions can penetrate into materials (see chapter 14). At MeV energies, electrons penetrate deeper than ions by at least two orders of magnitude. The electrons and ions deposited inside cannot recombine or conduct away. In the radiation belts, there are high-energy (MeV or above) electrons and ions, although their fluxes are low. The fluxes are higher during severe geomagnetic storms, such as those induced by solar coronal mass ejections. When the internal electric field in the dielectric material exceeds a critical value (typically 10^6 to 10^8 V/cm, depending on the material), a sudden discharge may occur. Depending on how thin a dielectric material is, a discharge may penetrate and conduct a pulse of electric current into some sensitive electronics unexpectedly. An example of thin dielectric material is the insulating sheaths of cables. An example of sensitive electronics is the popularly used PN junctions (appendix 7). The electromagnetic pulses generated in a discharge may also interfere with electronics and telemetry.

Compared with surface charging, deep dielectric charging is difficult to be detected. Spacecraft charging can easily be detected by means of voltage monitors mounted on the surface. When the surface potential reaches kilovolts in magnitude, a possible hazard due to surface charging is clearly present. In contrast, when the internal electric fields due to a layer of electrons deposited deeply inside reach high values (even millions of volts per cm, for example), the surface monitor can hardly detect them.[1] This is because there is often a layer of positive ions deposited at a shallower depth, neutralizing to some extent the electric fields on the surface but not between the layers.[2] Even so, a sudden discharge may occur if the internal electric field reaches a critical value. The location where the discharge is initiated is probably at a sharp point, or along a fault line, inside an imperfect dielectric material. The electric field is usually higher at a sharp point and less uniform along a fault line. An external factor, such as a meteoroid hypervelocity impact[3] or a hit by a very energetic ion, may also generate an ionization track and induce a sudden discharge.

More of the various aspects outlined here will be discussed in this chapter.

16.3 High-Energy Electron and Ion Fluxes

In the geosynchronous environment, measurements show that the fluxes of high-energy (MeV) electrons and ions are lower than those at space plasmas (10 keV or less) by orders of magnitudes.*

Mathematically, the Maxwellian model of electrons illustrates the property that the velocity distribution $f(E)$, in which velocity is expressed in terms of energy E, decreases as the electron energy E increases:

$$f(E) = n(m/2\pi kT)^{3/2} \exp(-E/kT) \tag{16.1}$$

In equation (16.1), the function $f(E)$ decreases exponentially as energy E increases. The number of electrons in an energy interval, E to $E + dE$, is given by the energy integral of $f(E)$ over the interval. The number at equal energy intervals decreases as E increases. The total flux J is defined by

$$J = \int_0^\infty dE \frac{dJ(E)}{dE} \tag{16.2}$$

where the kernel $dJ(E)/dE$ is called the *differential flux* (which has unit of energy in the denominator). As discussed in chapter 4, the total flux J [equation (16.2)] can be written in terms of $f(E)$ as follows:

$$J = \int_0^\infty dE E f(E) \tag{16.3}$$

with a multiplicative constant. In the integrand $E f(E)$, the factor E increases while $\exp(-E/kT)$ decreases with E. One can easily show that the exponential $\exp(-E/kT)$ decreases faster than E as E increases. Therefore, the differential flux in equal energy intervals, $E \pm \Delta E$, decreases with E.

Sometimes, the space plasma distribution $f(E)$ deviates from being Maxwellian, especially during and after large disturbances. In such events, a double Maxwellian distribution or a kappa distribution[4,5] is often a better description of the space plasma. These deviations are transient[6,7] [figure 16.1(A)], but the MeV electron and keV electron fluxes differ by orders of magnitude [figure 16.1(B)]. As a consequence of the flux difference, the spacecraft surface potential responds much more to the variations of the low energy (keV) electrons and much less to those of high-energy (MeV) electrons.

*It is customary to refer to the MeV electrons and ions in the magnetosphere as *radiation*, because they are so fast in velocity and so low in density that they do not interact with each other collectively. They fill the Van Allen radiation belts, or simply the radiation belts, which include the geosynchronous region.

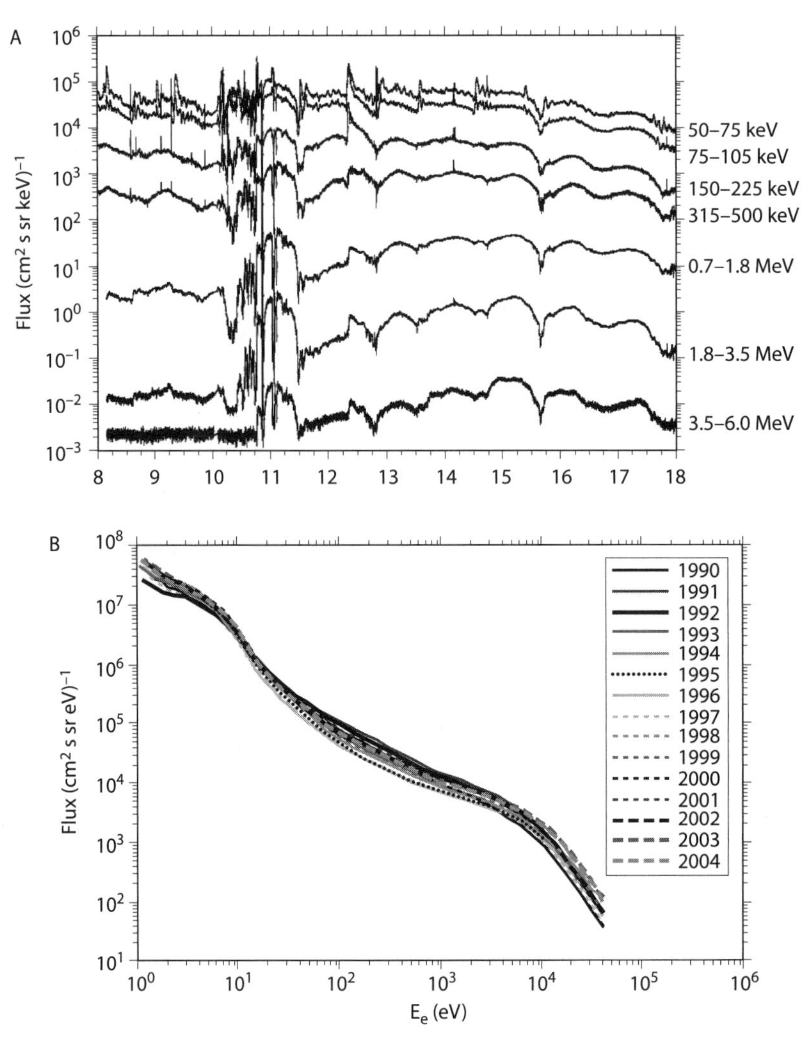

Figure 16.1 (A) Differential fluxes of electrons measured at different energy channels on Los Alamos National Laboratory (LANL) satellite 1991-080 at geosynchronous altitudes. As a general feature, the fluxes are higher at lower energies. Even during the rare event (10–11 January 1997), when the relativistic enhancement greatly disturbed the equilibrium condition in the magnetosphere, the general feature persisted. (Adapted from reference 6.) (B) Mean differential flux obtained on LANL geosynchronous satellites over more than a solar cycle. The flux is lower at higher energy values. (Adapted from reference 7.)

However, the high-energy electrons can remain deep inside a dielectric material, which has very low conductivity, for days and months. The accumulation of electrons or ions in dielectrics can slowly build up the electrostatic potential inside. The buildup, i.e., deep dielectric charging, may be hazardous to spacecraft electronics, as discussed in section 16.2.

16.4 Penetration of High-Energy Charges into Materials

High-energy electrons and ions impacting a material penetrate to different depths, depending on the particle energies (figure 16.2). At keV energies or below, electrons and ions can

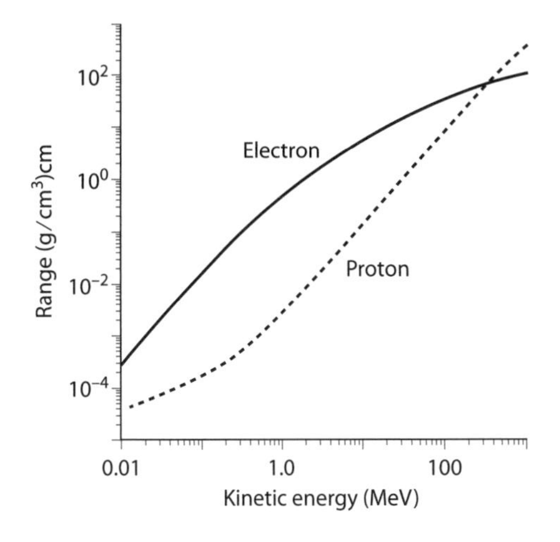

Figure 16.2 Penetration depths of electrons and protons in a dielectric material, kapton polymide (density = 1.42 g/cm^3).

penetrate only to shallow depths near the surface; there is no significant difference between their depths. At MeV energy, the electrons penetrate deeper than the ions and the difference between their penetration depths becomes significant. At multiple hundreds of MeV and above, ions penetrate to deeper depths than electrons; meson production and nuclear reactions may become possible. Chapter 14 discussed some fundamentals of the Bethe-Bloch theory of charged particle penetration into solids and provided some simple approximate formulae for practical use (see chapter 14 and the references therein).

16.5 Properties of Dielectrics

This section describes some general effects of temperature, electric field, and radiation on dielectric materials. Some references are scattered in the literature.[8–15]

16.5.1 Electric Field in a Sheet Charge Model

This section discusses some basic properties of charged dielectric materials. Electrons deposited inside the material can generate high electric fields. Depending on the conductivity, leakage current reduces the electric field. The conductivity is affected by the temperature, electron deposition, and high electric fields.

Charge deposition. Consider high-energy electron and ion penetration into a dielectric. As discussed earlier, the electron flux is higher than the ion flux by two orders of magnitude, and, in the MeV energy range, electrons penetrate deeper in dielectrics than ions by two orders of magnitude. Suppose that as a result, an electron layer is formed deeply inside the dielectric. By the Gauss law, an electric field E is generated:

$$E = \frac{Q}{A\varepsilon} \tag{16.4}$$

where Q is the net charge deposited in the layer, A is the cross-sectional area, and ε is the permittivity of the material. $\varepsilon = \varepsilon_r \varepsilon_0$, where ε_r is the relative permittivity (also called the *dielectric constant*) and ε_0 the permittivity of vacuum. The deposition rate of charge density is the current density J_d:

$$\frac{dE}{dt} = \frac{J_d}{\varepsilon} \tag{16.5}$$

Leakage conduction. Consider a conduction current density J_c. By Ohm's law, the electric field E is related to the current density J_c and the conductivity σ as follows:

$$E = \frac{J_c}{\sigma} \tag{16.6}$$

Total current. The sum of J_c and J_d is the total current J:

$$J_c + J_d = J \tag{16.7}$$

which can be written as follows:

$$\varepsilon \frac{dE}{dt} + \sigma E = J \tag{16.8}$$

or

$$\tau \frac{dE}{dt} + E = \frac{J}{\sigma} \tag{16.9}$$

where

$$\tau = \frac{\varepsilon}{\sigma} \tag{16.10}$$

In equation (16.10), τ is the decay time constant. For low-conductivity σ, the decay time is long [equation (16.10)]. For constant parameters, the following equation (16.11) is a solution of equation (16.9):

$$E(t) = E_0 \exp\left(-\frac{t}{\tau}\right) + \frac{J}{\sigma}\left[1 - \exp\left(-\frac{t}{\tau}\right)\right] \tag{16.11}$$

As a physical interpretation, equation (16.11) represents two contributions to the internal electric field E, viz, the initial electric field $E_0(t = 0)$ and the contribution by the total current J. Both are affected by the decay time constant τ,, which is usually long for dielectrics. In time $t \ll \tau$, one expands the exponentials of equation (16.11) in a Taylor series and keeps the first-order terms as follows:

$$E(t) = E_0\left(1 - \frac{t}{\tau}\right) + \frac{Jt}{\sigma\tau} \tag{16.12}$$

In the limit, $t/\tau \rightarrow 0$, one recovers $E(t) = E_0$. Table 16.1 lists some examples of the parameters σ, ε, and τ. More on the time constant τ and its typical values is given later in sections 16.13 and 16.14 on fluence and leakage.

Although equation (16.12) offers basic physical insights and provides a simple mathematical model for one to work with, one must bear in mind that it is a simple approximation. In practice, the incoming electrons are not mono-energetic. Also, the dielectric conductivity σ and the time constant τ can change slowly as a result of the high-energy electron and ion radiation. For example, radiation-induced conductivity and temperature-dependent properties are relevant in deep dielectric charging research. This area needs further experimental studies, theoretical understanding, and computational simulations and analysis.

TABLE 16.1
Examples of Some Dielectric Parameters

Material	*Conductivity* σ $\Omega^{-1}\,m^{-1}$	*Relative Permittivity* ε_r	*Decay Time* τ
Kapton	1×10^{-15}	3.45	8.5 h
Mylar	1×10^{-16}	3.00	74.0 h
Epoxy	4×10^{-15}	3.60	2.2 h
Polythene	1×10^{-14}	2.26	0.6 h
Teflon	6×10^{-15}	2.15	0.9 h
ETFE fluoropolymer	1×10^{-16}	2.50	61.0 h
Glass (CMX)	1×10^{-13}	3.80	0.1 h
Glass (borosilicate)	1×10^{-12}	6.70	1.0 min

Data from reference 16.

16.5.2 Temperature-Dependent Conductivity

Physically, thermal energy helps increase the mobility of electrons over potential barriers. Indeed, the conductivity σ of a dielectric material increases with temperature T. For typical insulators, the temperature-dependent conductivity $\sigma(T)$ is given by the equation

$$\sigma(T) = \sigma_\infty \exp\left(-\frac{E_A}{kT}\right) \tag{16.13}$$

where σ_∞ is the maximum conductivity (asymptote) at high temperature, and E_A is the material-dependent activation energy, which is not the bandgap energy. From equation (16.13), one obtains the functional dependence: $\log\sigma$ is proportional to $-1/kT$. The European Space Agency (ESA) Internal Charging Engineering Tool (ICET)[12,13] has obtained data on the conductivity of several polymers at various temperatures. The results agree with the preceding equation (figure 16.3).

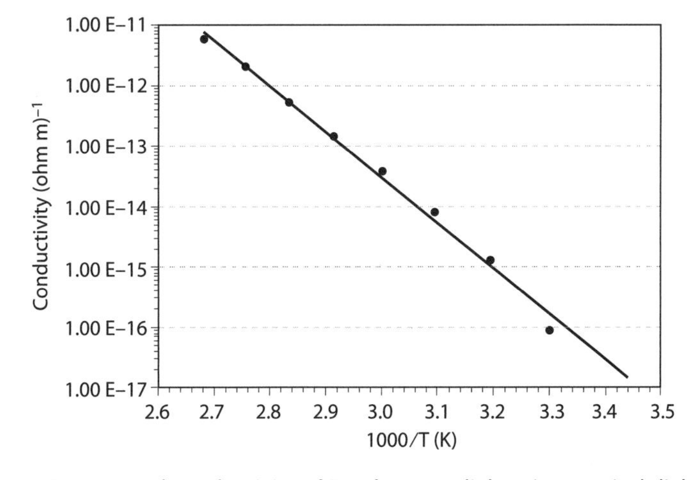

Figure 16.3 Measured conductivity of Raychem 44 dielectric, a typical dielectric material, as a function of temperature T. (Adapted from reference 13.) The log of the conductivity decreases linearly with $1/kT$.

This temperature property is significant, because equation (16.13) states that the conductivity of dielectric materials decreases as the temperature decreases. Therefore, the property implies that deep dielectric charging is more prominent at lower temperatures.

As an example of the temperature dependence of the conductivity of dielectric materials, let us take polythene ($E_A = 1$ eV). When the temperature T cools down from 25°C to 0°C, the conductivity of polythene decreases by a factor of 30 (reference 17).

16.5.3 Electric Field–Induced Conductivity

If one applies an external electric field E to a dielectric material, the conductivity changes. If the applied field exceeds a critical E^*, breakdown occurs. At electric fields below the critical value, the electric-field-induced conductivity $\sigma(E,T)$ at temperature T for dielectric materials is given by the Adamec and Calderwood equation[18,19]:

$$\sigma(E,T) = \sigma(T)\left[\frac{2 + \cosh(\beta E^{1/2}/2kT)}{3}\right]\left[\frac{2kT}{eE\delta}\sinh\left(\frac{eE\delta}{2kT}\right)\right] \tag{16.14}$$

where $\sigma(T)$ is given in equation (16.13), E is the electric field, $\beta = (e^3/\varepsilon)^{1/2}$, δ is a jump distance (typically 10 angstroms; see references 12 and 13). Sorensen[12,13] remarked that this equation, equation (16.14), has not been thoroughly validated but thought that it was in a suitable analytical form for the European Space Agency (ESA) deep dielectric charging software (reference 20).

16.5.4 Radiation-Induced Conductivity

Electron irradiation on dielectric materials can increase the conductivity. The basic behavior is governed by the Fowler equation[21]:

$$\sigma(D) = \sigma_0 + kD^{\Delta} \tag{16.15}$$

where $\sigma(D)$ is the radiation induced conductivity, and σ_0 is called the *dark conductivity*, which is the conductivity without the applied radiation. The symbol k is the coefficient of radiation-induced conductivity [$\Omega^{-1}\,\text{cm}^{-1}\,\text{rad}^{-1}\,\text{s}$], the dose rate is denoted by D [rad s^{-1}], and Δ is a dimensionless material-dependent exponent [$\Delta < 1$]. Sorensen[12,13] has reported tests using several typical polymer samples and found that the Fowler equation is valid (figure 16.4).

The following equation (16.16) is called *Frederickson's equation.*[12,13] It gives an estimate of the peak electric field E_p in irradiated dielectrics[10]:

$$E_p = (A/k)/(1 + \sigma_0/kD) \tag{16.16}$$

where $A = 10^{-12}$ sec-volt/ohm-rad-cm^2, k is the coefficient of radiation-induced conductivity (sec/cm-ohm-rd), σ_0 is the dark conductivity (ohm-cm)$^{-1}$, and D is the dose rate (rad/sec). If D is replaced by D^{Δ} in equation (16.16), the equation can be simplified to the following form:

$$E_p = AD^{\Delta}/\sigma(D) \tag{16.17}$$

where $\sigma(D)$ is given in equation (16.15). It should be cautioned[12,13] that Frederickson[10] gave this equation (16.16) without derivation, did not specify the dielectric material thickness, did not clarify the grounding arrangement, and assumed that the dose rate is the same at any depth. Careful measurements are needed to validate or improve Frederickson's equation.

16.5.5 Delayed Conductivity

Radiation and electric fields enhance the conductivity of dielectric materials. If the radiation or applied electric field is turned off after a period of application, the conductivity decreases

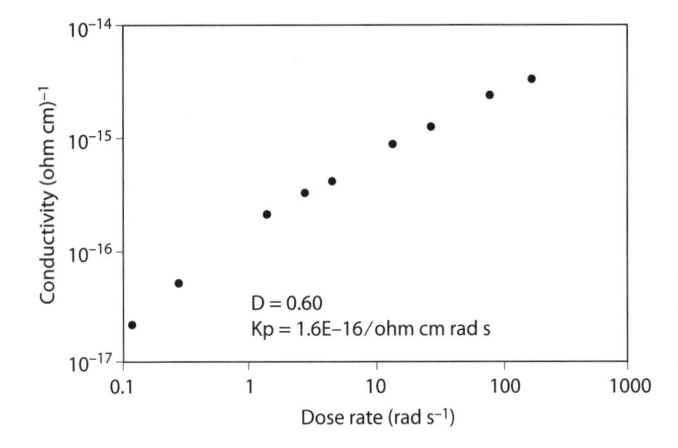

Figure 16.4 Radiation-induced conductivity measured by the European Space Agency. (Adapted from references 12 and 13.) The graph obtained agrees with the Fowler equation.

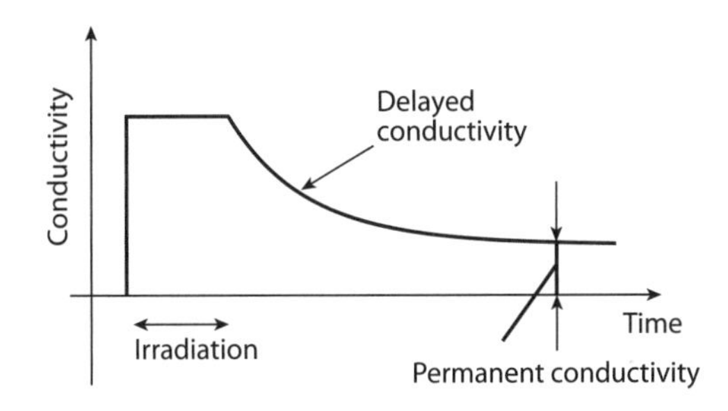

Figure 16.5 Delayed conductivity. Radiation enhances the dielectric material's conductivity. After the irradiation, the conductivity decreases slowly (typically, over more than an hour) to its normal value. (Adapted from reference 12.)

exponentially toward the permanent conductivity, which was the value in the absence of irradiation or electric fields (figure 16.5). Frederickson[10] gave the electric field decay equation in the following form:

$$E(t) \approx (r/\varepsilon)[1 - \exp(-t/r\varepsilon)] \quad (16.18)$$

where t is time, r the resistivity, and ε the dielectric constant. As Frederickson[10] had cautioned, the decay equation (16.18) is not meant to give accurate quantities but is helpful in understanding the decay time constant approximately. The exponential decay equation equation (16.18) was later used by others such as, for example, Green and Dennison.[15]

ESA measurements on polymers using beta radiation from cobalt 60 showed that the decay constant depends also on the temperature and the total dose.[12,13] Polar polymers such as plexiglass and PVC took hours to decay, whereas nonpolar polymers such as polythene and polystyrene decayed in shorter time. The polarization in the polar polymers requires hours to relax. Sorensen cautioned about using conductivity values quoted in the literature, because it is usual to measure the conductivity after only 60 s. For more on the polarization mechanism in polar polymers, see Adamec and Calderwood.[18,19]

16.6 Observations Attributed to Deep Dielectric Charging

16.6.1 Space Observations

Data observed on the CRRES satellite by Violet and Frederickson[1] have shown that the internal discharges correlate with the high-energy electron and ion fluences (the integral of flux over a period of time) but not with the surface potential (see figure 15.5). A physical interpretation[2] will be given as follows.

The high-energy electron and ion fluxes are responsible for the buildup of deep dielectric charging. The electrons and ions penetrate to different depths. For example, at MeV energies, electrons penetrate deeper than ions by two orders of magnitude. As a result, an internal negative charge layer builds up. The electric field due to the electron layer extends to outside the dielectric surface. Since nature prefers neutrality, ambient ions are attracted from outside toward the electron layer. However, the ions stop at shallower depths (see chapter 15) and therefore cannot reach the electron layer. At equilibrium, the electric field generated by the ion positive charge cancels that of the electron negative charge. But the cancellation is outside the double layers only. Since the ions cannot reach the electrons, the electric field between the layers is significant. The double layer mechanism[2] explains why, even when the internal electric field is very high, a surface monitor on a satellite cannot detect any significant electric field. In short, differential charging between different surfaces is important in surface charging, whereas differential charging between different depths is important in deep dielectric charging.

16.6.2 Laboratory Observations

Laboratory experiments have shown that when an applied electric field reaches a critical value, depending on the dielectric material properties, dielectric breakdown occurs. That is, at or above the critical electric field, the material becomes conducting. The value of the critical electric field E* is typically of the order of 10^6 V/m to 10^8 V/m, depending on the material. In practice, dielectric breakdown rarely starts globally throughout the entire material. Instead, it usually starts at some point of defect or some fault channel.

Since the electric charge buildup in a dielectric material creates the electrostatic potential energy, a sudden dielectric breakdown releases the energy as kinetic energy of the electrons and ions along the conduction channel. The kinetic energy, in turn, can contribute to the damages of the physical structure of the dielectric material by impact ionization and heat generation. As a result, when a discharge occurs, the discharge track becomes wider and wider downstream, resembling a river system. The discharge pattern (figure 16.6) is called a *Lichtenberg discharge pattern*. After irradiation of a dielectric sample by means of high-energy electrons for some time, a spontaneous discharge may or may not occur. One can induce a discharge by touching the sample with a grounded conducting wire.

Why should the critical electric field E^* be about 10^6 V/m to 10^8 V/m for various dielectric materials? Let us consider some reasons.

16.7 Avalanche Ionization in a High Electric Field

The following is a simple argument as to why the critical electric field E^* for dielectric breakdown is of the order of 10^6 V/m to 10^8 V/m. When an electron in a solid is accelerated along the direction of the applied electric field E, the electron gains kinetic energy. When it collides with a bound electron of an atom, some of the kinetic energy is spent in knocking loose an electron from the atom—i.e., ionization. The ionization energy of typical atoms is of the order of 10 Ev (reference 15), while the mean free path for a low-energy electron in an insulator is of the order of 10^{-6} m (reference 15). Therefore, in an external electric field of 10^7 V/m, the electron gains an energy $\Delta V = 10$ eV as it moves a distance $\lambda = 10^{-6}$ m. If the electron ionizes the atom upon collision, a new electron is released from a neutral atom, which itself becomes an ion. The two electrons start their journeys anew and will gain 10 eV in a mean free path and so on. In this manner, more and more electrons travel together downstream. They may also lose some energy in inelastic collisions, excitation of atoms or molecules, radiation, heating, electron-ion recombination, and lattice deformation, but if the energy gain exceeds the loss

Figure 16.6 A Lichtenberg discharge pattern in a dielectric sample. The pattern resembles that of a river system. (Adapted from http://en.wikipedia.org/wiki/Lichtenberg_figure. Copyright Bert Hickman.)

in every cycle, avalanche results. This theory, although admittedly crude, is consistent with the widening pathways downstream in Lichtenberg discharge patterns and the order of magnitude of the mysterious critical electric field $E^* \approx 10^7$ V/m.

In terms of the double layer framework, the electrons in the deep layer are repelled by the high electric field that they have piled up themselves and attracted by the positive ions near the dielectric surface. As a result, the electrons tend to move slowly toward the ion layer through the dielectric material. The electron mobility is low. The electrons gain energy as they move through the electric field. If impact ionization occurs and the gain exceeds the losses, avalanche ionization emerges.

16.8 Related Questions and Related Mechanisms

One might question the 10 eV ionization energy used in section 16.7. For some materials, it may take more energy to ionize. Collision with ionization would result in energy lost to excitation, radiation, and heating, and so the effective ionization energy would be higher. Suppose that one uses 100 eV instead of the 10 eV; the critical electric field would be $E^* = 10^8$ V/m.

One might ask another question. In solid state theory, the electrons form valence bands and conduction bands. For insulators, the bands are separated by a gap of typically 1 to few eV depending on the material (e.g., references 22 and 23). If the valence band is full and the conduction band is empty, the electrons in the valence band cannot jump over the gap. If so, we have an insulator. Thermal energy is too small to overcome the gap energy. Suppose that an applied electric field of $E^* = 3 \times 10^6$ V/m is applied and the mean free path λ equals 10^{-6}m, the electron gains an average energy ΔV of 3 eV in a collision. Therefore, the energy (3 eV) gained in the electric field along the mean free path may be sufficient to excite the electron from the valence band to the conduction band.

One might then ask "If the mean free path λ is inversely proportional to the material density n, how would the critical electric field vary?" For a shorter mean free path, the critical electric field E^* needs to be higher in order to achieve the same ionization energy per collision. Therefore, the critical electric field E^* is proportional to the material density n.

Summarizing the preceding results, we conclude that the critical electric field E^* is of the order of 10^6 to 10^8 V/m.

Another question might be: "Sir Neville Mott had a respectable theory that at a critical charge density n_e deposited in a dielectric, an insulator-conductor transition occurs. Does it apply to deep dielectric charging on spacecraft?" And another question: "Poole and Frenkel had a microscopic (atomic level) theory of insulator-conductor transition under a critical high electric field. How about it?" Yet another question: "What about Zener's breakdown?" We will try to address these three questions in the following sections.

16.9 The Mott Transition

The idea of the Mott transition is as follows. The electron in an atom is attracted by the nucleus via a Coulomb force [figure 16.7(A)]. If there are electrons deposited in the vicinity of the atom, they generate a Debye shielding, effectively shortening the range of the Coulomb force. If there are enough electrons deposited, the Debye shielding shuts off the Coulomb force by so much that the atom's electron becomes free [figure 16.7(B)].

The electrostatic potential $\phi(r)$ at a distance r from a nucleus is of the Coulomb form:

$$\phi(r) = \frac{e}{4\pi\varepsilon r} \tag{16.19}$$

The electron and the nucleus form an atom, with the electron at the distance $r = R$ from the nucleus, where R is called the *effective Bohr radius*. In the presence of the electrons deposited nearby, the potential, equation (16.19), at $r = R$ is screened by the electrons and becomes a screened Coulomb form:

$$\phi(r) = \frac{e}{4\pi\varepsilon R} \exp\left(-\frac{R}{\lambda}\right) \tag{16.20}$$

The screened potential, equation (16.20), is a short-ranged one compared with the Coulomb potential. In the Thomas-Fermi approximation (for example, reference 23), the Debye distance λ is given by

$$\lambda^2 = \frac{R}{4}\left(\frac{\pi}{3n}\right)^{1/3} \tag{16.21}$$

where n is the electron density. Note that the Debye distance λ decreases as the density n increases [equation (16.21)]. At low densities, the Debye distance λ is much longer than the Bohr radius. In the limit of $\lambda \to \infty$ in equation (16.20), one recovers the Coulomb form, equation (16.19). As the density n increases, the distance λ decreases. Eventually, the Debye distance λ equals the effective Bohr radius R, so that the potential at the Bohr radius is lowered to the value

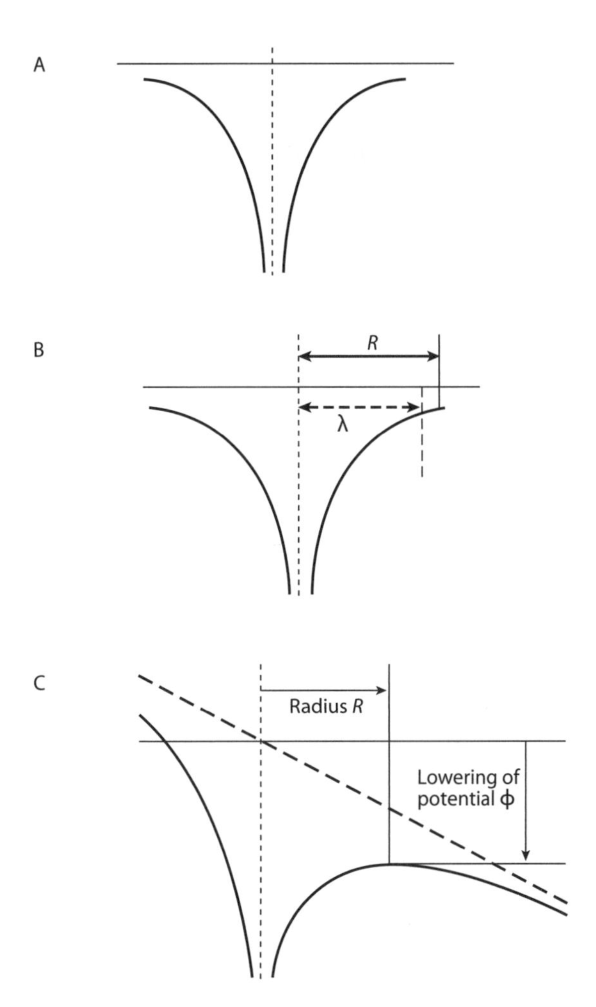

Figure 16.7 (A) Coulomb potential profile in an atom. (B) Short-range potential profile in an atom. Debye shielding shortens the range of the Coulomb force. If the Debye distance is shorter than the effective Bohr radius, the electron escapes. Mott transition is at work. (C) Tilting of the Coulomb potential profile by an applied electric field. With sufficient tilting, the electron escapes. Poole's mechanism is at work. (Adapted from reference 34.)

$$\phi(R) = \frac{e}{4\pi\varepsilon R} \exp(-1) \tag{16.22}$$

Mott's idea[24] is that when $\lambda = R$, the screened potential [equation (16.22)] is a critical value, which is too small to hold the electron at $r = R$. The electron and the nucleus no longer form a bound state. The electron then roams free away from the nucleus. Substituting $\lambda = R$ in equation (16.21), one obtains the critical electron density $n = n_*$ given by

$$n_*^{1/3} R \approx 0.25 \tag{16.23}$$

Equation (16.23) is the Mott equation for insulator-conductor transition. For a monograph on Mott transition and related topics, see, for example, reference 25. Laboratory measurements[25] using many materials have obtained a similar equation:

$$n_*^{1/3}R \approx 0.26 \pm 0.05 \qquad (16.24)$$

Equation (16.24) is very close to the Mott transition equation (16.23).

Anderson, using quantum theory for disordered systems, obtained an insulator-conductor transition equation very close to equation (16.23). Mott and Anderson shared a Nobel Prize for their many contributions in physics.

16.10 The Poole-Frenkel High Electric Field Effect

The Poole-Frenkel idea is as follows. When an electric field is applied to an atom, the potential gradient of the applied field adds to the Coulomb potential profile of the atom. The added gradient tilts the potential profile and therefore lowers the profile more on one side than on the other [figure 16.7(C)]. If the tilting lowers the potential profile so much that the electron on the atom becomes free, the electron can leave the atom and hop to another atom and so on.

When an external electric field E is applied to the atom, the potential $\phi(r)$ at $r = R$ in the direction of E becomes

$$\phi(R) = \left[\frac{e}{4\pi\varepsilon R} - ER\right]\exp\left(-\frac{R}{\lambda}\right) \qquad (16.25)$$

In the limit of electric field $E \to 0$, one recovers from equation (16.25) the screened Coulomb potential, equation (16.20). Therefore, one recovers the Mott transition equation, equation (16.23), if there is very high electron density but no external electric field applied.

If the applied electric field E is high but the electron density n is low, one can let $\lambda \to \infty$. The potential $\phi(R)$ becomes

$$\phi(R) = \left[\frac{e}{4\pi\varepsilon R} - ER\right] \qquad (16.26)$$

which is lower than that of the Coulomb potential, equation (16.21). When the applied electric field E is sufficiently high, it lowers the potential $\phi(R)$ to the critical value, equation (16.22):

$$\phi(R) = \left[\frac{e}{4\pi\varepsilon R} - ER\right] = \frac{e}{4\pi\varepsilon R}\exp(-1) \qquad (16.27)$$

With that lowered potential, the electron at R becomes loose again. The critical electric field E^* for the transition, or breakdown, is therefore given by equation (16.27) and is of the order of $E^* \approx 10^6$ to 10^8 V/m for most materials.

If one considers the combined effect[26] of high charge density and high electric fields in a solid, the breakdown voltage is lowered [equation (16.25); figure 16.8].

16.11 Zener Breakdown

In a high electric field, an electron in the valence band may tunnel through the gap to the conduction band. In quantum mechanics, the electrons in a solid form wave functions touching and overlapping the neighboring wave functions. Together they form bands separated by gaps.[22,23] In an insulator, the valence band is full but the conduction band is empty. The gap energies of typical insulators are 1 to 2 eV. The electrons in a fully occupied valence band cannot conduct because of Pauli's exclusion principle. There is no conduction unless the valence electrons can jump over the gap to the conduction band.

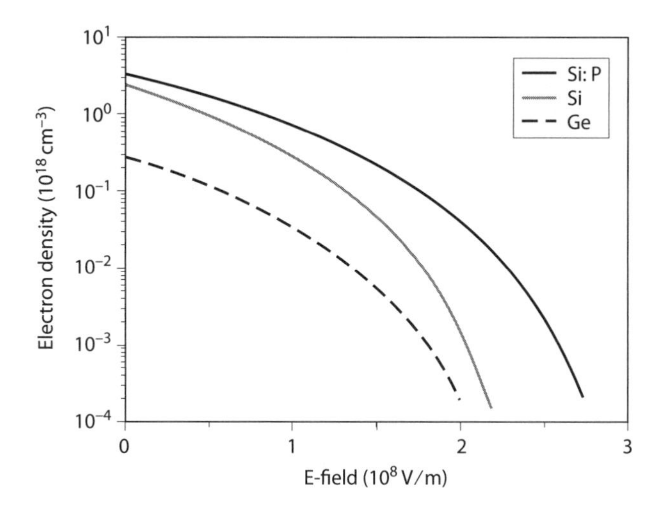

Figure 16.8 Mott's insulator-conductor transition with an applied electric field E. Without E, the electron density has to be high for the transition. At lower electron densities, the applied electric field has to be higher for the transition to occur. (Adapted from reference 26.)

In quantum mechanics, an electron can tunnel through a barrier, or a gap ε, with a finite probability. The probability increases as the barrier lowers or the gap narrows. Zener[27] suggested that applying a high electric field E may enhance the probability for the electron tunneling through the gap. Using Wentzel-Kramers-Brillouin (WKB) approximation as described in standard quantum texts (e.g., Merzbacher[28]), Zener[27] obtained the probability, or tunneling rate, γ, of tunneling through the energy gap:

$$\gamma = \frac{eEa}{h}\exp\left\{-\frac{\pi^2 m a \varepsilon^2}{h^2 eE}\right\} \tag{16.28}$$

where e is the electric charge, E the electric field, a the lattice spacing, eEa the electrostatic potential energy (in eV) across the lattice spacing, h the Planck's constant, m the electron mass, and ε the gap. For example, taking $a = 3 \times 10^{-8}$ cm, equation (16.28) gives the tunneling rate γ:

$$\gamma(\varepsilon, E) = 10^7 E \times 10^{-\frac{0.5\times10^7\times\varepsilon^2}{E}} \tag{16.29}$$

For a gap energy $\varepsilon = 2$ eV, equation (16.29) gives the Zener tunneling rate:

$$\gamma(2\text{ eV}, E) = E\ 10^{7-\frac{2\times10^7}{E}} \tag{16.30}$$

In equations (16.29) and (16.30), we have followed Zener's historical paper[27] using base 10. Instead, one can use the exponential formula, equation (16.28), directly. In figure 16.9, the Zener tunneling rate $\gamma(\varepsilon,E)$, calculated by using equation (16.30), is plotted as a function of the applied electric field E. In this figure, $\gamma(\varepsilon,E)$ remains low until a critical electric field E^* is reached. Above E^*, the tunneling rate rises rapidly as E increases, meaning that the electrons from the valence band can tunnel to the conduction band in rapidly increasing numbers. In

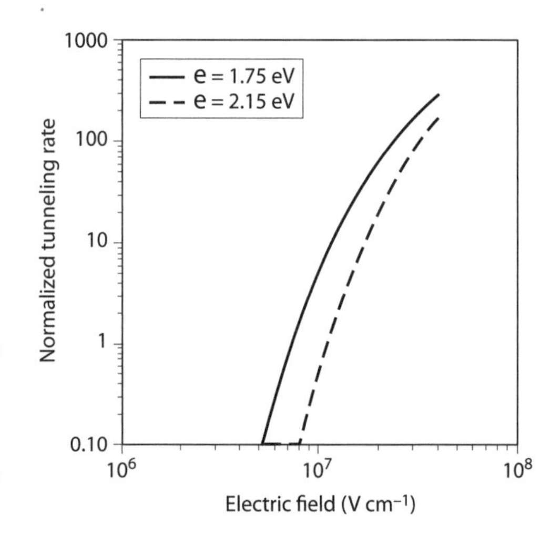

Figure 16.9 Zener breakdown as shown by the rapid rise in the tunneling rate as a function of the applied electric field E for various gap energies ε. A critical electric field E* exists [calculated by using equation (16.21)]. Below E*, the tunneling rate remains low. Above E*, the insulator becomes a conductor. (Adapted from reference 34.)

other words, the insulator becomes a conductor. The transition is abrupt as indicated by the rapid rise in tunneling rate when the applied electric field exceeds the critical value E^* (figure 16.9). This phenomenon is called Zener breakdown[27] of the insulating material (see, for example, reference 22).

16.12 Electron Fluence

Fluence F is the incoming flux accumulated over a period T. Flux J is the number of particles traveling to an unit area per unit time:

$$F(T) = \int_0^T dtJ(t) \tag{16.31}$$

The unit of flux J is the number of electrons per cm^2 per sec, and that of fluence F is the number of electrons per cm^2. If the material is very thin, only a fraction f of the flux is retained, and therefore one replaces J by fJ in equation (16.31). The complementary fraction $(1-f)$ of the flux passes through the material without staying inside and therefore does not contribute to the deep dielectric charging.

Some pedagogical notes are given as follows. For simplicity, consider the fraction $f = 1$. The fluence F received in area A equals the charge Q deposited in a deep layer of area A after a period T. The electric field E generated by the planar layer of charge Q is given by (for example, reference 29):

$$E = \frac{Q}{2\varepsilon A} \tag{16.32}$$

where

$$\varepsilon = \varepsilon_0 \kappa$$

In equation (16.32), ε is the permittivity of the dielectric material, ε_0 the permittivity of vacuum, and κ the dielectric constant (also called the *relative dielectric permittivity*). If a double layer is formed, multiply the electric field of equation (16.32) by 2 because the electric fields generated by each layer conspire[29] in the region between the layers.

Suppose that there is an electron layer of charge Q, area A, and thickness Δx, in a dielectric material. The charge density ρ is given by

$$\rho = \frac{Q}{A\Delta x} \tag{16.33}$$

The electric field E is the voltage difference $\Delta\phi$ per unit distance Δs. Suppose that an electron layer is formed at a depth Δs. The voltage difference between the layer and the dielectric surface is given as follows:

$$\Delta\phi = E\Delta s \tag{16.34}$$

16.13 Critical Fluence for Deep Dielectric Charging

Since there exists a critical electric field E^* for dielectric breakdown, there must exist a critical fluence F^*. Fluence is the accumulation of flux over a period. The critical fluence depends on the material properties, the fraction f of the incoming flux retained in the material, the charge leakage rate, and the geometry of the material considered.

Vampola[30] quoted Frederickson[31] on the critical fluence F^* for CRRES spacecraft discharges as follows: "The critical fluence F^* equals 10^{12} e^- cm^{-2} for an average flux $\langle J \rangle$ of 10^8 e^- cm^{-2} s^{-1} in a period T greater than the leakage time τ." As an exercise, let the leakage be negligible in this period, neglect geometrical effects, etc., and suppose that the initial value $F(0) = 0$. Equation (16.31) gives

$$F^*(T) = \int_0^T dt J(t) = T \times 10^8 e^- cm^{-2} s^{-1} = 10^{12} e^- cm^{-2} \tag{16.35}$$

which yields a time $T^* = 10^4$s—i.e., about 3 hours. This is the time needed for the flux accumulation to reach the critical value F^* for the given $\langle J \rangle$. The preceding estimate assumes an average $\langle J \rangle$. In reality, $J(t)$ varies with time.

If the initial value $F(0)$ is finite, the time needed becomes shorter accordingly. In a dielectric material with finite leakage, the charge deposited would be less.

16.14 Charge Density with Leakage

With leakage, the charge $Q(t)$ deposited decays exponentially with time t. Let the leakage time constant be τ. Effectively, the charge $Q(T)$ at the end of a period T is given by the sum of the incoming charge flux accumulating and decaying in that period, T_0 to T, and the initial charge $Q(T_0)$ decreasing exponentially:

$$Q(T) = \int_{T_0}^T dt J(t) \exp\left(-\frac{t - T_0}{\tau}\right) + Q(T_0) \exp\left(-\frac{t - T_0}{\tau}\right) \tag{16.36}$$

According to the CRRES results,[32] the leakage time constant τ was about one-half hour to a few hours. For the Double Star satellites,[33] the leakage time constant τ used was one hour.

16.15 A Remark on Spacecraft Anomalies

We have mentioned (chapter 15) the high correlation[17] between the spacecraft anomaly events and the high-energy (a few MeV) electron fluence observed on several spacecrafts.[20, 34] The high fluence of such electrons can cause deep dielectric charging. When some critical condition, such as critical electric field E^* for dielectric breakdown, is reached, deep dielectric discharging occurs. In general, the discharge mechanism is less well understood than the charging mechanism.[17,35] A few brief remarks on deep dielectric discharging are given as follows.

When deep dielectric discharging occurs, it may occur partially. A partial discharge does not form a complete Lichtenberg figure, unlike the full glory in figure 16.6. Partial discharges may occur along short paths only. Small discharges cause fewer effects or damages than large ones. With less damage, a spacecraft may survive without termination of its mission. The discharge signals monitored on the CRRES satellite, figure 15.5, were probably due to small discharges because CRRES survived after the period shown in the figure. With each small discharge, the likelihood of subsequent discharge occurring along the same path increases. This is because of ionization and lattice damage along the path. Subsequently, a large discharge may occur there.

It appears that the critical electric field observed in space[17] is usually lower than those quoted in materials data books. A spacecraft orbiting in radiation environments for months to years experiences many small discharges, thus rendering likely a large discharge subsequently. Furthermore, a spacecraft receives high-energy electrons penetrating deep in the dielectrics. On the other hand, the laboratory data are usually obtained by turning up the applied electric field quickly without waiting for months and without simultaneously bombarding the material with high-energy electrons.

When a discharge occurs between the double layers, a conduction path exists, at least temporarily. The electrons and ions shoot toward the layer of the opposite sign. As a conjecture, they can overshoot the layers that are not metallic electrodes. The overshoot is reminiscent of a Kaufman thruster (see, for example, figure 3 in reference 36) and may explain why buried electrons are sometimes ejected from charged dielectrics (see, for example, figure 10 of reference 35).

If a conductor ground is near the buried charge layers—for example, as a substrate near the deep layer of electrons or even sideways—a discharge may occur between the conductor and the buried charges.

Certain dielectrics, such as mica and porous ceramic, may carry adsorbed gas. The gas ionizes before the solid reaches breakdown condition. Intermittent sparking may occur. In space, outgassing may occur in the first few weeks after launch but should gradually subside.

16.16 Effect of Electrons Deposited inside Electronics

PN junctions are commonly used in electronics (see appendix 7). In a PN junction, two pieces of silicon are touching each other. In the language of solid state electronics, one piece is doped with excess electrons (N), while the other is doped with excess holes (P). For normal operations, the junction acts as a rectifier. When a forward electric field is applied, conduction occurs because the holes and electrons are moving toward each other. When a reversed electric field is applied, no conduction is expected because the holes and electrons are moving away from each other, thus forming a depletion layer. In space, anomalies can occur when (1) high-energy (MeV) electrons penetrate deep into the solid and deposit in the junction, especially at or near the depletion layer, which is delicate for the functioning of the rectifier, or (2) high electric fields build up at or near the depletion layer. These two points are briefly discussed as follows.

1. As a result of the bombardment of a high fluence of high-energy electron, excess electrons or holes accumulated inside a PN junction can change the densities of the electrons and holes in the P and N silicon materials. As a result, the current-voltage characteristic of the PN junction changes, possibly causing abnormal behavior.

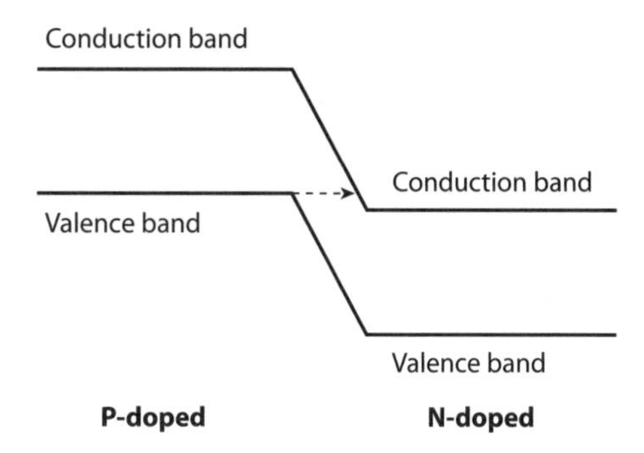

Figure 16.10 Zener breakdown in the depletion layer of a PN junction. (Adapted from reference 16.)

2. When an electric field is applied across the depletion layer of a PN junction, tunneling may occur. The tunneling rate abruptly rises when a critical electric field E^* is reached. This phenomenon is a Zener breakdown of the PN junction. The junction normally functions as a diode rectifier. With Zener breakdown, the PN junction diode (see, for example, reference 16) behaves abnormally.

PN junctions are common in space electronics. High-energy electron bombardment can cause not only (1) high-field breakdown deep in dielectrics such as cable sheaths but also (2) breakdowns in electronics. Mechanism 1 is more general, and much attention has been paid to it by the deep dielectric charging research community.[10,31,32] Mechanism 2, being more specific to the type of electronics used, may be responsible for many more spacecraft anomalies than previously thought. We suggest that charging in semiconductors and "breakdowns in electronics" such as Zener breakdown in PN junction diodes, are important areas for spacecraft anomalies studies.

16.17 Exercises

1. Suppose that the flux J is 2×10^8 $e^- cm^{-2} s^{-1}$ in a period of 10^5 s. What is the flux J in units of ampere per unit area per second? What is the fluence F for this period?

2. In the scenario of problem 1, suppose that the fraction f of electron flux retained in a thin dielectric material is 0.6. What is the total charge Q deposited in the period? What is the electric field E generated (in volts per cm) if the dielectric constant K equals 4.2?

3. Suppose that the electrons of certain energies penetrate to an approximate depth of 2 mm and the deposition is in the final one-tenth of the electron path approximately. Assuming that the area A is 1 cm^2, what is the approximate electron deposition volume? What is the charge density there?

4. Suppose that an electron layer is formed at a depth of 5 mm inside a dielectric material but an ion layer is not yet formed. What is the voltage difference between the layer and the dielectric surface?

5. Calculate the Coulomb potential $\phi(r)$ at $r = R$ for Si:P, Si, and Ge with $R = 17$, 19, and 39Å, respectively.[25]

6. Calculate the screened potential $\phi(r)$ for Si:P, Si and Ge, with $\lambda = R$.

7. Calculate the breakdown electric field E^* for Si:P, Si, and Ge at negligible charge densities ($\lambda \rightarrow \infty$). (Answers: 3.2×10^8 V/m, 2.5×10^8 V/m, and 6×10^7 V/m, respectively.)

8. Consider the Zener tunneling rate γ (equation 16.20). Plot the graph $d\gamma/dE$ versus the applied electric field E for various bandgap energies, 1.0, 1.5, and 2.0 eV. At what electric field E does the rate γ increases suddenly in each case?

16.18 References

1. Violet, M. D., and A. R. Frederickson, "Spacecraft anomalies on the CRRES satellite correlated with the environment and insulator samples," *IEEE Trans. Nucl. Sci.* 40, no. 6: 1512–1521 (1993).
2. Lai, S. T., "A mechanism of deep dielectric charging and discharging on space vehicles," presented at IEEE International Conference on Plasma Sciences, Raleigh, NC (1998). Conference Record of 1998 IEEE International Conference on Plasma Science, p. 154, doi: 10.1109/PLASMA.1998.677564, ISSN: 0730-9244 (1998).
3. Lai, S. T., E. Murad, and W.,J. McNeil, "Hazards of hypervelocity impacts on spacecraft," *J. Spacecraft and Rockets*, 39, no. 1, 106–114. (2002).
4. Vasyliunas, V. M., "A survey of low-energy electrons in the evening sector of the magnetosphere with Ogo 1 and Ogo 3," *J. Geophys. Res.* 73: 2339–2385 (1968).
5. Meyer-Vernet, N., "How does the solar wind blow? A simple kinetic model," *Eur. J. Phys.* 20: 167–176 (1999).
6. Reeves, G. D., R.H.W. Friedel, R. D. Belian, M. M. Meier, M. G. Henderson, T. Onsager, H. J. Singer, D. N. Baker, X. Li, and J. B. Blake, "The relativistic electron response at geosynchronous orbit during the January 1997 magnetic storm," *J. Geophys. Res.* 103, no. A8: 17559–17570 (1998).
7. Thomsen, M. F., M. H. Denton, B. Lavaud, and M. Bodeau, "Statistics of plasma fluxes at geosynchronous orbit over more than a full solar cycle," *Space Weather* 5: S03004 (2007).
8. Gross, B., "Radiation-induced charge storage and polarization effects," in *Electrets*, ed. G. J. Sessler, pp. 217–279, Springer-Verlag, Berlin and New York (1980).
9. Sessler, G. M., Ed., *Electrets*, Springer-Verlag, Berlin and New York (1980).
10. Frederickson, A. R., "Partial discharge phenomena in space applications," in *Fourth European Symposium on Spacecraft Materials in Space Environment*, pp. 221–232, *ESA*, CERT/ONERA, Toulouse, France (1989).
11. Frederickson, A. R., "Progress in high-energy electron and x-irradiation of insulating dielectrics," *Brazil. J. Phys.* 29, no. 2: 241–253 (1999).
12. Sorensen, J., "Engineering tools for internal charging," Final Report, DERA/CIS/CIS2/CR990401, Issue 1.0, J. Sorensen, Tech. Manager, European Space Agency, The Netherlands (1999).
13. Sorensen, J., "Engineering tools for internal charging," Final Report, DERA/CIS2/CR000277, Issue 1.0, J. Sorensen, Tech. Manager, European Space Agency, The Netherlands (2000).
14. Dennison, J. R., J. Gillespie, J. Hodges, R. C. Hoffman, J. Abbott, and A. Hunt, "Radiation induced conductivity of highly insulating spacecraft materials," presented at 10th Spacecraft Charging and Technology Conference, Biarritz, France (2007).
15. Green, N. W., and J. R. Dennison, "Deep dielectric charging of spacecraft polymers by energetic protons," *IEEE Trans. Plasma Sci.* 36: 2482–2490 (2009).
16. Rudden, M. N., and J. Wilson, *Elements of Solid State Physics*, 2nd ed., Wiley, New York (1993).
17. Rodgers, D. J., and K. A. Ryden, "Internal charging in space, spacecraft charging technology," in *Proceedings of the Seventh Spacecraft Charging Technology Conference, ESTEC*, ed. R. A. Harris, ESA SP-476, pp. 25–32, European Space Agency, Noordwijk, The Netherlands (2001).

18. Adamec, V., and J. H. Calderwood, "Electrical conduction in dielectrics at high fields," *J. Phys. D* 8: 551–560 (1975).
19. Adamec, V., and J. H. Calderwood, "Electrical conduction and polarization phenomena in polymer dielectrics at low fields," *J. Phys. D. Appl. Phys.* 11: 781–790 (1978).
20. Sorensen, J., D. J. Rodgers, K. A. Ryden, P. M. Latham, G. L. Wrenn, L. Levy, and G. Panabiere, "ESA's tools for internal charging," *IEEE Trans. Nucl. Sci.* 47, no. 3: 491–497 (2000).
21. Fowler, J. F., "X-ray induced conductivity in insulating materials," *Proc. Roy. Soc.* 236A: 464–480 (1956).
22. Ziman, J. M., *Principles of the Theory of Solids*, Cambridge University Press, Cambridge, UK (1964).
23. Kittel, C., *Introduction to Solid State Physics*, 8th ed., John Wiley and Sons, New York (1986).
24. Mott, N. F., *Metal-Insulator Transitions*, Taylor and Francis, London, UK (1974).
25. Kamimua, H., and H. Aoki, *The Physics of Interacting Electrons in Disordered Systems*, Clarendon Press, Oxford, UK (1989).
26. Lai, S. T., "Mott transition as a cause of spacecraft anomalies," *IEEE Trans. Plasma Sci.* 28, no. 6: 2097–2102 (2000).
27. Zener, C., "A theory of the electrical breakdown of solid dielectrics, *Proc. Royal Soc. (London), Series A* 145: 524–529 (1934).
28. Merzbacher, E., *Quantum Mechanics*, 3rd ed., Wiley, New York (1998).
29. Griffiths, D. J., *Introduction to Electrodynamics*, Prentice Hall, Englewood Cliffs, NJ (1981).
30. Vampola, A. L., "Thick dielectric charging on high-altitude spacecraft," *J. Electrostatics* 20, no. 1, 21–30 (1987).
31. Frederickson, A. R., "Bulk charging and breakdown in electron-irradiated polymers," in *Spacecraft Charging Technology—1980*, eds. N. J. Stevens and C. P. Pike, NASA CP-2182, pp. 33–51, NASA Lewis Research Center, Cleveland, OH (1981).
32. Frederickson, A. R., E. G. Holman, and E. G. Mullen, "Characteristics of spontaneous electrical discharging of various insulators in space radiation," *IEEE Trans. Nucl. Sci.* 39, no. 6: 1773–1782 (1992).
33. Han, J., J. Huang, Z. Liu, and S. Wang, "Correlation of double star anomalies with space environment," *J. Spacecraft and Rockets* 42, no. 6: 1061–1065 (2005).
34. Wrenn, G. L., and R.J.K. Smith, "Probability factors governing ESD effects in geosynchronous orbit," *IEEE Trans. Nucl. Sci.* 43, no. 6: 2783–2789 (1996).
35. Catani, J. P., "Electrostatic discharges and spacecraft anomalies," in *Proc. Seventh Spacecraft Charging Technology Conference*, ed. R. A. Harris, ESA SP-476, p. 33, European Space Agency, Noordwijk, The Netherlands (2001).
36. Wilbur, P. J., V. K. Rawlin, and J. R. Beattle, "Ion thruster development trends and status in the United States," *J. Propulsion and Power* 14, no. 5: 708–715 (1998).

17

Charging Mitigation Methods

Various methods to mitigate spacecraft charging have been, proposed, discussed, or tested in the past decade. They all have advantages and disadvantages. In this chapter, we discuss and critique the various mitigation methods. In the last part of this chapter (section 17.10), we discuss briefly mitigation of deep dielectric charging, a recent development.

17.1 Introduction

In general, there are two types of spacecraft charging mitigation methods: active and passive. The active type is controlled by commands; the passive type is automatic without control. The main methods are listed in table 17.1.

Alternatively, the mitigation methods can be arranged into two types: (1) electron ejection and (2) ion reception. In method 1, a device draws electrons from the spacecraft ground and ejects them into space.[1] This method is effective for reducing the negative charge of the spacecraft ground but is ineffective for mitigating the dielectric surface potential. As a result, differential charging between the dielectric surfaces and the conducting ground ensues. The resulting differential charging may pose a worse situation than before. In addition, if a high-current, high-energy electron beam is emitted, the potential sheath of the spacecraft may engulf the neighboring surfaces, such as those on booms, resulting in current flows between surfaces.[2,3]

In method 2, positive ions arrive at a spacecraft that is charged negatively. This method is effective in mitigating any negatively charged surface, regardless of dielectric or conductor. The ions neutralize the negative charges. The ions may preferentially land on the *hot spots*, where the negative potential is higher.[4] Furthermore, if the ions are energetic enough, they may act as secondary electron generators. The secondary electrons are repelled by the negative surface potentials and therefore leave, carrying away negative charges.

Thus, method 2 is effective for reducing differential charging. A disadvantage is that prolong use may end up electroplating the entire spacecraft. A combination of both types (1 and 2) is recommended. We will now discuss each of the methods listed in table 17.1.

17.2. Sharp Spike Method

Sharp spikes protruding from charged surfaces generate very high electric field E (figure 17.1). The E field at the spike tip is proportional to r^{-2}, where r is the radius of curvature of the tip. At sufficiently high fields, field emission of electrons occurs reducing the negative potential of the conducting surfaces connected to the spike. The current density J of field emission is given by the Fowler-Nordheim equation[5]:

$$J = AE \exp(-BW^{3/2}/E) \tag{17.1}$$

where A,B are constants, and W is the work function.

This is a convenient, passive method requiring no command/control. Its disadvantage is that the electron emission draws electrons from the conducting ground only. Thus,

TABLE 17.1
Critical Overview of Spacecraft Charging Mitigation Method

Method	*Type*	*Physics*	*Comment*
Sharp spike	Passive	Field emission	Requires high E field. Ion sputtering of the sharp points. Mitigates charging of conducting ground surface but not dielectrics. Differential charging ensues.
Conducting grids	Passive	Prevention of high E field	Periodic surface potential.
Semiconducting paint	Passive	Increase of conductivity on dielectric surfaces	Mitigates dielectric surface charging. Paint conductivity may change gradually.
High secondary electron yield material	Passive	Secondary electron emission	Mitigates for primary electrons at energies between the $[\delta(E) = 1]$ crossing points.
Hot filament	Active	Thermal electron emission	Space charge current limitation. Mitigates conducting ground charging only. Differential charging ensues.
Electron beam or emission	Active	Emission of electrons	Mitigates conducting ground charging only. Differential charging ensues.
Ion beam or emission	Active	Return of low-energy ions	Neutralizes the "hot" spots. Effective for both conducting and dielectric surfaces. The ions may act as secondary electron generators. Cannot reduce potential below the emitted ion energy unless charge exchange occurs.
Plasma emission	Active	Emission of electrons and ions	More effective than electron or ion emission alone.
Evaporation	Active	Evaporation of polar molecules that attach electrons	Mitigates conducting and dielectric surface charging. Not intended for deep charging. May cause contamination.
Metal-based dielectrics	Passive	Increase of conductivity in dielectrics	Mitigates deep dielectric charging. Metal-based material needs to be homogeneous to be useful. Conductivity change and control need to be studied.

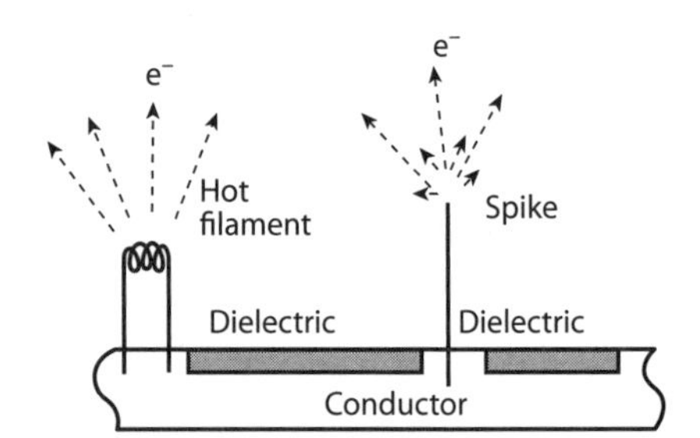

Figure 17.1 Electron emission from a sharp spike and a hot filament. (Adapted from reference 4.)

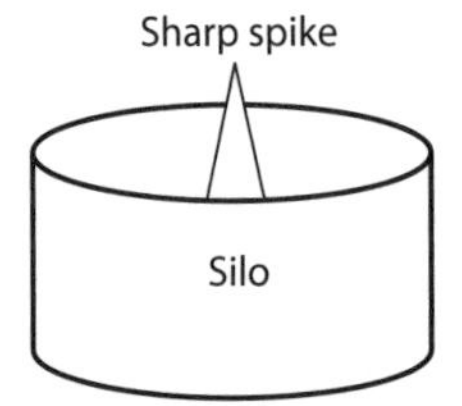

Figure 17.2 A sharp spike housed in a silo. The sharp spike, connected to the spacecraft ground, is negatively charged. The positive ambient ions, being heavier than electrons, come in with helical trajectories in the presence of ambient magnetic fields and, as a result, may hit the silo instead of the sharp spike tip. In this way, the sharp spike may have a longer lifetime.

differential charging may ensue, as discussed in the previous section. There is another disadvantage, viz, ion sputtering of the tips can blunt them, reducing the efficiency of field emission. This is because ambient positive ions attracted by the high E field of the tip may knock out some atoms on impact.

There are ways to mitigate sputtering. One way is to protect a spike tip by means of ceramic coating. Such a coating would help prevent ions from sputtering the tip inside, because the ion collisional cross section in the coating is larger that of electrons. The ions have slowed down when they reach the tip inside. Another way is to house the spike inside a silo (figure 17.2). Electrons and ions gyrate in ambient magnetic fields. In the ionosphere, the geomagnetic field is significant. Ions gyrate with larger gyroradii than electrons. As a result, some ions may hit the silo structure instead of the spike tip.[6–8] As a result, the silo helps maintain the sharpness of the tip for a longer time. Some examples of recent progress in field emission of electrons for spacecraft charging mitigation are found in references 9 to 11.

17.3 Hot Filament Emission Method

In this method, electrons are emitted from hot filaments. The filament materials used are of high melting points. The current J density emitted is given by the thermionic emission equation[12]:

$$J = AT^2 \exp(-W/kT) \tag{17.2}$$

where A is a constant, W the work function, and kT the thermal energy.

Near or above the melting points of the materials, both neutrals and ions are *evaporated*, and the ion current density J^+ is given by an equation of the same form as J but the constants are different.

For charging mitigation using hot filaments, electrons are emitted from hot filaments that do not melt. (The use of melting filaments would fall into a different category, viz, ion or plasma emission.) Since electron emission can reduce the charging level of the spacecraft ground but not the dielectric surfaces, differential charging may ensue (see section 17.2).

Furthermore, the current emitted may be limited by space charge saturation very near the filament, because the energy of thermal electrons is low.

17.4 Conducting Grid Method

One often-discussed method is to cover a nonconducting surface with a mesh of conducting wires. Although the wire mesh provides a uniform potential along the wires throughout the area, periodic potential differences between the wires and the surface area may develop. This method is convenient and passive. It may be adequate for some applications but is not recommended for most cases.

17.5 Partially Conducting Paint/Surface Method

The use of partially conducting paint eliminates the periodic potential problem described section 17.4 and is often effective and convenient. Examples of partially conducting paints are zinc ortho-titanate, alodyne, and indium oxide.[13] Frederickson et al.[14] have discussed the properties of a number of spacecraft polymer materials.

Two comments are offered: (1) Under bombardment by electrons, ions, and atoms (especially oxygen atoms), the surface material properties, including conductivity, change gradually in time. More measurements and research are needed in this area. (2) Introducing metal atoms into the interstitial lattice sites of polymers would produce metalized polymers not homogeneous enough for many purposes. The recent techniques of introducing metal atoms at the molecule level deserve significant attention, and this topic will be discussed later in section 17.10.

17.6 High Secondary Electron Yield Method

The use of coatings of high secondary electron emission ($\delta_{max} \gg 1$) would work for a certain primary electron energy range (typically up to about 1 keV) only. Beyond that range, the secondary emission decreases to below unity ($\delta(E) < 1$) and therefore offers no protection against charging. A case in point is the copper-beryllium surface of the SC10 boom[15] on SCATHA. The material has a $\delta_{max} = 4$. When the space plasma became stormy ($kT \gg$ keVs), on day 114, the boom suddenly jumped, in a triple-root fashion, from nearly zero V to a high potential of the order of kV negative.[16]

17.7 Electron and Ion Emission Method

Electron emission alone is not effective in reducing the negative potentials of a spacecraft as a whole, because the surfaces electrically isolated from the beam emission device are unaffected. Paradoxically, emission of low-energy positive ions from highly negatively charged spacecraft can reduce the potential effectively. This method has been observed (figure 17.3) on SCATHA[3,17] and simulated on a computer.[18]

A physical interpretation[3] of this apparent paradox is that the low-energy ions cannot go very far and have to return to the spacecraft (figure 17.4). This method is effective in mitigating differential charging. It is effective even without ion-induced secondary electrons, because the ions can automatically home in toward the more negatively charged surfaces.

As a corollary, the method of emitting ions alone is not expected to complete the mitigation of charging if there is no other player, or chess piece, in the system. The mitigation process would stop when the spacecraft potential energy $e\phi$ is equal to or less than the emission energy E_i of the ions.

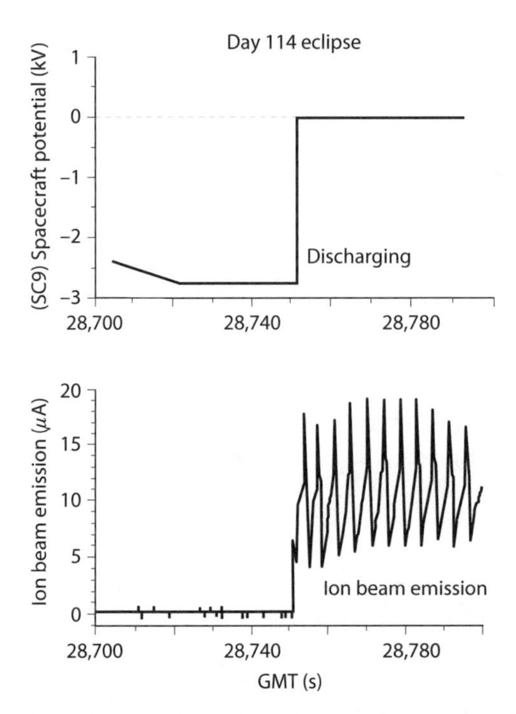

Figure 17.3 Xe+ (50 eV) ion emission from the SCATHA satellite charged to nearly −3 kV. The emission mitigated the charging. The potential was measured at 16-s intervals. (Adapted from reference 3.)

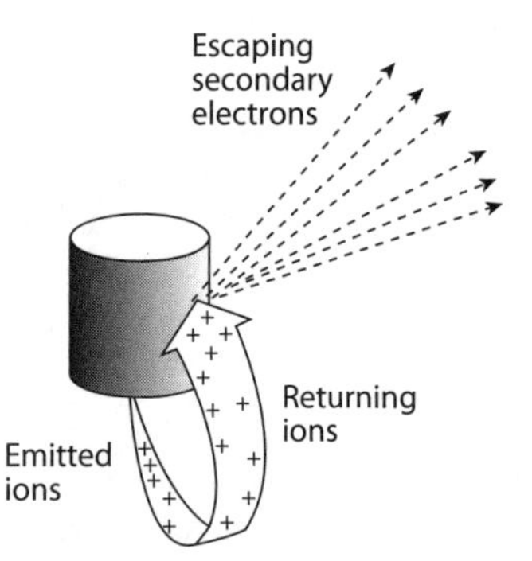

Figure 17.4 Positive ion emission from a highly negatively charged spacecraft. The ions are returning. (Adapted from reference 3.)

$$e\phi \leq E_i \tag{17.3}$$

For example, if positive ions of higher energy (for example, 200 eV) are emitted from a spacecraft of lower potential (say, −100 V), the ions will escape without returning to the negatively charged spacecraft surfaces. In this case, the spacecraft potential cannot be reduced to below −100 V in magnitude. Such a phenomenon has been observed.

However, if there is another chess piece in the system, the game would change. Suppose that neutral gas molecules are abundant in the vicinity of the ion exit point. Positive ions A^+

can exchange charge (see chapter 20 and the references therein) with low-energy (thermal) neutral molecules M to form low-energy ions:

$$A^+ + M \rightarrow A + M^+ \tag{17.4}$$

where M^+ is the low-energy ion generated from the thermal-neutral molecule M. The charge exchange cross section depends on the species and the energy involved. Usually, the cross section decreases with the ion energy. The charge exchange rate depends also on the densities (see chapter 20).

17.8 The DSCS Charge Control Experiment

Emission of a mixture of low-energy ions and electrons—i.e., plasma—would be a reasonable method for active charge mitigation. It would combine the advantages of both the electron and the ion emission methods. The charge control experiment on DSCS is for demonstrating this method. Early results[19] showed that it worked. The next section will discuss some case studies. The DSCS satellite[19] is at geosynchronous altitudes. Two dielectric samples, kapton and quartz, are both on the same side of the spacecraft. A field-mill device behind each sample measures the potential difference between the sample and the spacecraft ground. The spacecraft is often in sunlight, and therefore the ground is charged slightly positively near 0 V. When the kapton reaches −1.5 kV, a device onboard would automatically trigger the release of an ionized xenon gas (plasma) of energy below 10 eV.

Figure 17.5 shows a mitigation experiment on the DSCS satellite. The upper panel shows xenon gas released at about 9500 s. Some xenon atoms are promptly ionized in sunlight to form xenon ions. The lower panel shows the charging potentials of two small surfaces, kapton and quartz, which are not facing the Sun. The potentials measured are relative to the spacecraft frame (metallic ground), which is practically uncharged in sunlight. The xenon release lowers the charging potentials almost immediately at nearly 10,000 s (from reference 4) ,

Figure 6.7 showed a sequence of events on the DSCS satellite. At about 1900 s, the two dielectric surfaces (kapton and quartz) charge to over −2 kV relative to the spacecraft frame (metallic ground). Low-energy xenon plasma is released at about 1950 s, reducing the magnitudes of charging immediately. The plasma release ended at about 5000 s, thereby moving any control to the natural charging of the two dielectric surfaces. From about 5000 to 12,000 s the charging levels relative to the spacecraft frame vary in time in response to the

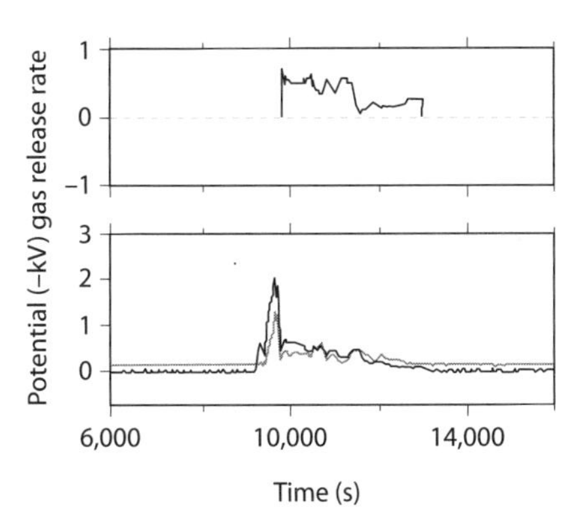

Figure 17.5 Mitigation experiment on the DSCS satellite. The upper panel shows xenon gas released at about 9500 s. Some xenon atoms are ionized in sunlight to form xenon ions. The lower panel shows the charging potentials of two small surfaces, kapton and quartz. The potentials are relative to the spacecraft frame (metallic ground). The xenon release lowers the charging potentials almost immediately. (Adapted from reference 4.)

natural plasma. At about 12,000 s, the spacecraft frame itself charges to high negative voltages (at about −2 kV, as indicated in the top panel of positive ion flux peaks), and as a result, the relative voltage drops to 0 V. The 0 V levels of the quartz and kapton surfaces in the lower panel are inexact because of an instrument handicap.

17.9 Vaporization Method

Polar molecules, such as water, attach electrons readily. This is why touching a door knob after walking over a carpet on a dry winter day may generate an electrostatic spark, whereas no spark occurs on a humid day. Some polar molecule species, SF_6, attach electrons more readily than water.[20] CCl_4 molecules also attach electrons readily. During evaporation, highly charged droplets may disrupt into several smaller droplets[21] when the Coulomb force of the electrons in the droplet exceeds the surface tension:

$$CCl_4 + e^- > CCl_3 + Cl^- + \Delta E$$

Lai and Murad[22-24] suggested a charge control method by spraying polar molecule liquid droplets all over a spacecraft. The polar liquid droplets attach the electrons on the spacecraft surfaces, evaporate, are repelled by the surface potential, and take away the excess electrons, and therefore reduce the surface potential. This method has an advantage that it mitigates metals and dielectric surfaces alike, thereby reducing differential charging. Unlike the ion or plasma release methods, prolonged use of this method does not end up electroplating the entire spacecraft. This is because the charged droplets evaporate away. It is not meant for deep dielectric charging. It should not be used if contamination is a concern.

17.10 Deep Dielectric Charging

Deep dielectric charging can occur when high-energy electrons and ions are deposited inside dielectric materials. Charge accumulation in dielectrics can build up high electric fields.[25,26] To mitigate deep charging inside dielectrics, metalized dielectrics can be useful. Although introducing metal atoms into random interstitial lattice sites of a dielectric material can alter the conductivity, the spatial distribution of the resultant conductivity inside the material would be inhomogeneous. For many purposes in highly delicate electronics, pure homogeneous conditions may be needed. The recent success[27,28] of introducing metal atoms into the molecular level instead of the lattice level gives a promising method for mitigating deep dielectric charging. By opening the rings of dielectric polymer molecules, metal atoms can be inserted, resulting in pure homogeneous metallized dielectrics. Preliminary laboratory results[29] on discharges in irradiated metal-based polymer are encouraging. The conductivity change and control in space needs further study.

Hardening of electronics is helpful for protection. Using very thin electronics can also reduce deep charging in the electronic devices. Shielding can also be used. Depending on the thickness of the shield, electrons and ions below certain energies cannot pass through. The higher-energy electrons and ions can still pass through but are slowed down. A small opening in the shield, however, would let electrons and ions pass through. One obvious disadvantage of shielding is that it adds weight.

17.11 Exercises

1. Suggest some situations under which electrons emitted from a spacecraft may not be able to leave completely.

2. Why is low-energy electron emission not effective for mitigation of spacecraft charging?
3. Why is high-energy electron emission not effective for mitigation of spacecraft charging?
4. Why is positive ion emission from a high-level negatively charged spacecraft able to reduce differential charging?
5. Why is low-energy plasma emission effective in reducing both differential charging and absolute charging?
6. If a spacecraft is charged to 200 V negative, can emission of 2 keV positive ions mitigate the charging?

17.12 References

1. Grard, R.J.L., "Spacecraft potential control and plasma diagnostic using electron field emission probes," *Space Sci. Instrum.* 1: 363–376 (1975).
2. Lai, S. T., H. A. Cohen, T. L. Aggson, and W. J. McNeil, "The effect of photoelectrons on boom-satellite potential differences during electron beam ejections," *J. Geophys. Res.* 92, no. A11: 12319–12325 (1987).
3. Lai, S. T., "An overview of electron and ion beam effects in charging and discharging of spacecraft," *IEEE Trans. Nuclear Sci.* 36, no. 6: 2027–2032 (1989).
4. Lai, S. T., "A critical overview of spacecraft charging mitigation methods," *IEEE Trans. Plasma Sci.* 31, no. 6: 1118–1124 (2003).
5. Fowler, R. H., and L. W. Nordheim, "Electron emission in intense electric fields," *Proc. Roy. Soc.* A119: 173–181 (1928).
6. Spindt, C. A., "A thin-film field emission cathode," *J. Appl, Phys.* 39, no. 7: 3504 (1968).
7. Spindt, C. A., C. E. Holland, A. Rosengreen, and I. Brodie, "Field-emitter arrays for vacuum microelectronics," *IEEE Trans. Electron. Dev.* 38, no. 10: 2355–2363 (1991).
8. Aguero, V. M., and R. Adamo, "Space applications of Spindt cathode field emission arrays," in *Proc. 6th Spacecraft Charging Technology Conference*, Hanscom AFB, MA (1998).
9. Mandell, M. J., V. A. Davis, B. M. Gardner, F. K. Wong, R. C. Adamo, D. L. Cooke, and A. T. Wheelock, "Charge control of geosynchronous spacecraft using field effect emitters," Paper AIAA 2007-284, presented at 45th AIAA Aerospace Science Meeting, Reno, NV (2007).
10. Iwata, M., T. Sumida, H. Igawa, Y. Fujiwara, T. Okumura, M.A.R. Khan, K. Toyoda, S. Hatta, T. Sato, and T. Fujita, "Development of electron-emitting film for spacecraft charging mitigation: observation, endurance and simulations," Paper AIAA 2009-560, presented at 47th AIAA Aerospace Science Meeting, Orlando, FL (2009).
11. Khan, A. R., T. Sumida, M. Iwata, K. Toyota, M. Cho, and T. Fujita, "Development of electron-emitting film for spacecraft charging mitigation: environment exposure tests," Paper AIAA-2010-76, presented at 48th AIAA Aerospace Science Meeting, Orlando, FL (2010).
12. Richardson, O. W., "On the negative radiation from hot platimium," *Proc. Camb. Phil. Soc.* 11: 286–295 (1901).
13. Purvis, C. K., H. B. Garrett, A. C. Whittlesey, and N. J. Stevens, "Design guidelines for assessing and controlling spacecraft charging effects," NASA Tech. Paper 2361, NASA Lewis Research Center, Cleveland, OH (1984).
14. Frederickson, A. R., D. B. Cotts, J. A. Wall, and F. L. Bouquet, *Spacecraft Dielectric Material Properties and Spacecraft Charging*, AIAA, New York (1986).
15. Lai, S. T., "Theory and observation of triple-root jump in spacecraft charging," *J. Geophys. Res.* 96, no. A11: 19269–19282 (1991).

16. Lai, S. T., "Spacecraft charging thresholds in single and couble Maxwellian space environments," *IEEE Trans. Nucl. Sci.* 19: 1629–1634 (1991).
17. Cohen, H. A., and S. T. Lai, "Discharging the P78-2 satellite using ions and electrons," presented at AIAA 20th Aerospace Sci. Mtg., Paper AIAA-82-0266, AFGL-TR-83-0139, Air Force Geophysics Laboratory, Hanscom AFB, MA (1983).
18. Wang, J., and S. T. Lai, "Numerical simulations on virtual anodes in ion beam emissions in space," *J. Spacecraft and Rockets* 34, no. 6: 829–836 (1997).
19. Mullen, E. G., A. R. Frederickson, G. P. Murphy, K. P. Ray, E. G. Holeman, D. E. Delorey, R. Robinson, and M. Farar, "An automatic charge control system at geosynchronous altitude: flight results, for spacecraft design consideration," *IEEE Trans. Nucl. Sci.* 44: 2188–2194 (1997).
20. Christophorous, L. G., *Electron-Molecule Interactions and Their Applications*, vol. 2, Academic Press, New York (1984).
21. Roth, D. G., and A. J. Kelley, "Analysis of the disruption of evaporating charged droplets," *IEEE Trans. Ind. App.*: 771–775 (Sept./Oct. 1983).
22. Lai, S. T., and E. Murad, "Spacecraft charging using vapor of polar molecules," in *Proc. Plasmadynamics Lasers Conf.*, AIAA-95-1941, pp. 1–11 (1995).
23. Lai, S. T., and E. Murad, "Mitigation of spacecraft charging by means of ionized water," U.S. Patent No. 6,463,672 B1 (2002).
24. Lai, S. T., and E. Murad, "Mitigation of spacecraft charging by means of polar molecules," U.S. Patent No. 6,500,275 B1 (2002).
25. Violet, M. D., and A. R. Frederickson, "Spacecraft anomalies on the CRRES satellite correlated with the environment and insulator samples," *IEEE Trans. Nucl. Sci.* 40, no. 6: 1512–1520 (1993).
26. Wrenn, G. L., "Conclusive evidence for internal dielectric charging anomalies on geosynchronous communications spacecraft," *J. Spacecraft and Rockets* 32: 514–520 (1995).
27. Manners, I., "Ring-opening polymerization of a silaferroceno-phane within the channels of mesoporous silica: poly (ferrocenylsilane)-mcm-41 precursors to magnetic iron manostructures," *Adv. Materials* 10: 144–149 (1998).
28. Manners, I., "Ring-opening polymerization of metallocenphanes: a new route to transition metal-based polymers," *Adv. Organometallic Chem.* 37: 131–168 (1995).
29. Manners, I., and K. Balmain, "Materials and methods of charge dissipation in space vehicles," U.S. Patent No. 6,361,869 B1 (2002).

18

Introduction to Meteors

The purpose of this chapter is to introduce the some fundamentals of meteor science. Meteors form an important part of the space environment. Both of the two standard textbooks on space environment[1, 2] include a very brief chapter on meteors, pertaining to spacecraft interactions. It is often a popular question[3] whether meteor impact can cause spacecraft charging, spacecraft discharging, or physical damage to spacecraft. The next chapter will discuss these effects. For pedagogical purposes, in this chapter we introduce the basic features of meteors, before the effects of meteoric impact on spacecraft surfaces are discussed.

Meteors originate from comets or asteroids. They come down into the Earth's atmosphere everyday. Before they enter the atmosphere, meteors are called *meteoroids*. After entry, they are called *meteors*. Atmospheric friction heats them to high temperatures. When they *ablate*[4] (that is, vaporize as a result of the high temperature of friction) in the atmosphere, they become visible to the naked eye or radars. The visible streak of light is also called a meteor. Ablation of very small meteors is negligible. Larger ones can survive ablation. Their remnants land on the ground and are called *meteorites*. In this chapter, we will examine (1) meteor size distribution, (2) meteor characteristics depending on their origin, (3) meteor velocity distribution, and (4) meteor composition. In the next chapter, we will discuss meteor hypervelocity impact.

18.1 Size Distribution

Meteor sizes[5] range from very tiny to very large (See figures 18.1 and 18.2 for quantitative information). The very large ones are rare but massive. Some meteors, becoming very bright for a short time, are called *fireballs* and occur often well below 100 km altitude. Most visible meteors occur around 90 to 120 km altitude. They are the *visual meteors* (figure 18.1). The small ones are invisible but can be still detected by radar. They are called *radio meteors*. The small ones, though invisible to the eye, form the major part of the distribution (figure 18.2) and are called *micrometeorites*. These cannot be observed even by the radars of today, but their accumulated effects in the ionosphere can be observed.

18.2 Meteor Showers

Meteor showers[6,7] occur at predictable times every year (table 18.1). Looking up, an observer on Earth sees shower meteors streaming toward the Earth from almost the same spot in the sky. Meteor showers are mostly of cometary origin. Most comets have elliptical orbits around the Sun. If an orbit is hyperbolic, the comet would leave the solar system and never return. As a comet comes near the Sun, sunlight heats the comet, which then ejects gas and particles. The ejected material continues to follow the comet's orbit approximately. Orbiting the Sun once a year, the Earth experiences a closest approach to the comet's orbit once a year (figure 18.3). The falling of the ejected particles into the Earth's atmosphere forms a meteor shower.

A meteor shower associated with a comet occurs once a year and may last for a few days or weeks. Perseids and Leonids are perhaps the most well-known meteor showers. Perseids last for weeks, and the peak (maximum number of visible meteors per hour) is on August 11

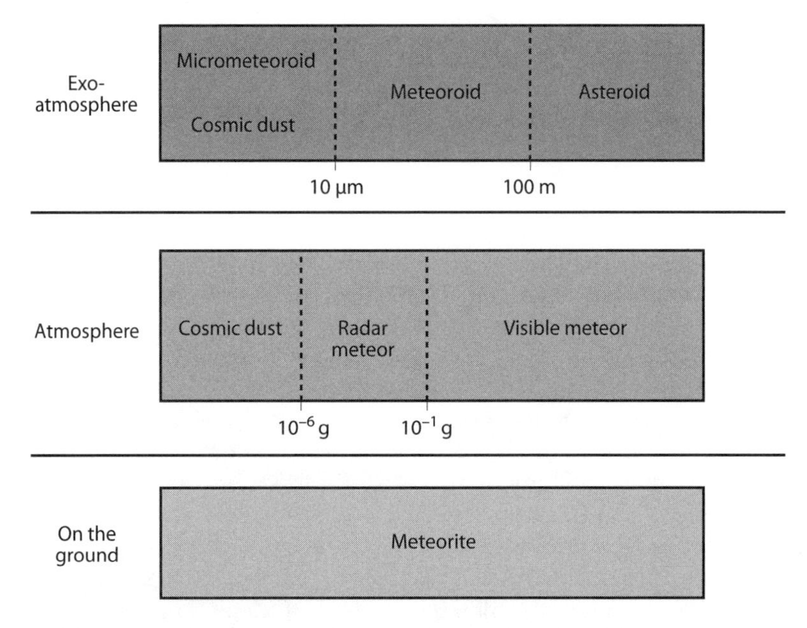

Figure 18.1 Types of meteor particles according to their sizes. The exoatmosphere particles and bodies are cosmic dust, micrometeoroids, meteoroids, and asteroids. When they enter the atmosphere, they are called meteors and are observable at altitudes below about 120 km. When they land on the ground, they are called meteorites, including the small meteor particles, which are difficult to observe.

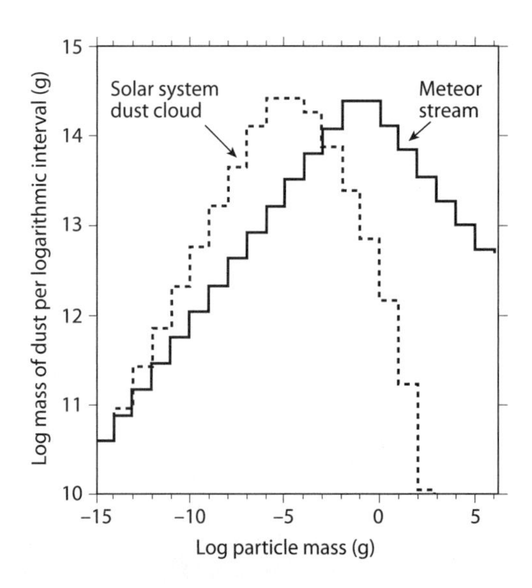

Figure 18.2 Mass distributions of meteors. (Adapted from reference 5.)

usually. Leonids last for a few days. The zenith hourly rate (ZHR) of a meteor shower is defined as the number of meteors an observer would see in one hour under a clear, dark sky, if the limiting magnitude is 6.5 mag and the radiant (the direction of the meteor stream) is in the zenith. The Leonids peak is usually on November 17. Due to irregularity of the comet ejections and the gravitational pulls of Jupiter and other planets, a meteor shower peak may arrive hours earlier or later than predicted. Time tables of annual meteors are available at some websites. Table 18.1 provides a handy compilation of annual meteor peak dates, peak hours.

TABLE 18.1
Visible Meteor Shower Calendar 2010

Name	*Dates*	*Peak date*	*Right ascension*	*Declination*	*Velocity km/s*	*ZHR*
Quadrantids	Dec 28–Jan 12	Jan 03	230°	+49°	41	120
Lyrids	Apr 16–Apr 25	Apr 22	271°	+34°	49	18
Eta Aquariids	Apr 19–May 28	May 06	338°	-01°	66	85
June Bootids	Jun 22–Jul 02	Jun 27	224°	+48°	18	Variable
Delta Aquariids	Jul 12–Aug 19	Jul 28	339°	–16°	41	16
Sigma Capricornids	Jul 03–Aug15	Jul 30	307°	–10°	23	5
Perseids	Jul 17–Aug 24	Aug 12	48°	+58°	59	100
Kappa Cygnids	Aug 03–Aug 25	Aug 18	286°	+59°	25	3
Alpha Aurigids	Aug 25–Sep 08	Sep 01	84°	+42°	66	6
September Perseids	Sep 05–Sep 17	Sep 09	60°	+47°	64	5
Delta Aurigids	Sep 18–Oct 10	Sep 29	82°	+49°	64	2
Draconids	Oct 06–Oct 10	Oct 08	262°	+54°	20	Var
Orionids	Oct 02–Nov 07	Oct 21	95°	+16°	66	30
Southern Taurids	Sep 25–Nov 25	Nov 05	52°	+15°	27	5
Northern Taurids	Sep 25–Nov 25	Nov 12	58°	+22°	29	5
Leonids	Nov 10–Nov 23	Nov 17	152°	+22°	71	20
Alpha Monocerotids	Nov 15–Nov 25	Nov 21	117°	+01°	65	Var
Geminids	Dec 07–Dec 17	Dec 14	112°	+33°	35	120
Comae Berenicids	Dec 12–Jan 23	Dec 20	161°	+30°	65	5
Ursids	Dec 17–Dec 26	Dec 22	217°	+76°	33	10

Source: International Meteor Organization website (http://www.imo.net/calendar/2010) and http://www.popastro.com/sections/meteor/showers.htm, both compiled by Alastair McBeath. Most of the shower calendar data varies slightly every year. Some meteor showers have vastly different ZHR from year to year. For example, the Leonids's ZHR was 100+ in the IMO calendar 2000. Historically, the Leonids in 1833 and 1966 featured peak ZHR of the order 10^4.

18.3 Meteor Velocity Limits

Meteors are very fast. This is the main reason why they can penetrate, or cause damage to, spacecraft surfaces. The purpose of this section is to derive the limits of meteor velocities. The fastest meteors travel at about 72 km/s in the Earth's atmosphere and the slowest at about 11 km/s. Faster meteors would escape from the solar system and never return.

To calculate the escape velocity V_S from a location R in the solar system, where R is the distance from the center of the Sun, consider the reverse motion. A particle, initially at rest just outside the solar system, is falling toward the Sun. The kinetic energy gained equals the potential energy lost:

$$\frac{1}{2}mV_S^2 = \frac{GM_S m}{R} \tag{18.1}$$

where M_S is the mass of the Sun, and m the mass of the particle. At the Earth's position, the solution of equation (18.1) gives $V_S \approx 42$ km/s approximately.

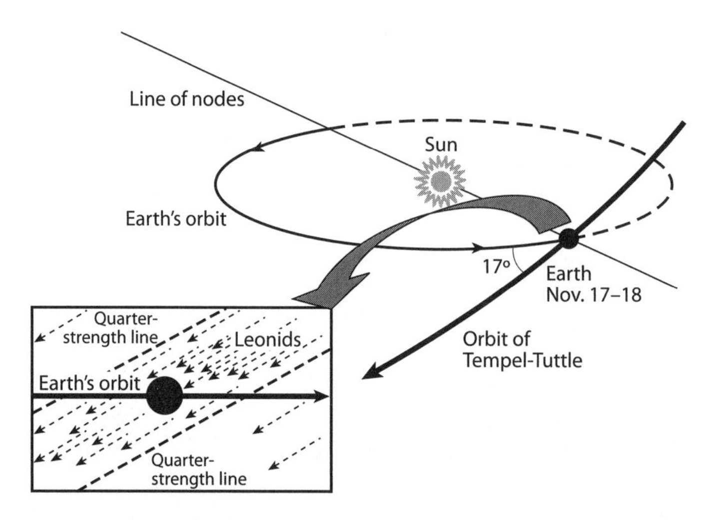

Figure 18.3 Intersection of orbits of the Earth and the comet Tempel-Tuttle. The Sun's orbital plane is at an angle with that of the comet and its remaining debris. The comet debris shed by the comet when it passed near the Sun many years ago forms the annual Leonids meteor shower. The shower strength varies every year. Typical width between the quarter-strength lines is about 35,000 km. (Adapted from http://www.palomar.edu/astronomy/astronomy/meteor_shower_information.htm.)

The Earth's orbital velocity V_o is calculated by equating the orbital centrifugal force to the attractive force by the Sun:

$$\frac{mV_o^2}{R} = \frac{GM_S m}{R^2} \tag{18.2}$$

Equation (18.2) gives $V_o \approx 29.7$ km/s. From equations (18.1) and (18.2), one obtains a relation [equation (18.3)] between the particle's escape velocity V_S from the Earth's location in the solar system and the Earth's orbital velocity V_o around the Sun (figure 18.4):

$$V_S = \sqrt{2}\,V_O \tag{18.3}$$

Ignoring the Earth's attraction, the fastest velocity V_{max} is obtained when the Earth with its orbital motion has a head-on collision with a particle of velocity V_S. Thus, the fastest velocity of a solar system particle in the Earth's orbit is $V_{max} = |V_o + V_S| = 71.7$ km/s.

To calculate the slowest velocity $V(r)$, consider the reverse motion of a particle initially at rest outside the Earth. The particle is falling toward the Earth. Equating the energies, we have

$$\frac{1}{2}mV^2 = \frac{GM_E m}{r} \tag{18.4}$$

where M_E is the mass of the Earth, and r is the distance from the center of the Earth. The radius of the Earth equals 6370 km approximately. Therefore, at 110 km altitude, $r = 6370 + 110 \approx 6480$ km. Equation (18.4) gives the slowest velocity $V \approx 11.2$ km/s approximately. Below 110 km, the atmospheric density is significant and, as a result, the meteor particle ablates with the atmosphere, heats up, evaporates, and slows down as it descends.

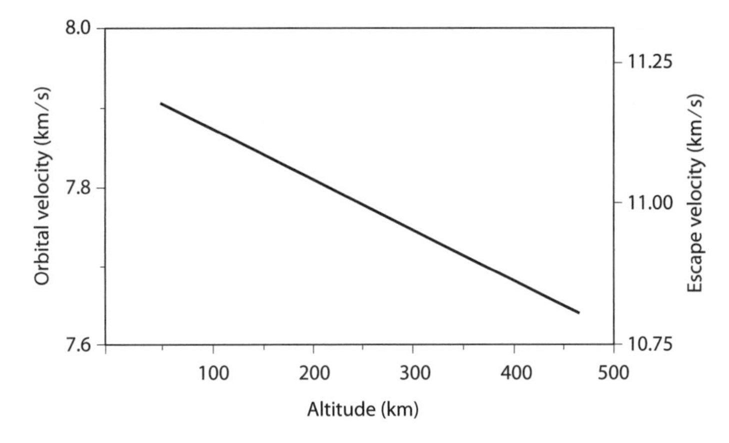

Figure 18.4 Relation between escape and orbital velocities. (Adapted from reference 3.)

If a particle, with an initial velocity of 71.7 km/s outside the Earth, is falling toward the Earth, the final velocity V_f at r would be given by the conservation of energy:

$$\frac{1}{2}mV_f^2 = \frac{1}{2}m(71.7)^2 + \frac{1}{2}m(11.2)^2 \tag{18.5}$$

which gives the maximum velocity $V_f \approx 72.6$ km/s.

To summarize this section, the slowest and fastest meteoroid velocities above about 110 km altitude are 11.2 km/s and 72.6 km/s. Below about 110 km altitude, atmospheric density becomes important. The meteoroid can ablate, heat up, slow down, and evaporate partially or completely.

18.4 Nonshower Meteors

Meteors of asteroidal origin are not as predictable as cometary meteors. Asteroids are of various sizes. Unlike comets that eject visible tails, the asteroids of small size cannot be seen. Occasionally, one may see a visible meteor during periods when no meteor shower is expected.

Interplanetary dust particles are tiny and have almost zero streaming velocity in the interplanetary space. When they happen to come near the Earth, they are attracted by the Earth's gravitation. They accelerate to a velocity $V_0(h)$ at altitude h. At about 120 km altitude, the atmospheric density becomes significantly high, and ablation of the falling particle begins. The value of V_0 at $h = 120$ km equals approximately the escape velocity (≈ 11 km/s) from the Earth.

Since the main bulk of the annual meteoric mass falling on Earth is due to the tiny dust particles, the mean velocity of all meteors is dominated by the interplanetary dust particles. The mean velocity $\langle V \rangle$ after gravitational attraction but before atmospheric deceleration is $\langle V \rangle \approx 17$ km/s (reference 8).

18.5 Debris

Debris poses hazards to spacecraft. Large debris are observable by radar and catalogued. Both debris and spacecraft are increasing in numbers. Debris orbits tend to peak around 600–900 km altitude and also around the geosynchronous altitude range. Outside the lower atmosphere, debris orbital velocity can be calculated with a centrifugal force equation analogous

TABLE 18.2
Composition of Typical Meteors

Element	*Percentage*
H	2.00%
C	3.50%
O	46.50%
Na	0.50%
Mg	9.70%
Si	10.60%
Ca	0.95%
Fe	18.50%

Data from reference 10.

to equation (18.2). Since the origin of debris is manmade satellites, their velocity depends on the orbital velocity of the debris-maker. If the debris comes off its satellite by breakup due to explosion or collision, the debris velocity is the vector sum of its original velocity plus that caused by the explosion or collision. The altitude and period of debris orbits resulted from breakup are usually described by using a Gabbard diagram.[9] Debris at high altitudes eventually falls down to low altitudes due to orbital decay caused by atmospheric drag and gravitation.

18.6 Meteor Composition

For completeness, we mention that there are many types of meteors. The most abundant type is chondrite. The elemental abundance[10] in chondrites is approximately similar to that of the Sun. There are variations[10] among types of chondrites, but we will not discuss the details here because the composition is not relevant for our topic, viz, meteor impacts on spacecraft surfaces. Table 18.2 lists the composition of typical meteors. The spectral lines observed following a meteor impact depends on the compositions of the meteor and the surface.

For more details on meteors, see, for example, references 11 to 14.

18.7 Exercises

1. By substituting numbers for M_S, M_E, R, etc., verify the numerical values of the meteor velocity limits.
2. Almost every orbiting object in the solar system is prograde. Suppose, however, that there are dust particles coorbiting with the Earth but traveling in the opposite direction; the head-on relative velocity would be $2 \times 29.7 \approx 59$ km/s for a massless Earth. What would the velocity be for a massive Earth?

18.8 References

1. Tribble, A. C., *The Space Environment*, rev. ed., Princeton University Press, Princeton, NJ (2003).
2. Hastings, D. E., and H. B. Garrett, *Spacecraft-Environment Interactions,* Cambridge University Press, Cambridge, UK (1996).

3. Lai, S. T., E. Murad, and W. J. McNeil, "Hazards of hypervelocity impacts on spacecraft," *J. Spacecraft and Rockets* 39, no. 1: 106–114 (2002).
4. Bronshten, V. A., *Physics of Meteoric Phenomena,* Reidel Publishing Co., New York (1983).
5. Hughes, D. W., "Meteors," in *Cosmic Dust*, ed. J.A.M. McDonnell, pp. 123–185, John Wiley & Sons, New York (1978).
6. Brown, P., "The Leonid meteor shower: historical visual observations," *Icarus* 138: 287–308 (1999).
7. Jenniskens, P., "Meteor stream activity I. The annual streams," *Astron. Astrophys.* 287: 990–1013 (1994).
8. Kessler, D. J., "Average relative velocity of sporadic meteoroids in interplanetary space," *AIAA J.* 7: 2337–2338 (1969).
9. Chobotov, V. A., and D. B. Spencer, "Debris evolution and lifetime following an orbital breakup," *J. Spacecraft* 28, no. 6: 670–676 (1991).
10. Lodders, K., and B. Fegley, *The Planetary Scientist's Companion*, Oxford University Press, Oxford, UK (1998).
11. Norton, O. R., *The Cambridge Encyclopedia of Meteorites*, Cambridge University Press, Cambridge, UK (2002).
12. Murad, E., and I. P. Williams, *Meteors in the Earth's Atmosphere: Meteoroids and Cosmic Dust and Their Interactions with the Earth's Upper Atmosphere*, Cambridge University Press, Cambridge, UK. (2002).
13. Jenniskens, P., *Meteor Showers and Their Parent Comets*, Cambridge University Press, Cambridge, UK (2006).
14. Lewis, J. S., *Physics and Chemistry of the Solar System*, Academic Press, San Diego, CA (1997).

19

Meteor Impacts

When a particle impacts on a surface, one or more processes may occur. Depending on the particle velocity, mass, and size, and on the target material density and thickness, the particulate may bounce back, make a dent, or penetrate through. If an impact is energetic enough, it may generate electrons and ions. Meteors pose hazards to spacecraft. Koons et al.[1] reported several cases of satellite failure or termination of mission by meteor or debris impact on spacecraft (table 19.1).

In this chapter, we will discuss meteoric impacts, mitigation techniques, and the meteoric impact effects on spacecraft charging and spacecraft discharging. Before we discuss the meteor impact effects on spacecraft charging and discharging, we discuss the impacts and the mitigation techniques. To begin, the kinetic energy of a meteoroid is estimated in section 19.1. Section 19.2 discusses meteoroid impacting on a surface and forming craters on the surface upon impact and presents some empirical formulae of crater depths. Possible mitigation methods are discussed in section 19.3. Whipple shields are discussed in section 19.4, which is followed by discussions on meteor impact probability in section 19.5 and the perturbation of angular momentum in section 19.6. The important question of whether spacecraft charging can occur is studied in section 19.7 and 19.8. Spacecraft charging is determined by the balance of currents. It is necessary to know the electron temperature of the plasma generated on impact, because the temperature very much controls the onset of charging. Last, the hazard of spacecraft discharging induced by meteor impact is discussed in section 19.9.

19.1 Kinetic Energy of Meteoric Particles

The Leonids meteor particles travel at about 71 km/s. Such a particle of 0.1 g, say, impacting on a spacecraft solar cell may have similar effects as a bullet hitting it. For a graph of the distribution of mass of the meteoric particles in an average meteor stream, see figure 18 of reference 2.*

Returning to the discussion on Leonids, which are fast, we would like to emphasize that even a calcium atom,[3] for example, released from a Leonids meteor particle, ablating in the atmosphere, would have such a high velocity that its kinetic energy may reach about 1 keV (figure 19.1). Impacts by debris and meteoric particles are ranked[4] as the most energetic interactions compared with all other spacecraft-environmental interactions, such as impacts by neutral oxygen atoms in the ionosphere, hot plasmas at geosynchronous orbits, energetic electrons and ions in the Van Allen radiation belts, and coronal mass ejection particles (figure 19.2).

19.2 Depth of Penetration

Laboratory measurements show that, for a given target, the penetration depth p depends on the particle velocity V and the particle diameter d. Figure 19.3 shows that experimental data

*Meteoric dust particles are smaller and slower. The most probable mass of meteoric dust particles falling into the atmosphere is about six orders of magnitude smaller. Most of them are nearly stationary, compared with meteoric streams, before they enter the Earth's atmosphere. They are much slower. Their average velocity is about 11–18 km/s at 150 km altitude.

TABLE 19.1
Spacecraft Failed and Damaged Due to Meteoroid or Debris Impact

Vehicle	*Date*	*Effect*	*Possible cause*
ISEE-1	October 1977	Detector window punctured.	Meteoroid
ISEE-1	August 1978	All isobutane gas lost in 5 days.	Meteoroid
Kosmos-1275	July 1981	Fragmented into 200 pieces.	Debris
HST	April 1990	5000 impacts in 4 years; solar cell punctured.	Meteoroid/debris
Solar-A	August 1991	Telescope punctured.	Meteoroid
STS-45	March 1992	Gouges on wing edge.	Debris
STS-49	May 1992	Chip in window pane.	Debris
Olympus	August 1993	Satellite failed to function.	Meteoroid
SEDS 2	March 1994	Mission terminated.	Meteoroid/debris
MSTI 2	March 1994	Satellite communication lost.	Meteoroid/debris

Note: Data adapted from reference 1.
HST, Hubble Space Telescope; ISEE, International Sun-Earth Explorer; and STS, Space Transportation System.

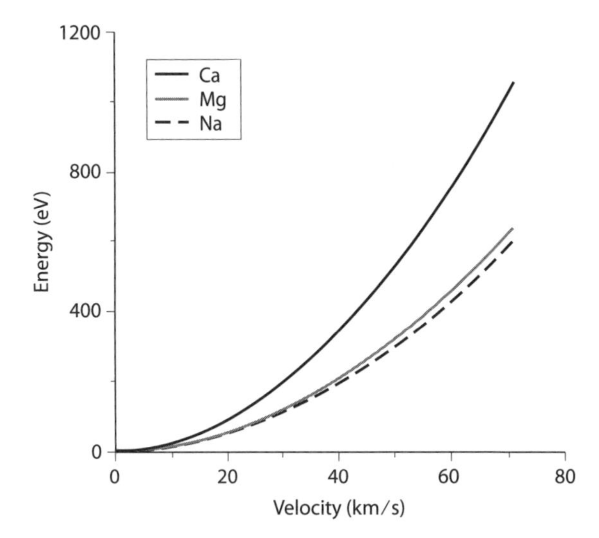

Figure 19.1 Kinetic energy of calcium, magnesium, and sodium atoms. (From reference 2.)

of log(I/I$^{1.056}$) plotted against log(I) gives a straight line for aluminum projectiles on an aluminium target.[5] Naturally, particles of faster velocities and smaller diameters penetrate deeper.

The Swift et al. impact model[6,7] offers some simple results (see also reference 8). Assuming that the craters are hemispherical and that all of the kinetic energy goes into melting the impacting particle, the model gives

$$\frac{4}{3}\pi\left(\frac{d}{2}\right)^3\frac{\rho_m V^2}{2}=\frac{1}{2}\frac{4}{3}\pi\left(\frac{p}{2}\right)^3 H \qquad (19.1)$$

where d is the diameter of the particle, ρ_m the particle density, p the penetration depth, and H the heat of fusion. This model assumes that the particle diameter d equals the diameter of the crater. The Swift et al. equation (19.1) can be written as

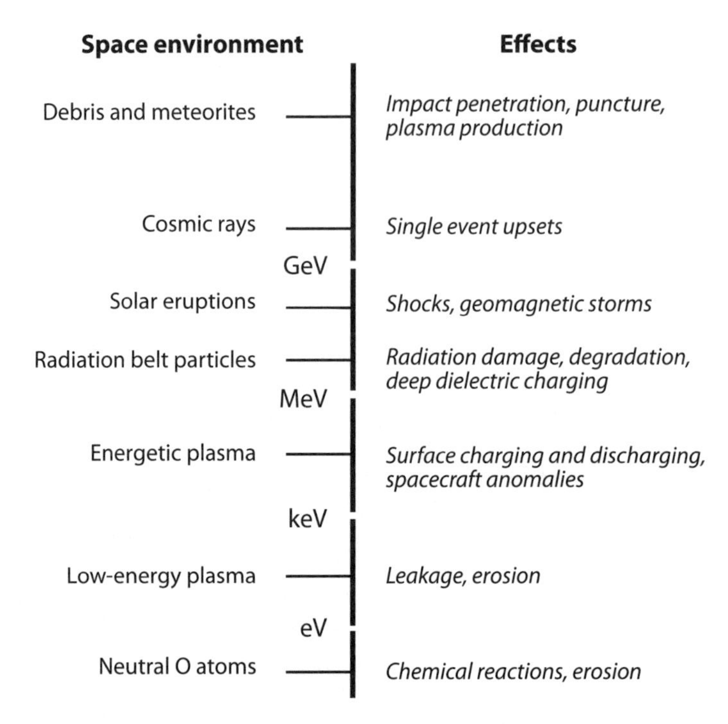

Figure 19.2 Energy scale of various spacecraft-environment interactions. (Adapted from reference 3.)

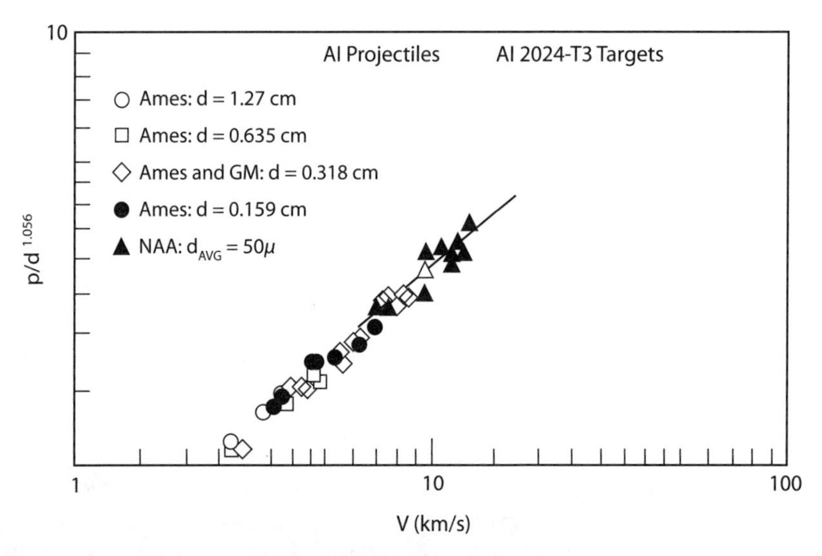

Figure 19.3 Ratio of penetration depth *p* and particle diameter *d* to the power 1.056 as a function of particle impact velocity. (Adapted from reference 31.)

$$\frac{p}{d} = \rho_m^{0.333} V^{0.667} H^{-0.333} \tag{19.2}$$

The Swift et al. equation (19.2) can be modified by including the heat of vaporization and by considering other crater shapes. The equation is simple and approximate but based on a physical model. It yields a straight line resembling the experimental data trend in figure 19.3.

Fitting s straight line to the experiemental data of figure 19.3 gives the Cour-Palais equation[5] as follows:

$$\frac{p}{d^{1.056}} = C\rho_m^{0.519} V^{2/3} \tag{19.3}$$

where ρ_m is in g cm^{-3}, V in km/s, and C is a constant that is characteristic of the target material and its condition.

McDonnell et al.[9] performed extensive laboratory experiments extending the Cour-Palais result by including the dependence on the particle tensile strength σ_m, target tensile strengths σ_T, particle density ρ_m, and target density ρ_T. The McDonnell et al. empirical formula[9] is more up-to-date than equations (19.2) to (19.3) . It is given as follows:

$$\frac{p}{d} = 0.7658d^{0.056}\left(\frac{\rho_m}{\rho_T}\right)^{0.476}\left(\frac{\sigma_{Al}}{\sigma_m}\right)^{0.134} V^{0.806} \tag{19.4}$$

where σ_{Al} is the tensile strength of aluminum target. V is in km/s. Table 19.2 lists the tensile strength and density of some typical materials. Using the McDonnell et al. empirical formula [equation (19.4)], a 1.3×10^{-5} g Leonids particle would penetrate 0.35 mm into typical solar panel mylar surface, and a 2.9×10^{-4} g Leonids particle would penetrate 1 cm into aluminum.

Another formula in the literature is based on observations from the LDEF (Long Duration Exposure Facility) space experiment. The LDEF penetration depth p is approximated by the following formula[10]:

$$p = Km^{0.352}\rho_T^{1.667} V^{0.875} \tag{19.5}$$

where m (g) is the mass of the particle, ρ_T the target density (g/cm^3), V the normal component of the particle velocity (km/s) relative to the surface, and K a material constant.

TABLE 19.2
Density and Tensile Strength of Some Surface Materials

Surface material	*Density (g cm^{-3})*	*Tensile strength (MegaPascals)*
	ρ_T	σ_T
Aluminum	2.71	90
Copper	8.90	150
Copper-beryllium	8.20	490
Gold	19.30	120
Iron	7.87	300
Mylar	1.395	40
Platinum	21.45	140
Silver	10.50	150
Stainless steel	7.80	460
Titanium	4.54	620

Note: Tensile strength, defined as the maximum force per unit area that a material can withstand before it fails, has units of Newtons per square meter (N/m^2 or Pascal). A Newton is 1 kg m s^{-2}.

19.3 Mitigation of Meteoric Impacts

There are some common mitigation methods for meteor impacts on spacecraft. A common practice is minimize the probability of meteor impacts. If an impact occurs, one tries to minimize the risk of a short-circuit or a discharge. If the impact is penetrating, one tries to protect the electronics behind some shields. More details about short-circuits or discharge will be given in section 19.9.

In anticipation of an intense meteor shower, a common practice is to orient the solar panels to a direction parallel to the meteor stream. This minimizes the effective area, A, normal to the meteor stream velocity so that the probability of impact is minimized. Another common practice is to shut off all the nonessential electrical power on a spacecraft during an anticipated intense meteor shower, in order to minimize the probability of a short-circuit or a discharge. Moving a spacecraft to the meteor eclipse side of the Earth temporarily would be helpful, but moving a spacecraft is often a difficult task. We discuss meteor shields, the most popular mitigation method, in the next section.

19.4 Meteor Shields

While a thick and heavy shield can most probably stop meteor penetration, it is important to consider a shield that is adequate to stop a meteor penetration and yet light in weight. There are two common designs: single-wall and double-wall designs.

When a particle penetrates through a wall, the particle may not only create a crater on the front side of the wall but it may also create a spall when the particle exits from the back side of the wall (figure 19.4).

A single-wall shield needs to have a minimum thickness t_c to accommodate a front crater and a back spall. The radii of the front crater ($d_c/2$) and the back spall ($d_s/2$) are of about the same order of magnitude. Thus, approximately,

$$t_c \geq d_c/2 + d_s/2 \tag{19.6}$$

Swift et al.[6] suggested that a minimum thickness $t_c = (3/2) \times (d_c/2)$ may be sufficient. Cour-Palais[5] obtained an empirical thickness formula from experimental observations:

$$t_c = C\rho_m^{1/6} m^{0.352} V_m^{0.857} \tag{19.7}$$

where C is a material constant (= 0.351 for aluminum), ρ_m is the density of the impacting particle and m is its mass in g.

Whipple[11] suggested the concept of a double-wall shield (figure 19.5). The first wall is the bumper shield, and the second wall the primary shield. When a meteor or impacting particle penetrates through the bumper shield, the particle is partially molten or vaporized. The

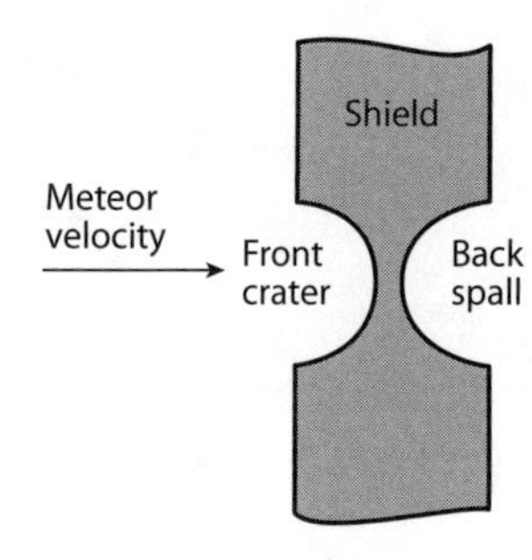

Figure 19.4 A meteor shield.

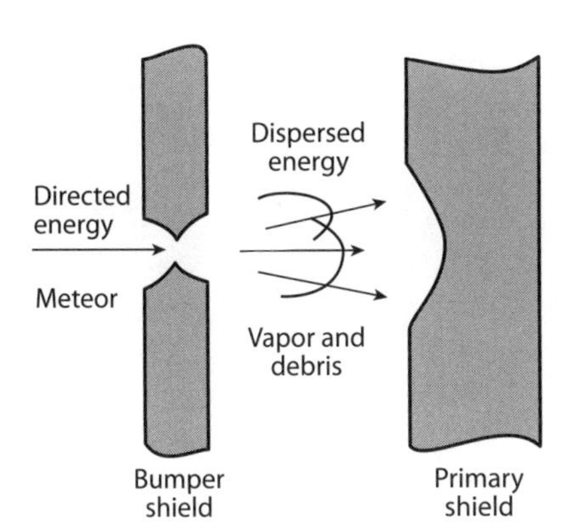

Figure 19.5 A Whipple meteor shield. (Adapted from reference 3.)

molten particle or vapor, together with the small debris from the bumper, will impact the primary shield. However, the impact on the primary shield will have less velocity and less focused pressure. Thus, the primary shield has a good chance to stop the entire mess of molten meteor, vapor, and dispersing small debris. The meteor mitigation methods using single or multiple walls have been well studied.[12–17]

19.5 Impact Probability of Meteors

The probability P of a hit by a particle on a surface is given by

$$P = \int_{t}^{t+\Delta t} dt F(t) A \tag{19.8}$$

where Δt is the duration, $F(t)$ the normal component of the flux of particles, and A the surface area. From elementary probability theory (see appendix 8), the probability P_N of N impacts, each of which is uninfluenced by the other, in a duration Δt is given by the Poisson distribution[18]:

$$P_N = \frac{P^N \exp(-P)}{N!} \tag{19.9}$$

If $P \ll 1$, one recovers $P_1 = P$ from equation (19.9) by using Taylor expansion of the exponential. By experience, the probability P for one or more meteoric impacts is very low.

The meteor flux is expected to be unattenuated from the top of the atmosphere down to about 120 km, where meteoric ablation with the atmosphere begins. Practically all spacecraft orbits are well above 120 km. Therefore, to calculate impact probability, there is no need to consider the attenuation of meteor flux as a function of altitude. McNeil et al.[3] have applied equation (19.8) to Leonids calculations. For a major Leonids storm (such as that of 1966) of flux $F = 8.63 \times 10^{-8}\ \mathrm{m^{-2}\ s^{-1}}$ lasting for $\Delta t \approx 2.5$ hours, a spacecraft of a typical size $A = 20\ \mathrm{m}^2$ (such as DSCS) would risk a 1.3% probability of a meteor hit. For a normal Leonids shower of $F = 4.69 \times 10^{-10}\ \mathrm{m^{-2}\ s^{-2}}$, the probability is 0.008%. In general, the probability P increases with the meteor particle flux $F(t)$, the shower duration Δt, and the satellite surface area A [see equation (19.8)]. There are many meteor showers each year. If the satellite lifetime is expected to be several years, the probability of getting a meteor hit increases with the expected life span. As already shown in table 19.1, Koons et al.[1] found that meteor impact is a significant hazard of satellite failures and damages.

For further study of probability of meteor impact on spacecraft, we refer the reader to the literature.[3, 17, 19–22]

19.6 Perturbation of Angular Momentum

To estimate the effect on angular momentum, let us take, for example, a spherical spacecraft of mass $M = 10^5$ gm, radius $R = 2$ m, with a light instrument panel, such as an antenna, extending to 10 m. A Leonids meteor of velocity v = 72 km/s and mass = 0.1 g (for an explanation of the mass chosen, see figure18.2) colliding with the tip of the extended instrument panel would add an angular momentum given by

$$I\Delta\omega = mvr \tag{19.10}$$

where I is the moment of inertia of the spacecraft, ω the angular velocity, $m = 0.1$ gm, v = 72 km/s, and $r = R + 10$ m = 1200 cm. For a sphere, the moment of inertia is given by $I = (2/5)$ MR^2. The result is $\Delta\omega \approx 0.035$ Hz. The telemetry data transmitted to the ground may show a sudden jump, $\Delta\omega \approx 0.035$ Hz, in the frequency of oscillation of the spacecraft. Such a jump can probably be detected using Fourier transform of the time sequence of the telemetry data.

19.7 Secondary Electrons and Ions by Neutral Particle Impact

If a neutral particle impacting on a surface is energetic enough, secondary electrons and ions will be generated.[23] Although the yields Y of secondary electrons and ions in the laboratory have been well studied for energetic electron and ion impacts, the yields by neutral impact have not been studied as thoroughly. The number of electrons generated for every primary particle impacting on the surface is called the yield Y_e. While it is easy to generate energetic electron and ion beams by means of electrostatic accelerations, it is difficult to accelerate neutral particles to velocities of tens of km/s in the laboratory. Currently, it is done[24,25] by using compressed light gas guns.

Early results[26] showed that the yields, Y_e, of secondary electrons are typically 0.2 to 0.4 (see explanation of yield in section 3.1), relatively independent of the incident particle energy provided that the latter exceeds the ionization threshold. Their results on secondary ion yields, Y_i, seemed to suggest 10^{-6} to 0.01 for clean surfaces, but 1–2 for oxidized surfaces. Later experiments[26–28] confirmed Y_e to around 0.2 but revised Y_i to around 0.04.

It is questionable how the incident neutral particle size, structure, shape, species, density, kinetic energy, incident angle, surface material properties, smoothness, surface oxidation, and temperature would affect the yields. More laboratory experiments are needed.

19.8 Plasma Generation by Neutral Particle Impact

McDonnell et al.[9] have been able to study plasma generation by neutral particle impact for high velocities in the laboratory. Even up to Leonids velocities can be achieved, albeit the faster particles have to be smaller. Electrons and ions are not each measured separately. From extensive laboratory experiments,[27] McDonnell et al.[9] have obtained an empirical formula for the electron charge production Q from aluminum surfaces due to a particle impact:

$$Q = 0.1m\left(\frac{m}{10^{-11}}\right)^{0.02}\left(\frac{V}{5}\right)^{3.48} \tag{19.11}$$

where m is the particle mass and V its velocity. The preceding result, equation (19.11), on plasma generation by fast small neutral particle impacts is the most up-to-date one.

McNeil et al.[3,4] have used the preceding formula to calculate the charge production rates for minor, moderate and major Leonid shower of ZHR (zenith hourly rate = hourly rates in theoretical perfect conditions; see International Meteor Organization, http://www.imo.net/) = 100, 5000, and 150,000, respectively. Assuming a population index r of 1.8 for the Leonids,[29] the charge production rates calculated are 2×10^{-10}, 2×10^{-8}, and 3×10^{-7} $C\ m^{-2}\ s^{-1}$, respectively. The perfect conditions for ZHR are unobstructed skies with the radiant overhead and the magnitude of the faintest visible star in the field of view equals +6.5. Quantitatively, it is given by the following equation[30]:

$$ZHR = \frac{CNr^{6.5 - lm}}{\sin(\sigma)T} \tag{19.12}$$

where C is a correction factor for the perception of the observer relative to an average person, $C = 1$ is for the average observer, N is the number of meteors observed in T hours, σ is the elevation of the shower radiant, lm is the limiting stellar magnitude, and r is the population index, which is defined as the ratio of the number of meteors in magnitude $m + 1$ to those in magnitude m.)

A centrally important question is whether the plasma generated by meteoric impact can cause spacecraft charging. Theory and spacecraft measurements have shown that space plasma with a temperature exceeding a critical temperature (see chapter 4) can cause charging of spacecraft surfaces to negative voltages. A typical value of the critical temperature in a Maxwellian space plasma is about 1 to 3 keV, depending on the spacecraft surface material. If the plasma generated by meteoric impact has a high enough temperature, it is conceivable that charging to negative voltages occurs. However, laboratory experiments[31] mimicking meteor impact on surface materials have reported plasma temperatures of 10 to 120 eV only. Such a temperature is too low to cause spacecraft charging. Laboratory measurements of the energy distribution of the impact generated plasma are lacking but much needed for better understanding.

It is interesting that two sudden charging events[4] of kV magnitude (negative) have occurred briefly on DSCS satellite during a major Leonid meteor storm. The space weather was quiet on that day. However, the two events[4] did not occur during the peak hour of the Leonid storm. Why the events[4] occurred is an important question.

19.9 Sudden Spacecraft Discharge Hazards

While the exact causes of spacecraft mission terminations are different in each case and are usually engineering-specific (depending on the geometry and usage of the instruments, the design and insulation of the circuits, for example) in addition to the dynamic condition of the space environment, we merely want to illustrate here the idea of sudden discharge hazards due to meteor impact. Meteor impact is an important type of spacecraft interaction with the space environment (figure 19.2). A simple scenario will suffice for illustrating the idea. Suppose that a meteor impacts a differentially charged spacecraft. Electrons, ions, and neutral gas are generated from the surface and the meteor particle (see figure 13.8). The electric fields between the differentially charged surfaces may accelerate the electrons and ions. If the electrons become energetic enough, they ionize some of the neutral gas molecules (figure 13.8). The newly created electron and ion pairs are also energized by the electric fields and may ionize further, provided Townsend's criterion[32–34] is satisfied. If the neutral gas becomes ionized, the total charge produced would be much larger than that produced without the neutral gas.

The situation is similar to a critical ionization velocity (CIV) discharge[35,36] with an important difference. The energy source for spacecraft surface discharge is the external electric fields between the differentially charged surfaces whereas the energy source in CIV is the relative velocity between a magnetized plasma and a neutral gas traveling through each other. For more details, we refer the reader to references 35 and 37. Some fundamentals of CIV will be given in the next chapter.

Another hazardous discharge scenario concerns deep dielectric charging. Before we describe this scenario, an introduction is in order. In the radiation belts, and less severely in the geosynchronous environments, the ambient electrons and ions often reach MeVs and beyond. These energetic electrons and ions can penetrate deeply into materials and stay there for days or weeks depending on the conductivity. These events are often significant in a few days after a coronal mass ejection (CME), which the Sun sends out occasionally. A CME cloud is an energetic plasma cloud. When a CME cloud reaches the Earth's outer magnetosphere, it compresses, and changes the direction of, the magnetic field lines, accelerating electrons to very high energies. If the electrons deposited inside the dielectric materials reach an electric field of 10^6 to 10^8 V/m, dielectric breakdown would occur.[7] Due to the low conductivity, it is conceivable that the electrons accumulate inside dielectrics, resulting in the buildup of very high electric fields.

In space and in the laboratory, electron flux is orders of magnitude larger than ion flux because of the mass difference. Therefore, electrons deposited inside dielectrics are much more abundant than ions. In the MeV energy range, electrons penetrate much deeper than ions.[7] For more on high-energy charge penetration in dielectrics, see chapter 16.

During a CME passage, the "killer" electrons penetrate and accumulate inside dielectrics. After the passage, the ambient plasma becomes much less energetic but denser. Now the ambient low-energy ions are attracted toward the dielectric surface by the attractive electric fields generated by the electrons inside. As a result, a double layer[4,38,39] of different electrostatic potentials is formed inside the dielectric, with positive ions deposited near the dielectric surface while electrons deep inside. This has potential for a hazardous discharge. We suggest that such a discharge should be called *anodized discharge*. All one needs is a spark, so to speak, to ignite the discharge. A meteor hit, for example, can provide such a "spark" for ignition (figure 19.6).

The double layer suggested earlier, although obvious, has not been actually proved in space or the laboratory. At the time of writing, there has been no laboratory or space experiment conducted for testing this idea.

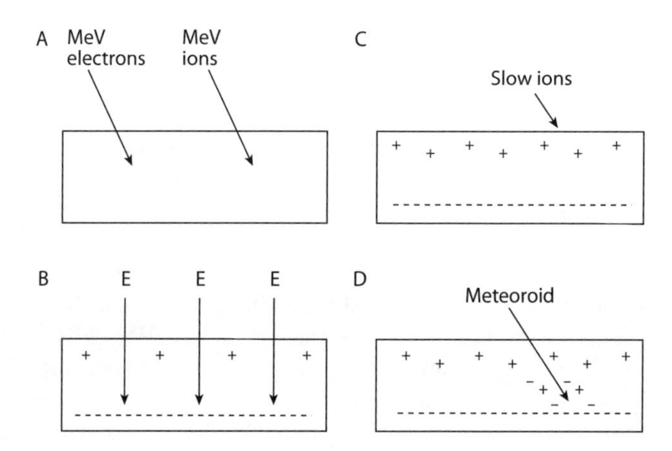

Figure 19.6 Double layer in deep dielectric charging. This is a potential hazard—a simple spark can ignite it. A meteoric hit can provide such a spark, initiating a discharge on a spacecraft. (Adapted from reference 35.)

As mentioned in the first paragraph of this chapter, Koons et al.[1] reported several cases of satellite failure, or termination of mission, by meteor impacts. Discharges can be generated by simple plasma production from the impact site on the surface (see section 19.8). Not all discharges initiated by meteor impact are due to the double layer mechanism suggested. The mechanism, if it occurs, would occur during, and after, the passage of a "killer electron" cloud from the Sun.

Last, we briefly point out that the generation of plasma by a sudden impact can produce electromagnetic waves,[40–42] which may interfere with the antennae system and electronics.

19.10 Summary

Since this chapter has discussed several topics, it is useful to include a summary here. A meteor hit can have effects similar to being struck by a bullet. The meteor particle can penetrate into material. The penetration can cause spacecraft failures and damages. The Whipple meteor shield is helpful for protection.

The probability of meteor impact on a spacecraft depends on the meteor flux, the duration of the meteor shower, the spacecraft size, and the spacecraft lifetime. Koons et al.[1] found several cases of meteor impact as the cause of satellite failures or damages.

Laboratory experiments using neutral particles and small pellets from air guns showed that the yields of electrons and ions are small. Since the temperature of the meteor-generated plasma is well below the critical temperature, spacecraft surface charging induced by meteor impact is practically ruled out.

It is important that meteor impact can induce satellite discharging. If a spacecraft is charged differentially on the surfaces or deeply inside the dielectrics a priori, a meteor impact can cause a sudden discharge. The discharge can occur even if the electric fields building up inside the dielectrics have not reached the critical value, which is about 10^6 to 10^9 V/m depending on the material. A sudden discharge can cause damages to the electronics onboard. Space debris can also induce such discharges and damages.

19.11 Exercises

1. Calculate the energy of (a) a bullet of mass 20 g traveling at 900 km per hour, and (b) a meteoroid of mass 0.1 g at 70 km/s.

2. An atom of atomic mass A weights 1.67×10^{-24}g multiplied by A approximately. Calculate the kinetic energy in keV for a 0.1 g meteoroid.

3. Using the empirical formula [equation (19.4)] of McDonnell, calculate the impact penetration depth p for various cases of projectile diameter d, mass m, and surface material tensile strength σ, density ρ, and meteor velocity V. Calculate the charge Q produced for each case. Assuming that the charge volume expands uniformly and spherically, calculate the charge density at a probe located at a distance x away from the meteor impact point.

19.12 References

1. Koons, H. C., J. E. Mazur, R. S. Selesnick, J. B. Blake, J. F. Fennell, J. L. Roeder, and P. C. Anderson, "The impact of space environment on space systems," Aerospace Report No. TR-99(1670)-1, Aerospace Corporation, El Segundo, Los Angeles, CA (1999).
2. Hughes, D. W., "Meteors," in *Cosmic Dust*, ed. J.A.M. McDonnell, pp. 123–185, John Wiley & Sons, New York (1978).

3. McNeil, W. J., S. T. Lai, and E. Murad, "Charge production due to Leonid meteor shower impact on spacecraft surfaces," in *Proc. 6th Spacecraft Charging Technology Conference*, pp. 187–191, Hanscom AFB, MA (2000).
4. Lai, S. T., E. Murad, and W. J. McNeil, "Hazards of hypervelocity impacts on spacecraft," *J. Spacecraft and Rockets* 39, no. 1: 106–114 (2002).
5. Cour-Palais, B. G., "Hypervelocity impacts in metals," *Int. J. Impact Eng.* 5: 221–238 (1987).
6. Swift, H. F., "Hypervelocity impact mechanics," in *Impact Dynamics*, ed. J. A. Zukas, pp. 215–239, Kreiger Publications, Malabar, FL (1992).
7. Swift, H. F., R. Bamford, and R. Chen, "Designing space vehicle shields for meteoroid protection: a new analysis," *Adv. Space Res.* 2, no. 12: 219–234 (1983).
8. Hastings, D. E., and H. B. Garrett, *Spacecraft-Environment Interactions*, Cambridge Univ. Press, Cambridge, UK (1996).
9. McDonnell, J.A.M., N. McBride, and D. J. Gardner, "The Leonid meteor stream: spacecraft interactions and effects," in *Proc. Second European Conference on Space Debris, Darmstadt, Germany*, ESA SP-393, pp. 391–396 European Space Agency, Paris, France (1997).
10. Tribble, A. C., *The Space Environment*, Princeton University Press, Princeton, NJ (1995).
11. Whipple, F. L., "Meteorites and space travel," *Astron. J.* 131: 1161 (1947).
12. Maiden, C. J., "Experimental and theoretical results concerning the protective ability of thin shields against hypervelocity projectiles," in *Sixth Symposium on Hypervelocity Impact*, pp. 69–156, Cleveland, OH (1963).
13. Maiden, C. J., and A. R. McMillan, "An investigation of protection afforded a spacecraft by a thin shield,"*J. AIAA* 2, no. 11: 1992–1998 (1964).
14. Swift, H. F., et al., "Characterization of debris clouds behind impacted meteoroid bumper plates," Paper AIAA-1969-380, in *Proc. AIAA Hypervelocity Impact. Conf.*, Cincinnati, OH (1967).
15. Swift, H. F., and A. K. Hopkins, "The effects of bumper material properties on operation of spaced hypervelocity particle shields," *J. Spacecraft and Rockets* 7, no. 1: 73–77 (1970).
16. Cour-Palais, B. G., "A career in applied physics: Apollo through space station," *Int. J. Impact Eng.* 23: 137–168 (1999).
17. Schonberg, W. P., "Protecting spacecraft against meteoroid/orbital debris impact damage: an overview," *Space Debris*, 1, no. 3: 194–210 (1999).
18. Smith, L. P., *Mathematical Methods for Scientists and Engineers*, Dover Publications, Inc., New York (1953).
19. Beech, M., and P. Brown, "Space-platform impact probabilities—the threat of the Leonids," *ESA J.* 18: 63–72 (1994).
20. Beech, M., P. Brown, and J. Jones, "The potential danger to space platforms from meteor storm activity," *Quart. J. Roy. Astron. Soc.* 36: 127–152 (1995).
21. Beech, M., P. Brown, J. Jones, and A. R. Webster, "The danger to satellites from meteor storms," *Adv. Space Res.* 20: 1509–1512 (1997).
22. Foschini, L., and G. Cevolani, "Impact probabilities of meteoroid streams with artificial satellites: an assessment," *Il Nuvo Cimento* 20 C: 211–215 (1997).
23. Pailer, N., and E. Grün, "Production of secondary particles by neutral and ionized cometary gas and dust impacting on the shield of the Giotto spaceccraft," *Eur. Space Agency Special Pub.* ESA SP 187: 1–4 (1982).
24. Kitta, K., E. Schneider, and A. Stilp, "Orbital-debris impact simulation experiments at the Ernst Mach Institut," in *Hypervelocity Impacts in Space*, ed. J.A.M. McDonnell, pp. 24–33, University of Kent, Canterbury, UK (1991).
25. Starks, M., personal communication (2007).

26. Schmidt, R., and H. Arends, “Laboratory measurement of impact ionization by neutrals and floating potential of a spacecraft during encounter with Halley's Comet,” *Planet. Space Sci.* 33, no. 6: 667–673 (1985).
27. Schmidt, R., and H. Arends, “Measurements of integral yields of charged secondary particles using neutral beams simulating a cometary fly-by,” in *The GIOTTO Spacecraft Impact Induced Plasma Environment*, ESA SP-224, pp. 15–19, European Space Agency Publ., Paris, France (1984).
28. Rudenauer, F. G., and W. Steiger, “Measurements of secondary charged particle emission under ion neutral impact,” in *The GIOTTO Spacecraft Impact Induced Plasma Environment*, ESA SP-224, pp. 1–10, European Space Agency Publ., Paris, France (1984).
29. Brown, P., “The Leonid meteor shower: historical visual observations,” *Icarus* 138: 287–308 (1999).
30. Brown, P., and R. Arlt, “Final results of the 1996 Leonid maximum,” *J. IMO* 25, no. 5: 210–214 (1997).
31. McDonnell, J.A.M., and K. Sullivan, “Hypervelocity impacts on space detectors, decoding the projectile parameters,” in *Proc. Workshop on Hypervelocity Impacts in Space,* ed. J.A.M. McDonnell, pp. 39–47, University of Kent, Canterbury, U.K (1991).
32. Ratcliff, P. R., and F. Alladadi, “Characteristics of the plasma from a 94 $km.s^{-1}$ microparticle impact,” *Adv. Space Res.* 17, no. 12: 87–91 (1996).
33. Von Engel, A., *Ionized Gases*, 2nd ed., Oxford University Press, London (1965).
34. Lai, S. T., and E. Murad, “Critical ionization velocity experiments in space,” *Planet. Space Sci.* 37, 865–872 (1989).
35. Lai, S. T., and E. Murad, “Inequality conditions for critical velocity ionization space experiments,” *IEEE Trans. Plasma Sci.* 20, no. 6: 770–777 (1992).
36. Lai, S. T., “A review of critical ionization velocity,” *Rev. Geophys.* 39, no. 4, 471–506 (2001).
37. Brenning, N., “Review of the CIV phenomenon,” *Space Sci. Rev.* 59: 209–314 (1992).
38. See, for example, the recent papers reporting on spacecraft discharging experiments conducted in the laboratory, *IEEE Trans. Plasma Sci. Special Issues*, Nov. 2006 and Oct. 2008.
39. Lai, S. T., “A mechanism of deep dielectric charging and discharging on space vehicles,” in *Proc. IEEE International Conf. Plasma Science*, p. 154, abstract (1998).
40. Meyer-Vernet, N., P. Couturier, S. Hoang, C. Perche, J. L. Steinberg, J. Fainberg, and C. Meetre, “Plasma diagnosis from thermal noise and limits on dust flux or mass in comet Giacobini-Zinner,” *Science* 232: 370–374 (1986).
41. Foshini, L., “Electromagnetic interferences from plasmas generated in meteoroid impacts,” *Europhys. Lett.* 43, no. 2: 226–229 (1998).
42. Swift, H. F., “The role of electromagnetic radiation in hypervelocity impact mechanics,” *Int. J. Impact Eng.* 26: 745–760 (2001).

20

Neutral Gas Release

Neutral gas release occurs in various situations. Some examples are neutral gas exhaust from spacecraft propulsion, plasma and neutral gas releases for mitigation of charging, emission of ion beams generated from ionization chambers, ion propulsion, outgassing from the surface materials, etc. Some spacecrafts use Vernier gas releases for attitude adjustment. In plasma releases, it is common to have neutral gas coming out from the ionization chamber. Since the efficiency of ionization in an ionization chamber is usually a few percent only and the ion extraction grids with high voltages do not "see" the neutral gas, the latter can wander out at will. Similarly, it is practically impossible to exclude neutral gas release in ion propulsion. Outgassing of neutral molecules can occur from the spacecraft surfaces during a transient period of days after launch.

Neutral gas does not, by itself, interact with spacecraft charging. If the neutral gas released from a spacecraft is ionized in the vicinity of the spacecraft, then the ionized gas can affect the ambient environment of the spacecraft. Chapter 10 discussed the phenomenon that the ions or electrons attracted back to the spacecraft can reduce the levels of spacecraft charging and differential charging. Electron impact ionization, charge exchange, photo-ionization, and stripping ionization are candidate mechanisms for ionization. There are other atomic and molecular processes. Alfvén's critical ionization velocity (CIV) has been suspected to be an efficient mechanism for ionizing neutral gas released from spacecraft. Active CIV experiments, however, have not clearly demonstrated CIV in space. High-energy neutral beams may also generate ionization by stripping.

20.1 Ionization and Recombination

A completely neutral gas released from spacecraft does not interact electrically. If the neutral gas turns into a partially ionized neutral gas, it interacts electrically. This chapter will discuss briefly some fundamentals of atomic and molecular processes relevant to gas released from spacecraft. More details on ionization mechanisms and chemical reactions can be found in comprehensive monographs such as references 1 to 4 or textbooks such as references 5 and 6.

20.1.1 Electron Impact Ionization

Electron impact ionization occurs when an electron impacting on a neutral atom or molecule ionizes the neutral atom or molecule:

$$A + e^- \rightarrow A^+ + 2e^- \tag{20.1}$$

where A is a neutral atom or molecule and e^- an electron (see, for example, references 1 and 2). The impact ionization cross section $\sigma(E)$ has a threshold at typically about 10 eV of electron energy E and a maximum at about $E = 110$ eV depending on the neutral species (see reference 3). The ionization rate dn_i/dt is given by

$$\frac{dn_i}{dt} = n\int_0^\infty dE E f(E)\sigma(E) \tag{20.2}$$

where n_i is the density of A^+ ions, n the density of neutrals A, $f(E)$ the velocity distribution of ambient electrons with the velocity expressed in terms of energy E, and σ the ionization cross section of the neutral species.

For neutral gas released in the ionosphere, the rate of electron impact ionization is low because the ambient electron energy is well below 1 eV. At the geosynchronous altitudes, the ambient electrons are energetic enough but their flux is low so that the rate is also low. The rate (equation 20.2) increases with the density n of the neutral gas released.

20.1.2 Charge Exchange

Charge exchange, also called *charge transfer*, occurs when a neutral atom A exchanges one outer electron with an ion B:

$$A + B^+ \rightarrow A^+ + B \tag{20.3}$$

The cross section peaks at zero energy for symmetric charge exchange, but if the energy of the reactants and products are different, there is a threshold. The peak is near zero energy when the energy of the reactants and products are similar. For details, see, for example, references 4–5. The charge exchange rate dn_x/dt is given by

$$\frac{dn_x}{dt} = nv\sigma n_+ \tag{20.4}$$

where n_x is the density of the ions generated by charge exchange, n the neutral gas density, v the relative velocity between the neutral gas atoms and the ambient ions, and n_+ the density of the ambient ions, which are mostly oxygen ions at about 200–400 km altitude in the nighttime ionosphere and at wider range extending to beyond 1000 km in the daytime. In the literature, $v\sigma$ is often written as $\langle v\sigma \rangle$, representing its average value, or simply as k ($\equiv \langle v\sigma \rangle$), which is also called the *rate coefficient*. The averaging is by means of the electron energy distribution only, the neutral gas being relatively slow with negligible energy spread.

Charge exchange is insignificant at geosynchronous altitudes because the density of the ambient neutral ions is very low and their energies too high. It is significant in the ionosphere where the ambient ion density is high while the ion energies are low.

Charge exchange does not produce a net increase in charge density but changes the species and velocity of the ions. For example, if one releases a neutral gas of barium atoms from a spacecraft orbiting at 6 km/s in the ionosphere where the ambient oxygen ions are at an average thermal velocity of 0.05 eV, the barium ions resulting from the charge exchange with the oxygen ions would be traveling at about 6 km/s, which translates into about 26 eV in energy.

As another example, suppose that one releases a xenon ion beam of 2 keV in space and the beam ions are generated in an ionization chamber of an ion beam device. With the present technology, the ionization efficiency is low, typically a few percent only (reference 7). The slow neutral xenon particles wandering out would charge exchange with the fast beam ions (2 keV), generating slow xenon ions and fast xenon neutrals. If the spacecraft is charged a priori to −500 V, say, the slow xenon ions would return, lowering the magnitude of the spacecraft potential as a result.

20.1.3 Photo-ionization

In photo-ionization, a neutral atom or molecule absorbs the energy from a photon and releases an electron:

$$A + h\nu \rightarrow A^+ + e^- \tag{20.5}$$

where A is a generic atom or molecule. The ionization rate dn_i/dt is given by

$$\frac{dn_i}{dt} = n\int_0^\infty d\omega I(\omega)\sigma_{ph}(\omega) \tag{20.6}$$

where ω is the photon frequency, $I(\omega)$ the number of photons per unit area per unit time, and σ_{ph} the photo-ionization cross section. The dominant UV line in sunlight is Lyman-Alpha, which has about 10 eV in energy, but there are higher frequency lines.

20.1.4 Ionization via Metastable States

Suppose that an electron impact excites an atom from its ground state to an excited state. When the excited state de-excites to the ground state, an amount of energy equal to the difference in the energy levels would escape as radiation:

$$A + e^- \rightarrow A^* + e^- \tag{20.7}$$

$$A^* \rightarrow A + h\nu \tag{20.8}$$

Suppose that the excited state has a relatively long lifetime. This is called a *metastable state.* While in the metastable state, a second electron can impact the atom. The impact supplies an amount of energy for the successive excitation from the excited state to the ionization state:

$$A^* + e^- \rightarrow A^+ + 2e^- \tag{20.9}$$

This process is called *ionization via metastable states* (figure 20.1). It requires a long metastable state lifetime together with a high electron impact rate.[8] To explain the effect of this process to a layperson, one could say that compared with direct ionization, the ionization via metastable states is a "democratic process" in which more little guys can participate.

20.1.5 Molecular Recombination

The electron-atomic ion recombination process is very slow:

$$A^+ + e^- \rightarrow A + h\nu \tag{20.10}$$

where A is a generic atom. The process, equation (20.10), is also called *dielectronic recombination*. The rate coefficient of equation (20.10) is of the order of 10^{-12} cm^3 atom^{-1} s^{-1} (references 9 and 10). On the other hand, molecular ion-electron recombination is very fast. Consider the case of a generic molecule, AB:

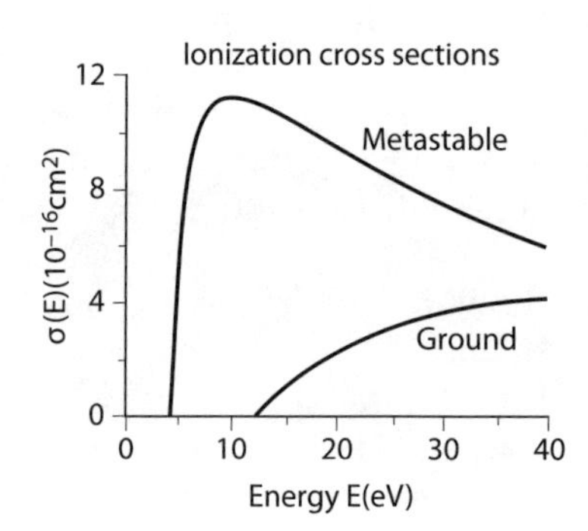

Figure 20.1 Electron impact ionization cross sections for xenon. (Adapted from reference 8.)

$$AB + e^- \rightarrow AB^+ + 2e^- \tag{20.11}$$

$$AB^+ + e^- \rightarrow A + B \tag{20.12}$$

The process, equations (20.11) and (20.12), is also *called dissociative recombination.* The rate coefficient of equation (20.12) is of the order of 10^{-6} and 10^{-7} cm^3 molecule^{-1} s^{-1} (references 11 and 12).

Gaseous molecules such as H_2O, CO_2, CO, N_2 and NO are present in the space shuttle exhaust[13] and launch vehicle exhaust. Since ion-molecular recombination is fast, it may deplete the electrons and quench an ionization discharge process in a neutral gas.[14]

20.2 Critical Ionization Velocity

This section gives a brief introduction to Alfvén's idea of critical ionization velocity (CIV).[15] We will not go into the mathematical details of plasma physics in this narration. At the time of this writing, CIV is not included yet in any plasma textbooks.

Alfvén suggested the idea of critical ionization velocity (CIV). When a neutral gas is traveling through a magnetized plasma with a relative velocity V, rapid ionization can occur if the perpendicular component of V exceeds a critical value V^*, which is given by

$$V^* = \sqrt{\frac{2e\phi}{M}} \tag{20.13}$$

Equation (20.13) represents the equality of the electron kinetic energy $(1/2)MV^*$ to the ionization energy $e\phi$of the neutral particle. The equality requires perfect efficiency in converting the kinetic energy to the ionization energy. *Perfect efficiency* means that there is no loss in the process of energy conversion. The critical ionization velocities V^* of the elements are shown in figure 20.2.

CIV is not a self-starting process. As an analog, an automobile gasoline engine needs an electric motor to start. For CIV to start, the neutral gas needs to have ions traveling with it. Ions have much larger gyroradii than electrons. As the ions travel across the magnetic field lines, the electrons are more or less tied down by the field lines. Beam-plasma instabilities,

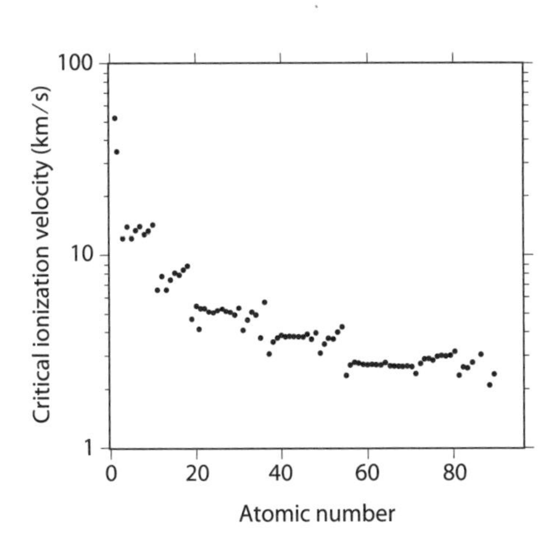

Figure 20.2 The critical ionization velocities of the elements. (Adapted from reference 25.)

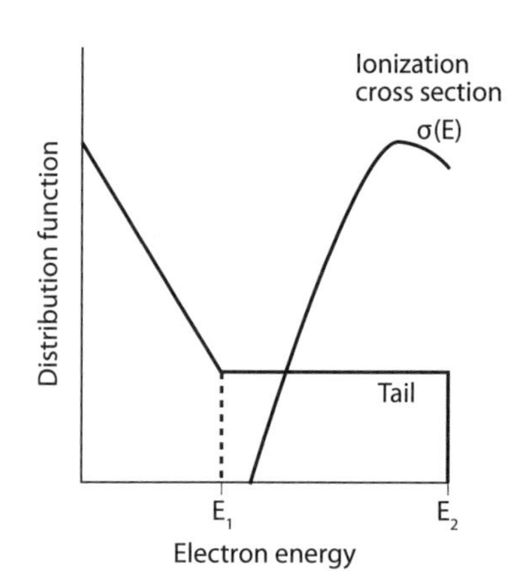

Figure 20.3 Idealized electron distribution and ionization cross section in CIV. (Adapted from reference 8.)

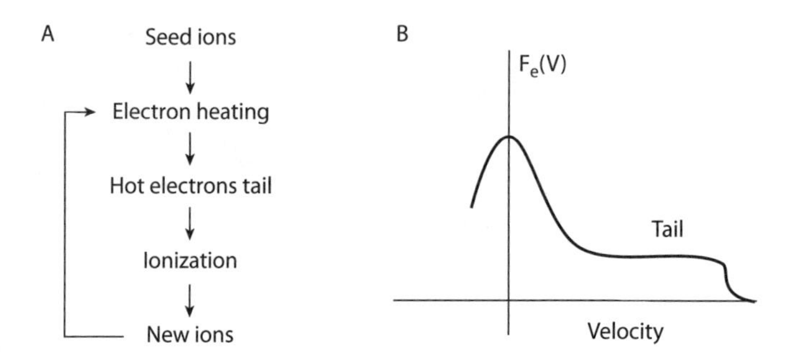

Figure 20.4 (A) Cyclic process of CIV. (B) Plateau tail distribution as a result of beam-plasma interaction. (Adapted from reference 25.)

such as the lower hybrid instability,* may occur, accelerating the electrons at the expense of the ion energies. As a result of the plasma instability, some electrons are accelerated, evolving into a plateau tail distribution.[16] If so, some of the electrons in the plateau tail may be energetic enough to ionize by impact (figure 20.3). The newborn ion travels with nearly the same velocity as the neutral particle and carries on the process of acceleration and ionization, etc., in a cyclic manner (figure 20.4).

Whether the cyclic process can be sustained depends on the competition between the gain and loss mechanisms (such as excitation or electron escapes without ionization). The cyclic process is analogous to a business cycle, in which one starts with some capital and accounts for the net gain or loss every season. If the number of ions increases after every cycle despite the losses, the cyclic process continues and may evolve into an avalanche ionization—i.e., a discharge.

Indeed, the first experiment on CIV using a homopolar device[15] reported that, when the relative velocity reached above the critical value, the neutral gas was promptly ionized to nearly 100% in 3 μs. Ever since, laboratory experiments using various devices have repeatedly demonstrated CIV with ease, although the critical value observed was always slightly higher than that given by equation (20.1), because one needs to account for the losses.

Naturally, one asks whether CIV works if one releases a neutral gas from spacecraft. For example, if one releases barium gas from a spacecraft in the ionosphere, will the gas be promptly

*For linear or nonlinear plasma instability theories in CIV, see references 16–19. There are debates about which plasma instability would work under what appropriate condition in CIV. These debates have their own merits. We will not digress further here.

ionized? The gas will have a velocity given by the vector sum of the gas velocity relative to the spacecraft and the spacecraft velocity in space. The critical ionization velocity, equation (20.13), of barium is $V^* = 2.708$ km/s, while a spacecraft orbiting at nearly 300 km altitude has a spacecraft velocity of $V_s = 6$ to 8 km/s. The gas velocity in the space frame equals the vector sum of the spacecraft velocity V_s in the space frame plus the release velocity V_R relative to the spacecraft. The critical ionization velocity, equation (20.13), is satisfied. Charge exchange between the barium atoms and the atmospheric atomic oxygen ions provides the seed barium ions traveling with the neutral gas and initiating beam-plasma interactions and perhaps CIV. Whether or not the CIV cyclic process can be sustained is a deciding factor for increasing ionization.

Theoretical models, laboratory experiments, and computer simulations have all shown CIV as feasible and reasonably well understood, although all CIV experiments in space have yielded negative results with perhaps three exceptions. There have been over 10 experiments to test CIV in space. The ionization yield (also called *percentage ionization*), observed in each CIV space experiment was low. Even in the few experiments in which the ionization observed was a few percent or slightly higher, the ionization could be explained by means of non-CIV processes.

It is not fully understood why the ionization yields have been so low in the CIV space experiments compared with those in the laboratory. There are many differences between the magnitudes of the parameters in space and in the laboratory.[20] Most importantly, the magnetic field is much lower in space, some atmospheric species may absorb energy by excitation without ionization, and the plasma density is much lower than that in the laboratory. There are inequality conditions to be satisfied for CIV to occur.[14,20] Papadopoulos[19] stressed that plasma turbulence is necessary for the CIV process to be efficient in space. In the laboratory, the plasma is preheated.

There are plausible reasons to explain partly the low efficiency of CIV in the space experiments. For example, the low efficiency of momentum coupling via Alfvén wings,[21] too much excitation energy loss, the lower hybrid wavelength exceeding the neutral gas dimension in the space experiments, ineffective beam-plasma interaction, the hot electrons escaping the neutral gas region with insufficient chance of collision and ionization, etc. are some plausible reasons. They are all worthy for further debate but will not be pursued in this brief introduction.

If CIV occurs in space, there are consequences. For example, the ions generated may cause contamination on the spacecraft. The ions may have characteristic spectral signatures. Of course, the ions or electrons returning to the spacecraft would reduce the spacecraft potential and mitigate differential charging as discussed in an earlier chapter.

Can the ionization electrons cause negative voltage spacecraft charging? Probably not, because the electrons are not energetic enough to cause surface charging, except perhaps at very low voltage which usually does no harm. To cause high-level negative voltage charging, the electrons must be hotter than the critical temperature. which is of the order of keV or more, depending on the surface material properties.

Last, electrons accelerated to about 100 eV have indeed been observed in CRIT II, a barium release experiment for testing CIV in the ionosphere (figure 20.5). Such electrons are energetic enough to ionize neutrals on impact. Their presence is crucial evidence that electron impact ionization, a central process in CIV, must be occurring in the experiment.

For reviews on CIV, see, for example, references 22 to 25.

20.3 Neutral Beam Stripping

For a high-energy neutral beam emitted to the atmosphere, stripping ionization can occur:

$$N + M \rightarrow N^+ + e^- + M \tag{20.14}$$

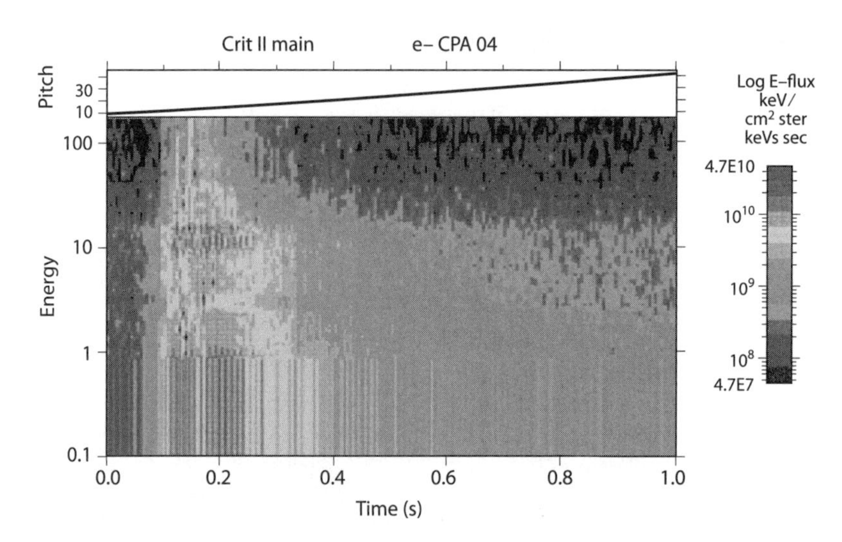

Figure 20.5 Electron energy distribution measured in CRIT II. Note that energetic electrons up to 100 eV or more are present. (Adapted from reference 23.)

where N is a generic neutral beam particle, and M a generic atmospheric particle. The reaction cross section peaks at multiple keVs typically, and is therefore unimportant at low energies. For hydrogen colliding with hydrogen, for example, the peak of the cross section is at about 10 keV in energy. Schematically, a way to get such a neutral beam is to let a high-energy ion beam undergo charge exchange with a neutral gas.

As a simple model, let the density $I(z)$ of a neutral beam undergoing stripping ionization at altitude z be given by

$$I(z) = I(z_0)\exp[-\beta/\cos\theta] \tag{20.15}$$

and

$$\beta = \sum_i \int n_i(z)\sigma_i(E(z))\,dz \tag{20.16}$$

where $I(z_0)$ is the initial density, θ is the angle between the beam and the vertical line (180° being downward), and E is the beam energy. The integral limits are z_0 and z. The densities n_i ($i = 1$ to 5) of the atmospheric species are given by the most abundant atmospheric species: O, O_2, N_2, He, and H. The survival probability P of the beam undergoing stripping is given by

$$P(E,z) = 1 - \exp[-\beta/\cos\theta] \tag{20.17}$$

For simplicity, one can take the Stein-Walker[26] neutral atmospheric density model for the atmospheric species densities n_i and the Green-McNeal stripping[27] model for the cross sections σ_i. Some results are shown in figure 20.6. For improved accuracy, one can use an improved neutral atmospheric density model[28] for n_i and some latest measurements of stripping ionization cross sections σ_I in equation (20.16).

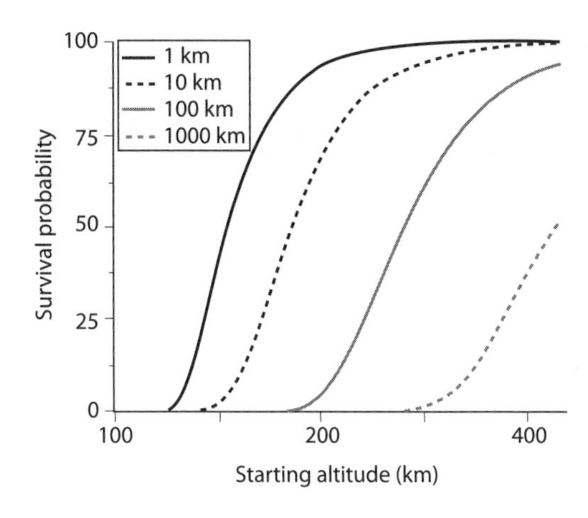

Figure 20.6 Survival probability of a 2 MeV neutral hydrogen beam propagating horizontally. (Adapted from reference 29.)

20.4 Exercises

1. The Skylab launch[30] reduced the electron density in its surrounding region in the ionosphere. Oxygen ions are abundant in the ionosphere above about 180 km altitude. The launch exhaust consisted of neutral water molecules and other chemicals. Write down the two chemical reactions that were probably responsible for the electron depletion?
2. Calculate the critical ionization velocities of hydrogen, barium, and xenon.
3. What is the main difference between a critical ionization velocity discharge and a Paschen discharge?

20.5 References

1. Kieffer, L. J., and G. H. Dunn, "Electron impact ionization cross-section data for atoms, atomic ions, and diatomic molecules, I. Experimental data," *Rev. Modern Phys.* 38: 1–135 (1966).
2. Maerk, T. D., and G. H. Dunn, *Electron Impact Ionization*, Springer Verlag, New York (1985).
3. Radzig, A. A., and B. M. Smirnov, *Reference Data on Atoms, Molecules, and Ions*, Springer Verlag, New York (1985).
4. Bates, D. R., Ed., *Atomic and Molecular Processes*, Academic Press, San Diego, CA (1962).
5. Bond, J. W., K. M. Watson, and J. A. Welch, Jr., *Atomic Theory of Gas Dynamics*, Addison-Wesley, Reading, MA (1965).
6. Von Engel, A., *Electric Plasmas: Their Nature and Uses*, Taylor and Francis, New York (1983).
7. Masek, T., "Rocket model satellite positive ion beam system," AFGL-TR-78-0179, ADA063253, U.S. Air Force Geophysics Laboratory, Hanscom AFB, MA (1978).
8. Lai, S. T., W. J. McNeil, and E. Murad, "The role of metastable states in critical velocity ionization process," *J. Geophys. Res.* 93, no. A6: 5871–5878 (1988).
9. Bates, D. R., and A. Dalgarno, "Electronic recombination," in *Atomic and Molecular Processes*, ed. D. R. Bates, pp. 245–271, Academic Press, Orlando, FL (1962).
10. Dunn, G. H., D. S. Belic, T. J. Morgan, D. W. Mueller, and C. Timmer, "Dielectronic recombination of some single-charge ions," in *Electronic and Atomic Collisions*, eds. J. Eichter, I. V. Hertel, and N. Stolterfoht, p. 809, Elsevier, Amsterdam (1984).
11. Bates, D. R. "Aspects of recombination," in *Atomic and Molecular Processes*, ed. D. R. Bates, pp. 235–259, Academic Press, Orlando, FL (1979).

12. McGowan, J. W., R. H. Kummler, and F. R. Gilmore, "Excitation and de-excitation process relevant to the upper atmosphere," *Adv. Chem. Phys.* 28, p. 379 (1975).
13. Wulf, E., and U. von Zahn, "The shuttle environment: effects of thruster firings on gas density and composition in the payload bay," *J. Geophys. Res.* 91, no. A3, 3270–3278 (1986).
14. Lai, S. T., and E. Murad, "Inequality conditions for critical velocity ionization space experiments," *IEEE Trans. Plasma Sci.* 20, no. 6: 770–777 (1992).
15. Alfvén, H., "Collision between a nonionized gas and a magnetized plasma," *Rev. Mod. Phys.* 32: 710–713 (1960).
16. Papadopoulos, K., "On the shuttle glow (the plasma alternative)," *Radio Sci.* 19, no. 2: 5712–5717 (1983).
17. Machida, S., and C. K. Goertz, "A simulation study of the critical ionization velocity process," *J. Geophys. Res.* 91: 11965–11976 (1986).
18. Moghaddam-Taaheri, E., and C. K. Goertz, "Numerical quasi-linear study of the critical ionization velocity phenomena," *J. Geophys. Res.* 98, 1443–1460 (1993).
19. Papadopoulos, K., "The CIV processes in the CRIT experiments," *Geophys. Res. Lett.* 19, 605–608 (1992).
20. Brenning, N., "A comparison between laboratory and space experiments on Alfvén's CIV effect," *IEEE Trans. Plasma Sci.* 20, no. 6: 778–786 (1992).
21. Kivelson, M. G., and C. T. Russell, *Introduction to Space Physics*, Cambridge University Press, Cambridge, UK (1995).
22. Torbert, R. B., "Review of critical velocity experiments in the ionosphere," *Adv. Space Res.* 10, no. 7: 47–58 (1990).
23. Torbert, R. B., C. Kletzing, K. Liou, and D. Rau, "Prompt ionization in the CRIT II barium release," *Geophys. Res. Lett.* 19: 973–976 (1992).
24. Brenning, N., "Review of the CIV phenomenon," *Space Sci. Rev.* 59: 209–314 (1992).
25. Lai, S. T., "A review of critical ionization velocity," *Rev. Geophys.* 39, no. 4: 471–506 (2001).
26. Stein, J. A., and J. C. Walker, "Models of the upper atmosphere for a wide range of boundary conditions," *J. Atmos. Sci.* 22**:** 11–17 (1965).
27. Green, A.E.S,. and R. J. McNeal, "Analytical cross sections for inelastic collisions of protons and hydrogen atoms with atomic and molecular gases," *J. Geophys. Res.* 76: 133–144 (1971).
28. Jacchia, L. G., "Thermospheric temperature, density, and composition: new models," Smithsonian Astrophysics Observatory Special Report No. 375, Cambridge, MA (1977).
29. Lai, S. T., "Transport of high energy neutral beams in the atmosphere," *Phys. Space Plasmas* 10: 455–460 (1991).
30. Mendillo, M., G. S. Hawkins, and J. A. Klobuchar, "A large-scale hole in the ionosphere caused by the launch of Skylab," *Science* 187: 343–346 (1975).

APPENDIXES AND ADDENDA

APPENDIX 1

Drift of Hot Electrons

A1.1 Relevance to Spacecraft Charging

Spacecraft charging and anomalies at or near geosynchronous orbits[1] occur more often in the morning sector, from midnight to the morning hours, than at other local time sectors (figure Ap.1.1). This effect of local time on spacecraft charging and anomalies is mainly due to the eastward drift of the hot electrons.[2] In contrast, the hot ions drift westward toward the dusk sector. This appendix discusses some fundamentals of this aspect of space physics.[3]

A1.2 Convection Drift

The nightside magnetospheric plasma sheet, or simply the nightside plasma sheet, in the equatorial plane of the magnetosphere, features a cross-tail electric field **E**. It points from the dawn side toward the dusk side and is also called the *dawn–dusk electric field*. The electric field strength is enhanced when the interplanetary magnetic field around the magnetosphere is southward or when a substorm occurs as a result of magnetic field reconnection somewhere in the magnetospheric tail. Since the Earth's magnetic field **B** in the equatorial plane points northward while the electric field **E** in the plasma sheet points from dawn to dusk, the electrons and ions in the plasma sheet drift sunward by the $\mathbf{E} \times \mathbf{B}$ convection (figure Ap.1.2).[3] Because of the convection drift, the electric field **E** is also called the *convection electric field*.

In cylindrical coordinates, let **r** be the radial vector from the Earth center to the point considered and φ be the eastward angle ($\varphi = 0°$ in the Sun direction). Figure Ap.1.2 is a view from high above the North Pole. In terms of the electric field components at the point (r, φ), the electric field **E** is given as follows:

$$\mathbf{E} = E \sin\varphi \, \hat{\mathbf{e}}_r + E \cos\varphi \, \hat{\mathbf{e}}_\varphi \tag{Ap.1.1}$$

where $\hat{\mathbf{e}}_r$ is the unit vector in the r (radial) direction, and $\hat{\mathbf{e}}_\theta$ the unit vector in the φ (eastward) direction. In cylindrical coordinates, the electric field **E** expressed as the negative gradient of a potential ϕ is given by

$$\mathbf{E} = -\nabla\phi = -\left(\hat{\mathbf{e}}_r \frac{\partial}{\partial r} + \hat{\mathbf{e}}_\varphi \frac{\partial}{r\partial\varphi} + \hat{\mathbf{e}}_z \frac{\partial}{\partial z}\right)\phi \tag{Ap.1.2}$$

where the electric potential ϕ is given as follows:

$$\phi = -Er\sin\varphi \tag{Ap.1.3}$$

As mentioned earlier in this section, both the electrons and ions in the plasma sheet are $\mathbf{E} \times \mathbf{B}$ drifting sunward, i.e., toward the Earth (figure Ap.1.2).

During periods of southward interplanetary magnetic field or substorms, energetic electrons—i.e., hot electrons—are abundant in the plasma sheet, often as a result of

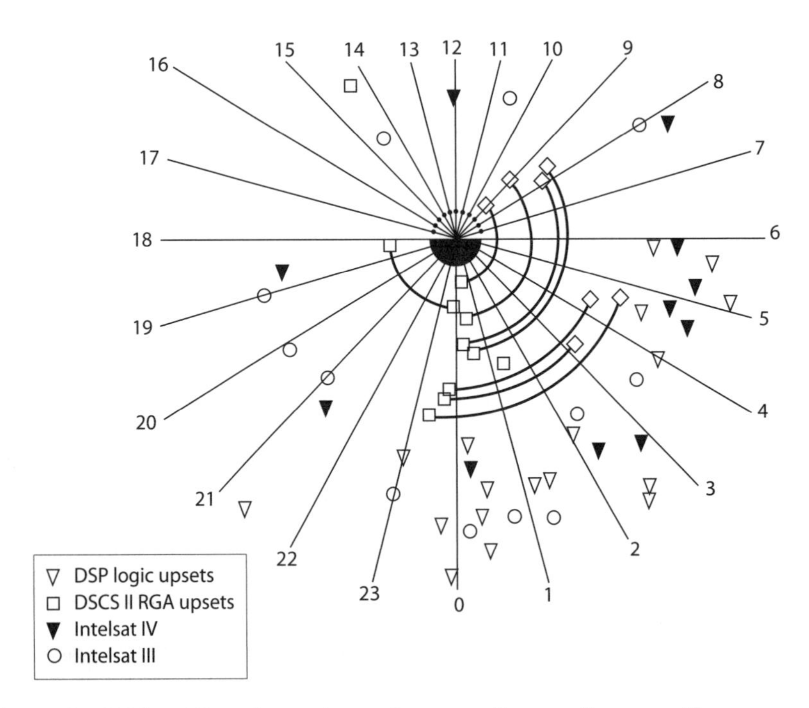

Figure Ap.1.1 Local time dependence of spacecraft anomalies caused by spacecraft charging. The Sun is to the top direction. Radial distance has no meaning. (Adapted from reference 1.)

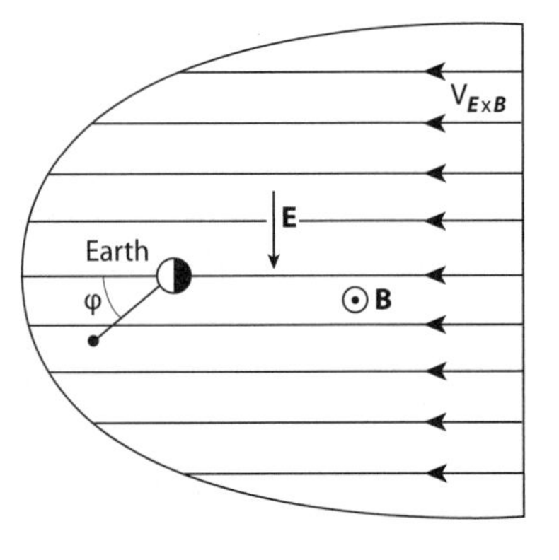

Figure Ap.1.2 E × B drift in the equatorial plane of the magnetosphere. The view is looking down from high above the North Pole of the Earth, half of which is in sunlight. The magnetic field **B** is pointing northward (that is, pointing out of the paper). The Sun is to the left. (Adapted from reference 3.)

reconnection somewhere in the magnetotail. The hot electrons undergoing convection drift toward the Earth would come near the geosynchronous orbit at midnight first, before they reach any other local time sectors in the orbit. However, as the electrons approach the geosynchronous orbit, and even before they reach it, they gradually feel the effects of the magnetic field corotating with the Earth, the gradient of the magnetic field, and the curvature of the magnetic field. There, they undergo corotation drift, gradient drift, and curvature drift. The total effect is that the hot electrons tend to drift eastward. This is why spacecraft charging and spacecraft anomalies at the geosynchronous altitudes occur mostly in the midnight to morning sector.

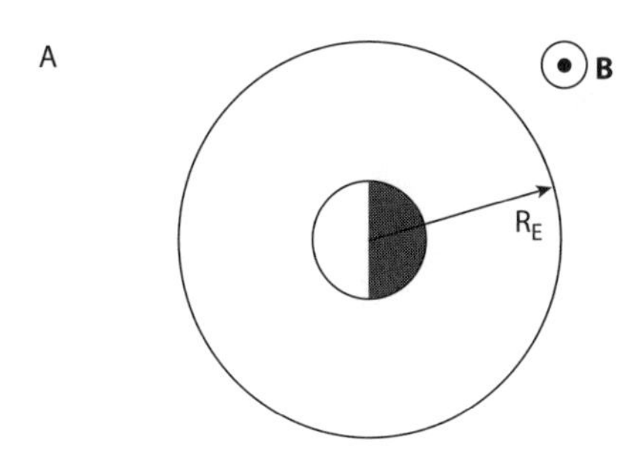

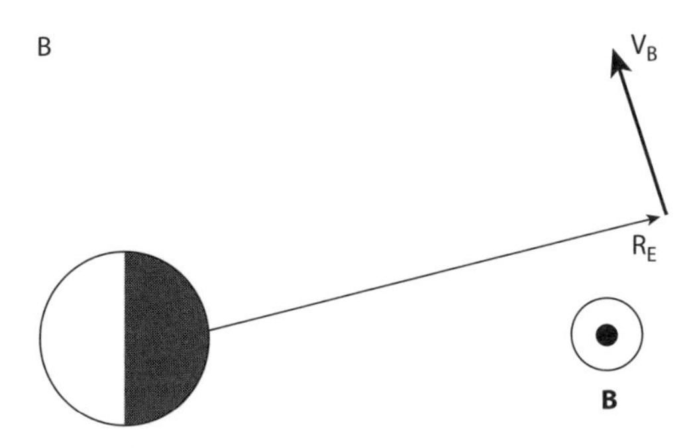

Figure Ap.1.3 Looking down from high above the North Pole, the view (A, B) is of the equatorial plane with the Sun to the left. The position vector R_E in the equatorial plane is radial from the Earth's center, and the magnetic field **B** is pointing northward (that is, pointing out of the paper). The corotation drift [equation (Ap.1.4)] is eastward.

A1.3 Corotation Drift

In figure Ap.1.3(A), the view is the equatorial plane, with the Earth denoted by a small half-shaded circle at the center and the near geosynchronous orbit denoted by a large circle. We are looking down from high above the North Pole. The magnetic field **B** in the orbit region is northward. Therefore, **B** is pointing toward the reader—that is, pointing out of the paper. When the electron, or ion, is approaching the geosynchronous orbit, the effect of the Earth's corotation appears gradually. The Earth is rotating eastward, carrying with it the corotating magnetic field **B**. Let the angular velocity of the rotating Earth be $\boldsymbol{\omega}$ (pointing north). The eastward velocity $\mathbf{V}_B$ of the corotating magnetic field at $\mathbf{r}$ is $\boldsymbol{\omega} \times \mathbf{r}$. Relative to the rotating system, the electron has a velocity opposite to $\mathbf{V}_B$. Therefore, the electron feels an electric field $\mathbf{E}_c$ due to the rotating Earth (and **B**):

$$\mathbf{E}_c = -\mathbf{V}_B \times \mathbf{B} = -(\boldsymbol{\omega} \times \mathbf{r}) \times \mathbf{B} \tag{Ap.1.4}$$

In figure Ap.1.3, $-\mathbf{V}_B$ is clockwise, and **B** is pointing north (out of the paper.) Therefore, the corotation electric field $\mathbf{E}_c$ is Earthward along $-\mathbf{r}$ and the corotation $\mathbf{E} \times \mathbf{B}$ drift direction is eastward along $\mathbf{V}_B$.*

*One might ask the question: An ion would feel an electric force in the opposite direction to that of an electron. The resultant $\mathbf{E} \times \mathbf{B}$ drift due to corotation would be westward! Correct?

Here is an answer: No, the motion-induced electric field $\mathbf{E}_C$ is $-\mathbf{V}_B \times \mathbf{B}$, irrespective of the sign of the charge. An electron and an ion drift in the same corotation electric field $\mathbf{E}_C$. There is a fundamental theorem that an $\mathbf{E} \times \mathbf{B}$ drift is independent of the sign of the charge (see, for example, reference 4). Therefore, the ion's $\mathbf{E} \times \mathbf{B}$ drift due to corotation is also eastward!

The magnitude of the Earth's magnetic field B in the dipole model is given by

$$B = B_E \frac{R_E^3}{r^3} \tag{Ap.1.5}$$

where r is measured from the Earth's center to the electron considered. Therefore, the magnitude of the corotation electric field, which is pointing Earthward, is given by

$$E_c = -\omega B_E \frac{R_E^3}{r^2} \tag{Ap.1.6}$$

where R_E is the Earth's radius (6370 km), and ω is the Earth's rotational angular velocity (radians/s). The electric field $\mathbf{E}_c$ is the negative gradient of the corotation electric potential ϕ_C. In cylindrical coordinates, we have

$$\mathbf{E}_C = -\nabla\phi_C = -\left(\hat{e}_r \frac{\partial}{\partial r} + \hat{e}_\varphi \frac{\partial}{r\partial\varphi} + \hat{e}_z \frac{\partial}{\partial z}\right)\phi_C \tag{Ap.1.7}$$

Since $\mathbf{E}_C$ is Earthward, only the $\partial/\partial r$ term in equation (Ap.1.7) is nonzero. Therefore, the corotation potential ϕ_C is obtained by integrating $\mathbf{E}_C$ of equation (Ap.1.6):

$$\phi_C = -\omega B_E \frac{R_E^3}{r} \tag{Ap.1.8}$$

A1.4 Combined (Convection plus Corotation) Drift

The combined potential ϕ_{CC} of convection and corotation is obtained by adding equations (Ap.1.3) and (Ap.1.8):

$$\phi_{CC} = \phi + \phi_C = -Er\sin\varphi - \omega B_E \frac{R_E^3}{r} \tag{Ap.1.9}$$

The combined electric field $\mathbf{E}_{CC}$ is

$$\mathbf{E}_{CC}(r,\varphi) = \left(E\sin\varphi - \frac{\omega B_E R_E^3}{r^2}\right)\hat{\mathbf{e}}_r + E\cos\varphi\,\hat{\mathbf{e}}_\varphi \tag{Ap.1.10}$$

Note the following:

1. At $\varphi = 0°$, the radial component of $\mathbf{E}_{CC} = \mathbf{E}_c$ and the φ component of $\mathbf{E}_{CC}$ is $\mathbf{E}$:

$$\mathbf{E}_{CC}(r,0) = \left(-\frac{\omega B_E R_E^3}{r^2}\right)\hat{\mathbf{e}}_r + E\,\hat{\mathbf{e}}_\varphi \tag{Ap.1.11}$$

2. At $\varphi = \pi/2$ (18:00 LT—i.e., dusk), $\mathbf{E}_{CC}$ has zero φ component:

$$\mathbf{E}_{CC}\left(r,\frac{\pi}{2}\right) = \left(E - \frac{\omega B_E R_E^3}{r^2}\right)\hat{\mathbf{e}}_r + 0\,\hat{\mathbf{e}}_\varphi \tag{Ap.1.12}$$

At $\varphi = \pi/2$, there is a critical radial distance r^* at which the total electric field is zero:

$$E_{CC}\left(r^*,\frac{\pi}{2}\right) = E - \frac{\omega B_E R_E^3}{r^{*2}} = 0 \qquad \text{(Ap.1.13)}$$

That is,

$$r^* = \sqrt{\frac{\omega B_E R_E^3}{E_0}} \quad \text{or} \quad \frac{r^*}{R_E} = \sqrt{\frac{\omega B_E R_E}{E_0}} \qquad \text{(Ap.1.14)}$$

The location r^* at $\varphi = \pi/2$ (i.e., LT 18:00) is called the *stagnation point*, where $\mathbf{E}_{CC} = 0$ in both components. Physically, at LT 18:00, the electric fields E and E_{cc} cancel each other. At that point, the convection drift is sunward, while the corotation drift is anti-sunward. They cancel each other at $(r^*, \pi/2)$, the stagnation point.

3. At $\varphi = 2\pi/3$ (LT 06:00), $\cos\varphi = 0$, but $\sin\varphi = -1$ in equation (Ap.1.10). There, unlike at $\varphi = \pi/2$, $\mathbf{E}_T$ does not vanish. Physically, both convection and corotation drifts are aligned (both are sunward) at $\varphi = 2\pi/3$ (LT 06:00), making the total drift maximum.

For diagrams showing the drift directions, see, for example, references 5 and 6.

Exercise 1: Calculate r^*, given $E = 5 \times 10^{-5}$ V/m, $\omega = 24$ H/day, $B_E = 3.1 \times 10^{-5}$T, and $R_E = 6370$ km.

Exercise 2: Plot the potentials φ_{CC} = constant, in x-y coordinates ($x = r\cos\varphi$, $y = r\sin\varphi$).

A1.5 Gradient Drift

The gradient drift is the drift due to the gradient of magnetic field. It is given by (reference 4; chapter 2):

$$\mathbf{V}_G = \pm v_\perp r_L \frac{\mathbf{B} \times \nabla \mathbf{B}}{2B^2} \qquad \text{(Ap.1.15)}$$

In equation (Ap.1.15), $\pm$ refers to the sign of the particle charge, and r_L is the Larmor radius. An explanation of gradient drift is given in figure Ap.1.4. Suppose that an electron is traveling with velocity v along the plasma sheet toward the Earth. The magnetic field is northward. The magnetic moment μ of the electron is given by

$$\mu = \frac{mv^2}{2B} \qquad \text{(Ap.1.16)}$$

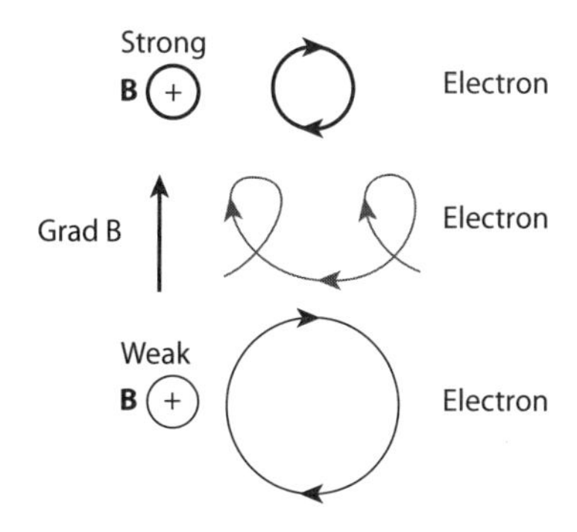

Figure Ap.1.4 The gradient drift. In a strong homogeneous magnetic field, the electron spirals in a tight circle (top). In a weak homogeneous magnetic field, the electron spirals in a wide circle (bottom). In a gradient magnetic field, the electron spirals tighter and wider alternately in every cycle, resulting in a drift toward the left (middle).

The Larmor radius of the electron is given by

$$r_L = \frac{mv}{|q|B} \tag{Ap.1.17}$$

Substituting equations (Ap.1.16) and (Ap.1.17) into equation (Ap.1.15), we have

$$V_G = \pm\frac{\mu}{|q|}\frac{\mathbf{B}\times\nabla\mathbf{B}}{B^2} = \frac{\mu}{q}\frac{\mathbf{B}\times\nabla\mathbf{B}}{B^2} \tag{Ap.1.18}$$

The Earth's magnetic field **B** increases as one travels toward the Earth. Therefore, the gradient of **B** (i.e., $\nabla\mathbf{B}$) is positive toward the Earth. Since **B** is northward at the equatorial plane, $\mathbf{B}\times\nabla\mathbf{B}$ is westward. For electrons, the minus sign in equation (Ap.1.15) renders the direction of the gradient drift $\mathbf{V}_G$ eastward. For ions, the gradient drift $\mathbf{V}_G$ is westward.

Since the magnetic moment μ [equation (Ap.1.16)] is a function of kinetic energy ($mv^2/2$), the energetic ions tend to "gradient drift" westward, winning the competition over the effect of the corotation drift. This behavior is in contrast to the electrons, which will drift eastward under the effects of corotation and gradient drifts.

A1.6 Curvature Drift

The effect of curvature drift is less important compared with the three drifts discussed earlier. The curvature drift velocity (reference 4; see chapter 2) will not be derived here. It is given as follows:

$$\mathbf{V}_U = \frac{1}{2}\frac{m}{q}(2v_\parallel^2 + v_\perp^2)\frac{\mathbf{R}_U\times\mathbf{B}}{R_U^2B^2} \tag{Ap.1.19}$$

where $\mathbf{R}_U$ is the radius of curvature of the curved magnetic field lines, and q is the charge. The radius of curvature of the Earth's magnetic field lines is outward, while the magnetic field lines in the equatorial plane are northward. Therefore, $\mathbf{R}_U \times \mathbf{B}$ is westward. The charge q in equation (Ap.1.19) renders the electron curvature drift eastward while that of ions is westward.

A1.7 Summary

Directions. In the magnetospheric equatorial plane (figure Ap.1.2), the convection drift is sunward from the tail toward the Earth. The corotation drift is eastward, regardless of electrons or ions. The gradient drift is eastward for electrons but westward for ions. The curvature drift is also eastward for electrons but westward for ions (figures Ap.1.5 and Ap.1.6).

Magnitudes. The convection and corotation drifts are not functions of particle energy. The gradient drift [equation (Ap.1.18)] is a function of particle energy (via the velocity term). For low-energy particles, corotation is more important, while gradient drift is more important for high-energy particles.

Physical picture. Figure Ap.1.7 shows the schematics of electrons and ions entering the geosynchronous orbit from the nightside of the magnetospheric equatorial plane. Such entrances are common during substorms and are called *substorm injections* from the plasmasheet in the magnetotail. As a result, spacecraft charging at the geosynchronous orbit occurs more likely in the sector from about midnight to about 6 o'clock in the morning. Figure

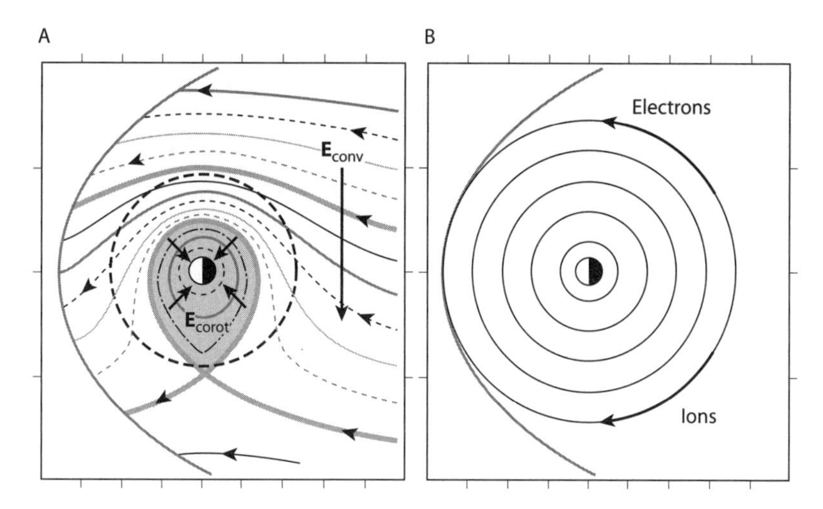

Figure Ap.1.5 (A) Combined convection and co-rotation electric fields. The dashed circle represents the geosynchronous orbit. The arrows indicate the resultant drift directions for electrons and ions. (B) Gradient and curvature drift directions. Electrons and ions drift in opposite directions. (Adapted from reference 5.)

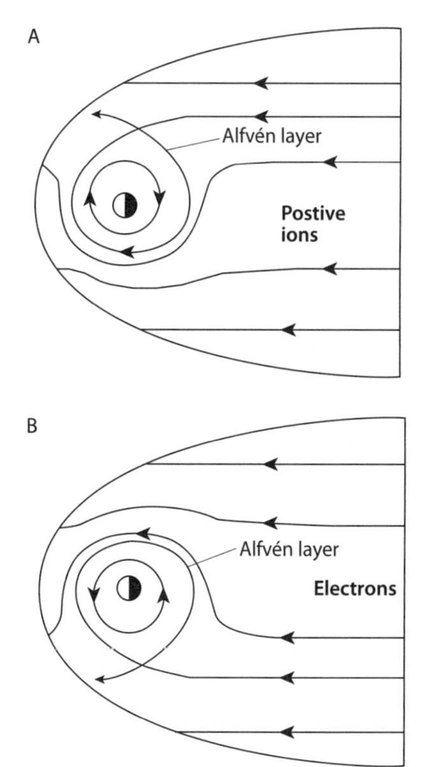

Figure Ap.1.6 Schematic of drift paths for typical plasma-sheet ions (A) and electrons (B). Shown is the magnetospheric equatorial plane, with the Sun to the left. (Adapted from reference 6.)

Ap.1.8 shows a superimposition of an electron flux map, the Freeman magnetospheric specification model (MSM)[7] for a given energy channel, on the spacecraft charging anomalies event map of Wilkinson (1991).[8] The features of figure Ap.1.8 are generally consistent with those of figure Ap.1.7.

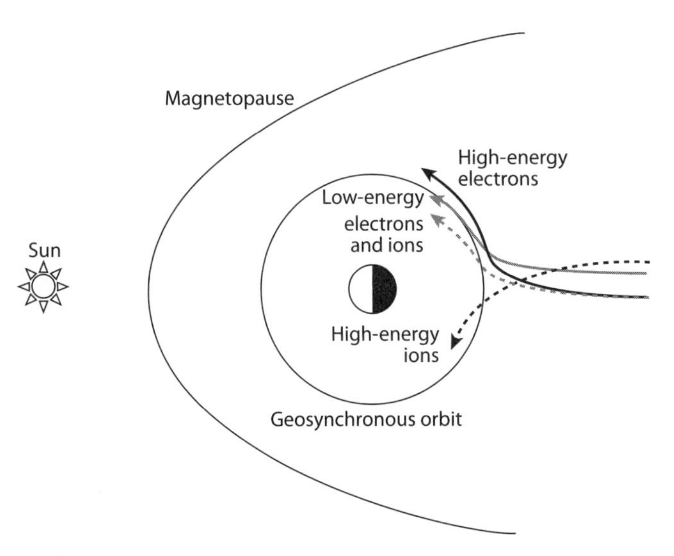

Figure Ap.1.7 Schematics of electrons and ions entering the geosynchronous orbit from the night side of the magnetospheric equatorial plane. Such entrances are common during substorms and are called substorm injections from the plasma sheet in the magnetotail. As a result, spacecraft charging at the geosynchronous orbit is more likely to occur in the sector from about midnight to about 6 A.M. (Adapted from figure 4 of reference 2.)

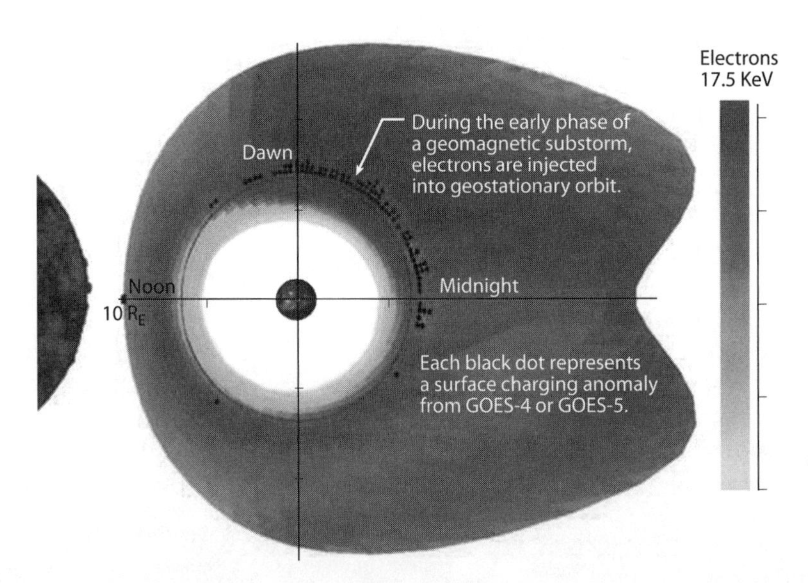

Figure Ap.1.8 Supposition of Wilkinson's spacecraft charging anomalies (dots) on Freeman's magnetospheric specification model. The view is on the magnetospheric equatorial plane. The Sun is to the left. The gray color illustrates the flux magnitude of the hot (17.5 keV) electrons. From the right-hand side, the electrons drift from the magnetotail toward the Sun. While approaching the geosynchronous region (faint ring), the hot electrons drift gradually eastward from the midnight sector. Spacecraft charging at or near geosynchronous orbits is most important in the morning sector.

1.8 References

1. McPherson, D. A., and W. R. Schober, "Spacecraft charging at high altitudes: the SCATHA satellite program," in *Spacecraft Charging by Magnetospheric Plasmas*, in *Progress in Aeronautics Astronautics*, ed. A. Rosen, vol. 47, pp. 237–246, MIT Press, Cambridge, MA (1976).
2. DeForest S. E., and C. E. McIlwain, "Plasma clouds in the magnetosphere," *J. Geophys. Res.* 76, no. 16: 3587–3611 (1971).
3. Kivelson, M. G., and C. T. Russell, *Introduction to Space Physics*, Cambridge University Press, Cambridge, UK (1995).
4. Chen, F. F., *An Introduction to Plasma Physics*, 2nd ed., Plenum, New York (1984).
5. Thomsen, M. F., Geospace Environment Modeling (GEM) Tutorial (2006), available at http://www-ssc.igpp.ucla.edu/gem/tutorial/2003MThomsen/.
6. Chen, A. J., "Penetration of low-energy protons deep into the magnetosphere," *J. Geophys. Res.* 75: 2458–2467 (1970).
7. Freeman, J. W., R. A. Wolf, R. W. Spiro, G. H. Voigt, and B. A. Hausman, "Development of a magnetospheric specification model. Volume 1. Final Report. 15 Nov. 1987–31 Mar. 1990," Air Force Research Contract F19628-87-K-0001, AD-A-244460/2/XAB, Rice University, Houston, TX (1990).
8. Wilkinson, D. C., S. C. Daughtridge, J. L. Stone, H. H. Sauer, and P. Darling, "TDRS-1 single event upsets and the effect of the space environment," *IEEE Trans. Nucl. Sci.* 38, no. 6: 1708–1712 (1991).

APPENDIX 2

Transformation of Coordinates

A2.1 Method 1

The Jacobian method is used for changing the integration variables from one coordinate system to another. For example, changing a volume integration in a rectangular coordinate system to that in a spherical one, the Jacobian method gives the following:

$$\iiint dx\,dy\,dz = \iiint dr\,d\theta\,d\varphi\,\frac{\partial(x,y,z)}{\partial(r,\theta,\varphi)} \tag{Ap.2.1}$$

where the Jacobian J is denoted by the ratio term on the right-hand side of equation (Ap.2.1). Explicitly, the Jacobian is given by a determinant:

$$\frac{\partial(x,y,z)}{\partial(r,\theta,\varphi)} = \begin{vmatrix} \frac{\partial x}{\partial r} & \frac{\partial x}{\partial \theta} & \frac{\partial x}{\partial \varphi} \\ \frac{\partial y}{\partial r} & \frac{\partial y}{\partial \theta} & \frac{\partial y}{\partial \varphi} \\ \frac{\partial z}{\partial r} & \frac{\partial z}{\partial \theta} & \frac{\partial z}{\partial \varphi} \end{vmatrix} = r^2 \sin\varphi \tag{Ap.2.2}$$

A2.2 Method 2

In method 2, one constructs the elemental volume of the new variables (figure Ap.2.1) as follows:

- As r increases at fixed θ and φ, it sweeps a length dr.
- As φ increases at fixed r and θ, it sweeps a length $r\,d\varphi$.
- As θ increases at fixed r and φ, it sweeps a length $r\sin\varphi\,d\theta$.

Multiplying the three lengths together, we have the new elemental volume: $r^2\sin\varphi\,d\theta\,d\varphi\,dr$.

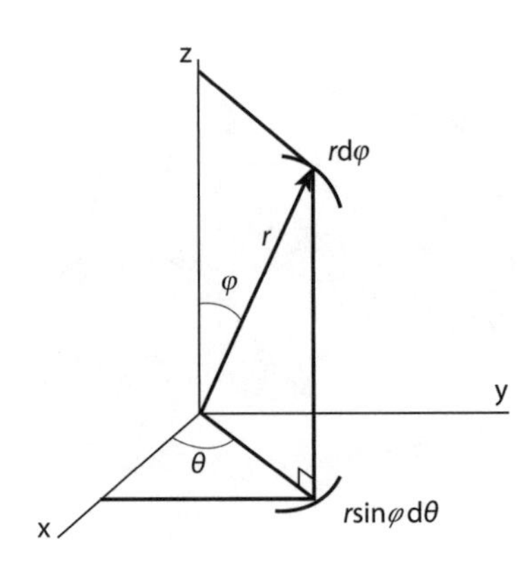

Figure Ap.2.1 Integration in polar coordinates. $\iiint dx\,dy\,dz = \iiint r^2 \sin\phi\, d\theta\, d\phi\, dr$.

APPENDIX 3

Normalization and Dimension of Maxwellian Distribution

A3.1 Normalization and Dimension

The Maxwellian velocity distribution $f(\mathbf{v})$ is defined as follows (for example, see reference 1):

$$f(\mathrm{v}) = n\left(\frac{m}{2\pi kT}\right)^{3/2} \exp\left(-\frac{\mathrm{v}^2}{\mathrm{v}_{th}^2}\right) \quad \text{(Ap.3.1)}$$

where n is the density, m the particle mass, k the Boltzmann constant, T the temperature, v the particle velocity, and v_{th} the particle thermal velocity. The velocities can be written as

$$\mathrm{v}^2 = \mathrm{v}_x^2 + \mathrm{v}_y^2 + \mathrm{v}_z^2 \quad \text{(Ap.3.2)}$$

and the thermal velocity v_{th} is given by

$$\mathrm{v}_{th}^2 = \frac{kT}{2m} \quad \text{(Ap.3.3)}$$

The integral of $f(\mathbf{v})$ over the three-dimensional velocity $\mathbf{v}$ is as follows:

$$\int_{-\infty}^{\infty} d^3\mathbf{v} f(\mathbf{v}) = n\int_{-\infty}^{\infty} d^3\mathbf{v}\left(\frac{m}{2\pi kT}\right)^{3/2} \exp\left(-\frac{\mathrm{v}^2}{\mathrm{v}_{th}^2}\right) \quad \text{(Ap.3.4)}$$

$$= n\left(\frac{m}{2\pi kT}\right)^{3/2} \int_{-\infty}^{\infty} d\mathrm{v}_x \exp\left(-\frac{\mathrm{v}_x^2}{\mathrm{v}_{th}^2}\right)\int_{-\infty}^{\infty} d\mathrm{v}_y \exp\left(-\frac{\mathrm{v}_y^2}{\mathrm{v}_{th}^2}\right)\int_{-\infty}^{\infty} d\mathrm{v}_z \exp\left(-\frac{\mathrm{v}_z^2}{\mathrm{v}_{th}^2}\right) \quad \text{(Ap.3.5)}$$

Using the property:

$$\int_{-\infty}^{\infty} d\mathrm{v} \exp(-a^2\mathrm{v}^2) = \frac{\sqrt{\pi}}{a} \quad \text{(Ap.3.6)}$$

one obtains the result:

$$\int_{-\infty}^{\infty} d^3\mathbf{v} f(\mathbf{v}) = n \quad \text{(Ap.3.7)}$$

From equation (Ap.3.7), is normalized so that

$$\frac{1}{n}\int_{-\infty}^{\infty} d^3\mathbf{v} f(\mathbf{v}) = 1 \quad \text{(Ap.3.8)}$$

From equation (Ap.3.8), one can see that the dimension of $f(\mathbf{v})$ is $(\mathrm{sec}^3/\mathrm{m}^6)$.

A3.2 Relating the Three-Dimensional Distribution to a One-Dimensional Distribution

For an isotropic plasma, the integral, equation (A3.7), of the Maxwellian distribution $f(\mathbf{v})$ integrated over the three-dimensional velocity $\mathbf{v}$ can be converted to an integral of another function $g(\mathrm{v})$ where v is a scalar:

$$\int_{-\infty}^{\infty} d^3\mathbf{v} f(\mathbf{v}) = \int_{-\infty}^{\infty} d\mathrm{v}\, g(\mathrm{v}) \tag{Ap.3.9}$$

where the one-dimensional (that is, scalar) function $g(\mathrm{v})$ is as follows (reference 1):

$$g(\mathrm{v}) = 4\pi n \left(\frac{m}{2\pi kT}\right)^{3/2} \mathrm{v}^2 \exp\left(-\frac{\mathrm{v}^2}{\mathrm{v}_{th}^2}\right) \tag{Ap.3.10}$$

Although the function $f(\mathrm{v})$ has its maximum at $\mathrm{v} = 0$, the function $g(\mathrm{v})$ equals zero at $\mathrm{v} = 0$ and has its maximum at a finite (nonzero) value of v. Using equations (Ap.3.8) to (Ap.3.10), one obtains

$$\frac{1}{n}\int_{-\infty}^{\infty} d\mathrm{v}\, g(\mathrm{v}) = 1 \tag{Ap.3.11}$$

From equation (Ap.3.11), one can see that the dimension of $g(\mathrm{v})$ is (sec/cm^4).

A3.3 Reference

1. Chen, F. F., *Plasma Physics*, 2nd ed., Plenum, New York (1984).

APPENDIX 4

Flux Integrals

The balance of currents can be written as the balance of fluxes, if the fluxes are uniform. Flux J is the amount of charge traveling across a surface area per unit time. For a single velocity v of electrons, the flux J is given by

$$J = nqv \tag{Ap.4.1}$$

where n is the electron density, and q the electron charge.

If the electron velocity distribution is $f(v)$, the flux J is given by

$$J = q\int_0^\infty d^3\mathbf{v}v_n f(v) \tag{Ap.4.2}$$

where v_n is the normal component of the incident electron velocity, and the electron density n is given by

$$n = \int_{-\infty}^\infty d^3\mathbf{v}f(v) \tag{Ap.4.3}$$

At equilibrium, the electron velocity distribution is a Maxwellian [equation (Ap.3.1)], which can be written in the following form:

$$f(v) = n\left(\frac{m}{2\pi kT}\right)^{3/2} \exp\left(-\frac{E}{kT}\right) \tag{Ap.4.4}$$

In spherical coordinates, equation (Ap.4.2) can be written as follows:

$$J = q\int_0^\infty dv v^2 \int_0^{\pi/2} d\varphi \int_0^{2\pi} d\theta \sin\varphi \, v_n f(v) \tag{Ap.4.5}$$

where we are considering the half space above the surface, which is assumed infinite and flat. For a surface element on a spacecraft, the ambient flux from below the half surface can not reach the surface. In equation (Ap.4.5), $v_n = v\cos\varphi$ if the incident electrons are coming in at angle φ. We assume isotropic ambient electron flux. Since the electron energy $E = (1/2)mv^2$, we have

$$dE = mv\,dv \tag{Ap.4.6}$$

Substituting equations (Ap.4.4) and (Ap.4.6) into equation (Ap.4.5), we obtain the flux:

$$J = n\left(\frac{1}{2m\pi}\right)^{1/2}\left(\frac{1}{kT}\right)^{3/2}\int_0^\infty dE\,E\exp(-E/kT) \tag{Ap.4.7}$$

In equations (Ap.4.7), the flux J is of the form

$$J = \int_0^\infty dE \frac{dJ(E)}{dE} = c\int_0^\infty dE\,E \exp(-E/kT) \qquad \text{(Ap.4.8)}$$

where $c = n(1/2m\pi)^{1/2}\,(1/kt)^{3/2}$.The integrand is called the *differential flux*. Thus, the differential flux is of the form

$$\frac{dJ(E)}{dE} = ncE\exp\left(-\frac{E}{kT}\right) \qquad \text{(Ap.4.9)}$$

Since the right-hand side of equation (Ap.4.4) is a function of energy E, let us change the notation to $f(E)$. That is,

$$f(E) = n\left(\frac{m}{2\pi kT}\right)^{3/2} \exp\left(-\frac{E}{kT}\right) \qquad \text{(Ap.4.10)}$$

Strictly speaking, the function $f(E)$ is a velocity distribution and not an energy distribution. The energy distribution $f_E(E)$ will be defined in appendix 5.

From equations (Ap.4.7) and (Ap.4.8), we have

$$J = \frac{2\pi}{m^2}\int_0^\infty dE\,E f(E) \qquad \text{(Ap.4.11)}$$

We must bear in mind that $f(E)$ in equations (Ap.4.10) to (Ap.4.11) is a velocity distribution, in which v is expressed in terms of E, where $E = (1/2)m\mathrm{v}^2$.

APPENDIX 5

Energy Distribution

One can define the electron energy distribution function $f_E(E)$ as follows:

$$n = \int_{-\infty}^{\infty} d^3\mathbf{v} f(\mathrm{v}) = \int_0^{\infty} dE f_E(E) \tag{Ap.5.1}$$

where the functional forms of $f(\mathrm{v})$ and $f_E(E)$ are different. The function $f(\mathrm{v})$ for a Maxwellian distribution has been defined in equation (Ap.4.4) of appendix 4:

$$f(\mathrm{v}) = n\left(\frac{m}{2\pi kT}\right)^{3/2} \exp\left(-\frac{E}{kT}\right) \tag{Ap.5.2}$$

where $E = (1/2)m\mathrm{v}^2$.

For isotropic electrons, the volume integral can be easily done in spherical coordinates. Since the volume element of each spherical shell is $4\pi\mathrm{v}^2 d\mathrm{v}$, we have

$$\int_{-\infty}^{\infty} d^3\mathbf{v} f(\mathrm{v}) = \int_0^{\infty} d\mathrm{v}\mathrm{v}^2 \int_0^{\pi} d\varphi \sin\varphi \int_0^{2\pi} d\theta f(\mathrm{v}) = 4\pi \int_0^{\infty} d\mathrm{v}\mathrm{v}^2 f(\mathrm{v}) \tag{Ap.5.3}$$

We can convert $d\mathrm{v}\mathrm{v}^2$ to an energy expression:

$$d\mathrm{v}\mathrm{v}^2 = \frac{(2E)^{1/2}}{m^{3/2}} dE \tag{Ap.5.4}$$

Thus, equation (Ap.5.3) becomes

$$\int_{-\infty}^{\infty} d^3\mathbf{v} f(\mathrm{v}) = \frac{4\pi 2^{1/2}}{m^{3/2}} \int_0^{\infty} dE E^{1/2} f(\mathrm{v}) \tag{Ap.5.5}$$

Equating the energy integrals of equation (Ap.5.1) with equation (Ap.5.5), we have

$$f_E(E) = \frac{4\pi 2^{1/2}}{m^{3/2}} E^{1/2} f(\mathrm{v}) \tag{Ap.5.6}$$

Substituting in the Maxwellian distribution $f(\mathrm{v})$, equation (Ap.5.2), into equation (Ap.5.6), we obtain

$$f_E(E) = n \frac{2}{\pi^{1/2}(kT)^{3/2}} E^{1/2} \exp(-E/kT) \tag{Ap.5.7}$$

which agrees with equation (5.28) of the classic text listed in reference 1.

A5.1 Reference

1. Mayer, J. E., and M. G. Mayer, *Statistical Mechanics*, John Wiley, New York (1963).

APPENDIX 6

Sheath Engulfment

A6.1 Introduction

This is a phenomenon of multibody interaction between differentially charged surfaces. When a surface charges to a high potential, the potential profile $\phi(r)$, which is a function of distance r from the surface, may extend to the neighboring surfaces. The neighboring surfaces may be partially or totally engulfed by the potential $\phi(r)$ generated by the highly charged surface. The dominating potential $\phi(r)$ forms a sheath region. *Sheath engulfment* is a term describing this phenomenon. As a result, currents may flow from one surface to the other. As a result of sheath engulfment, the surface potentials can be affected.

Sheath engulfment can occur naturally or artificially. It can occur naturally as a result of different charging properties, or different ambient conditions, of neighboring surfaces. As an example of sheath engulfment occurring naturally, we consider the monopole-dipole potential distribution around a sunlit dielectric spacecraft. In this example, the high negative potential of the dark side (without photoemission) extends to, or even wraps around, the low positive potential of the sunlit side (with photoemission). As a result, the potential of the sunlit side can swing negative (see chapter 8 and the references therein).

Significant different charging and sheath engulfment can be driven by artificial high current during nonneutral beam emissions from spacecraft with multiple surfaces, some of them being isolated.

A6.2 High-Current Beam Emission

High-current emission of a charged particle beam from a spacecraft surface induces charging to a high potential of the opposite sign, depending, of course, on the balance of all currents to and from the surface. If some neighboring surfaces are electrically isolated from the subject surface, their potentials would not be directly affected by the beam emission. However, the profile $\phi(r)$ (i.e., the sheath), of the high potential may extend significantly to the neighboring surfaces. To be concrete, let us consider a scenario as follows.

A satellite in the geosynchronous environment is emitting an electron current of about 1 to 2 mA at a beam energy of about 1 to 2 keV. The satellite can charge to hundreds of volts positive, because the incoming ambient electron current cannot compete with the outgoing electron beam current until the satellite voltage is high enough. The satellite body is assumed spherical and uniform. Following Whipple et al.[1] and Godard and Laframboise,[2] we model the sheath potential profile $\phi(r)$ as follows:

$$\phi(r) = \phi(0)\frac{R}{r+R}\exp\left(-\frac{r}{\lambda_D}\right) \tag{Ap.6.1}$$

where r is the distance from the satellite surface, R the radius of the satellite, and λ_D the Debye shielding distance of the plasma in the vicinity of the satellite. For a satellite in geosynchronous orbits, λ_D is of the order of 100 m as long as the satellite is not inside the plasmasphere. The positive potential $\phi(0)$ of the satellite surface ($r = 0$) is controlled by a high electron beam current. One notes that in equation (Ap.6.1), the ratio, $\phi(r)/\phi(0)$, between the potential at a distance r and the spacecraft potential (at $r = 0$) is independent of the spacecraft potential $\phi(0)$ but depends on the Debye distance λ_D only. As an exercise, one can take the SCATHA satellite[3] with radius R equal to about 0.8 m and calculate the sheath profile $\phi(r)$ at any distance r for any satellite potential $\phi(0)$ by using equation (Ap.6.1) for $\lambda_D = 20$ to 40 m. The results would show that the sheath potential $\phi(r)$ is substantial over tens of m in distance r.

Next, consider a pair of long booms extending, and electrically isolated, from the satellite body. The booms are in sunlight generating photoelectrons. Some photoelectrons will leave the boom surface and go away, but some photoelectrons will go to the satellite body depending on the engulfment by the sheath. Let us construct an analytical model[3] as follows.

Consider a point on a long boom. Its distance is r from the satellite body. The sheath potential at that point is $\phi(r)$ as given by equation (Ap.6.1). Let the Sun angle be θ (normal = 90°) at that point. Following Whipple,[4] we model the distribution of photoelectrons as Maxwellian with a temperature T_{ph}. The fraction f of photoelectron current going to the satellite body is given by the partition of the photoelectron distribution function (figures Ap.6.1 and Ap.6.2).

$$f[\phi(r)] = \frac{\int_0^{\phi(r)} dEE \exp(-E/kT_{ph})}{\int_0^{\infty} dEE \exp(-E/kT_{ph})} \tag{Ap.6.2}$$

As a simple case, the photoelectron current $I_{ph}(\theta,\phi)$ going to the satellite body is given approximately by equation (Ap.6.3):

$$I_{ph}(\theta,\phi) = D|\sin\theta| \int_0^L dr j_{ph}(\theta) f[\phi(r)] \tag{Ap.6.3}$$

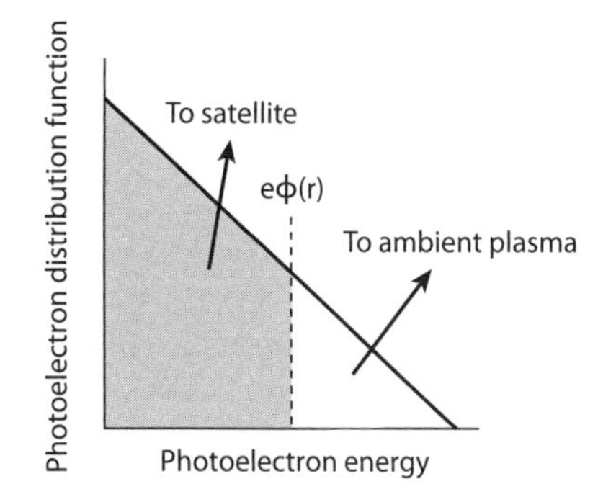

Figure Ap.6.1 Partition of photoelectron distribution function by the sheath potential energy.

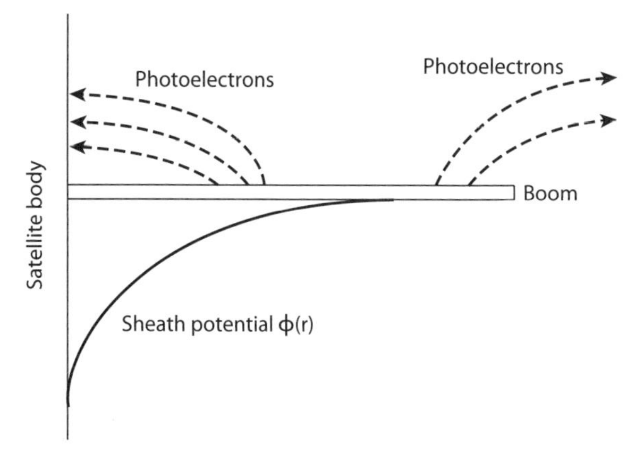

Figure Ap.6.2 Sheath engulfment. Photoelectrons in the sheath flow toward the satellite body.

where D is the diameter of the boom, L the length, j_{ph} the photoemissivity of the boom material, f the fraction given by equation (Ap.6.2), and the sheath potential ϕ at the point a function of the distance r and the satellite potential $\phi(0)$ as given by equation (Ap.6.1). If the satellite features two booms diametrically, then the right-hand side of equation (Ap.6.3) has to be doubled.

This photoelectron current, equation (Ap.6.3), flows toward the satellite body as a result of the sheath engulfment by the satellite body potential (figures Ap.6.3 and Ap.6.4). If the satellite body potential is driven by an artificial beam emission from the satellite, the degree of engulfment increases with the beam current. Consequently, a higher beam current emission gives rise to larger fraction of photoelectron current flowing toward the satellite body. In turn, the photoelectron current changes the satellite body potential, which is determined by the balance of all currents. Furthermore, since the photoelectron current is a function of Sun angle, the magnitude of the photoelectron current oscillates as the satellite rotates in sunlight. As a result, the satellite body potential modulates accordingly. For further details of this analytical method of modeling of sheath engulfment and its comparison with actual

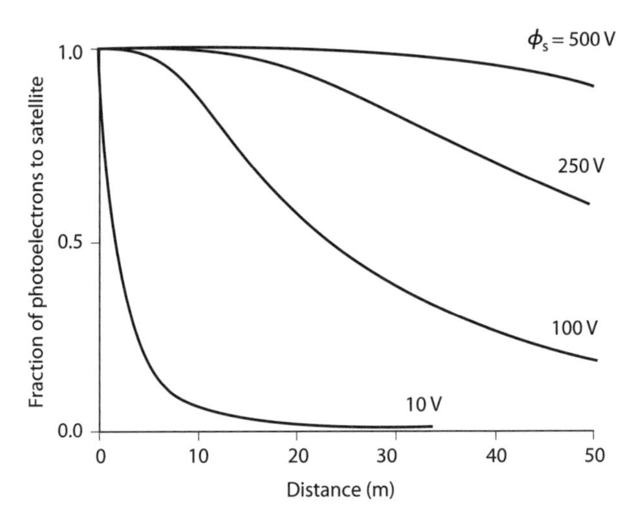

Figure Ap.6.3 Fraction of photoelectron current going toward the satellite as a function of distance from the satellite surface. (Adapted from reference 3.)

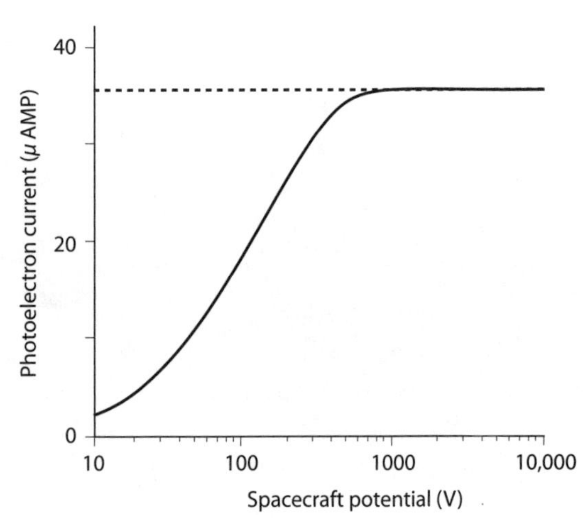

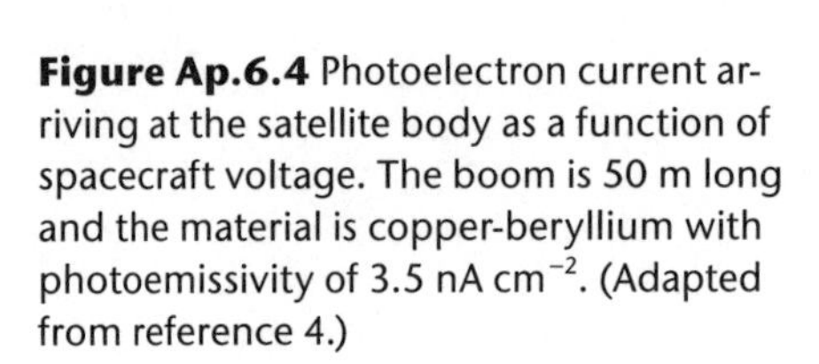

Figure Ap.6.4 Photoelectron current arriving at the satellite body as a function of spacecraft voltage. The boom is 50 m long and the material is copper-beryllium with photoemissivity of 3.5 nA cm^{-2}. (Adapted from reference 4.)

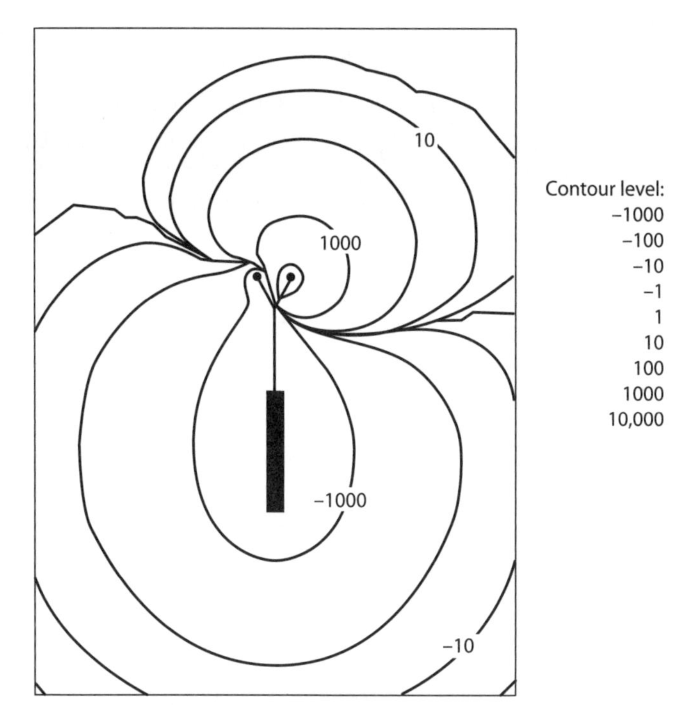

Figure Ap.6.5 NASCAP/LEO potential contours for one sphere at +46 kV with respect to spacecraft ground, which is at −6 kV. (Adapted from reference 8.)

data measured on the rotating SCATHA satellite during beam emissions in sunlight, see references 3 and 5. An improved formula for the photoelectron current from the boom is given in reference 6. It is instructional to study an application of this analytical method used in references 5 and 6.

Computer simulations can show sheath engulfment in vivid manner. Sheath engulfment is prominent when high beam currents are emitted from spacecraft. For example, the SPEAR-1 experiment, which featured high beam current emissions, yielded data showing sheath engulfment of neighboring surfaces. Computer simulations[7,8] have demonstrated vividly sheath engulfment in the SPEAR-1 experiment (figure Ap.6.5).

A6.3 References

1. Whipple, E. C., J. M. Warnock, and R. H. Winckler, "Effect of satellite potential in direct ion density measurements through the magnetosphere," *J. Geophys. Res.* 79: 179–186 (1974).
2. Godard, R.J.L., and J. G. Laframboise, "A symmetrical model of current collection of a spherical electrostatic probe in a flowing plasma,"*EOS Trans. AGU* 54: 392, abstract (1973).
3. Lai, S. T., H. A. Cohen, T. L. Aggson, and W. J. McNeil, "The effect of photoelectrons on boom-satellite potential differences during electron beam ejections," *J. Geophys. Res.* 92, no. A11: 12319–12325 (1987).
4. Whipple, E. C., Jr., "Potential of surface in space," *Rep. Prog. Phys.* 44: 1197–1250 (1981).
5. Lai, S. T., "Current collection and plasma measurements during electron beam emission at near geosynchronous altitudes," presented at 32nd AIAA Aerospace Sci. Mtg., Reno,

NV, Jan. 10–13, 1994, AIAA-94-0370 (1994). Available online at http://handle.dtic.mil/100.2/ADA277097.

6. Lai, S. T., "An improved Langmuir probe formula for modeling satellite interactions with near geostationary environment," *J. Geophys. Res* 99: 459–468 (1994).
7. Allred D. B., J. D. Benson, H. A. Cohen, W. J. Raitt, D. A. Burt, I. Katz, G. A. Jongeward, J. Antoniades, M. Alport, D. Boyd, W. C. Nunnally, W. Dillon, J. Pickett, and R. B. Torbert, R.B., "The SPEAR-1 experiment: high voltage effects on space charging in the ionosphere," *IEEE Trans. Nucl. Sci.* 35, no. 6: 1386–1393 (1988).
8. Katz, I., G. A. Jongeward, V. A. Davis, M. J. Mandell, R. A. Kuharski, J. R. Lilley Jr., W. J. Raitt, D. L. Cooke, R. B. Torbert, G. Larson, and D. Rau, "Structure of bipolar plasma sheath generated by SPEAR-1," *J. Geophys. Res.* 94: 1450–1458 (1989).

APPENDIX 7

PN Junctions

A7.1 Introduction

Microelectronics onboard spacecraft are subject to anomalies caused by interactions with energetic ambient electrons and ions. There are two types of anomalies: soft and hard. Soft anomalies are self-recoverable shortly afterward, but hard anomalies are nonrecoverable or hard to recover. Some symptoms of soft anomalies are phantom commands, spurious signals, and resets in logical circuits. Some examples of hard anomalies are damages or defects generated in the microelectronics. PN junctions (also called P-N junctions) are commonly present in electronic chips and are subject to anomalies induced by electron or ion bombardment in space. This appendix provides some fundamentals of PN junctions and explains why they are prone to anomalies.

A7.2 Electronic Bands

In quantum mechanics, electrons behave as waves. The periodic potential structure of a solid lattice allows only certain frequency bands and gaps of electron waves to exist (figure Ap.7.1). The bands and gaps apply to the outermost electrons (valence electrons) only. The electrons in the inner shells of an atom do not participate in electron wave propagation or conduction. A good explanation why such bands and gaps form in periodic potential structures is given by the Kronig-Penney model, which is available in standard solid-state textbooks.[1,2]

For example, a silicon atom has four valence electrons. The valence band is full, the conduction band is empty, and the energy gap[3,4] between them is about $E_G = 1.12$ eV. Silicon is an insulator at normal temperature, because the electron thermal energy (well below 1 eV) is less than the gap energy. One can increase the conductivity of silicon by doping (meaning adding impurities to it). Suppose that one dopes silicon with atoms (such as arsenic, As) of five valence electrons; the energy required by the extra electron to hop into the conduction band is much less than E_G. The doping atoms with extra electrons are *donors* (or N-type). Suppose that one dopes silicon with atoms (such as gallium Ga) of three valence electrons; the energy required to create a hole in the valence band is much less than E_G. Such doping atoms are *acceptors* (or P-type).

A7.3 PN Junction

A silicon device with a PN junction is shown in figure Ap.7.2. The silicon labeled P is silicon doped with acceptors such as Ga; that labeled N is silicon doped with donors such as As. The P-silicon has holes, while the N-silicon has conduction electrons. When an electric field **E** is applied downward (figure Ap.7.2), electric current flows because the holes move downward and the electrons upward. However, when the field **E** is upward (figure Ap.7.3), no continuous current flows because the holes and the electrons move away from each other. As they move away, a depletion layer (not shown) forms at the junction between the P-silicon and the N-silicon, thus creating an electric field opposite to the applied one. Since the applied electric field controls the on-off of the current flow in a rectifier, the PN junction acts as a rectifier (figures Ap.7.2 and Ap.7.3).

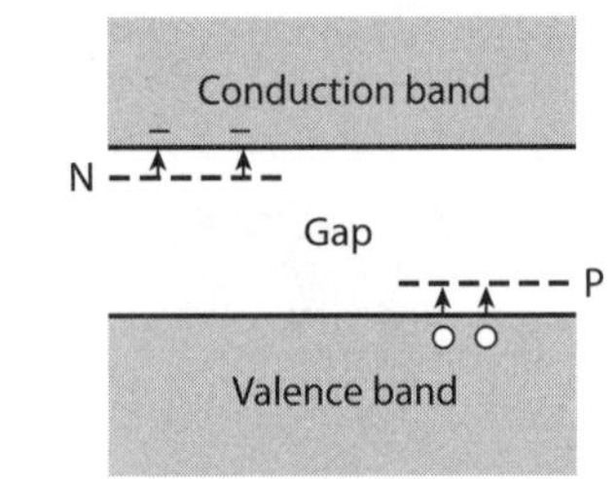

Figure Ap.7.1 Bands, gaps, and impurity doping.

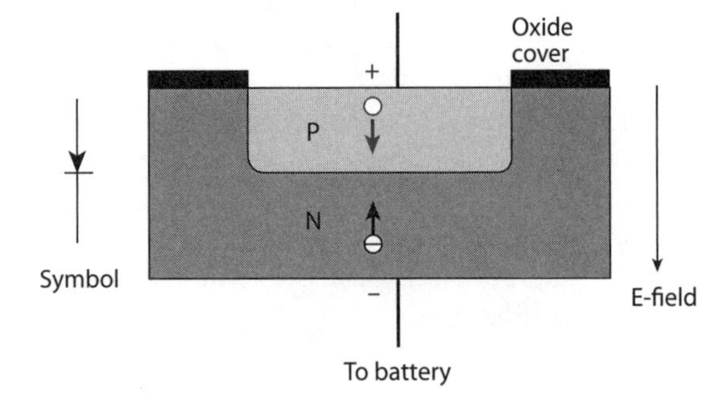

Figure Ap.7.2 A forward-biased PN junction. Current flows. The junction acts as a rectifier.

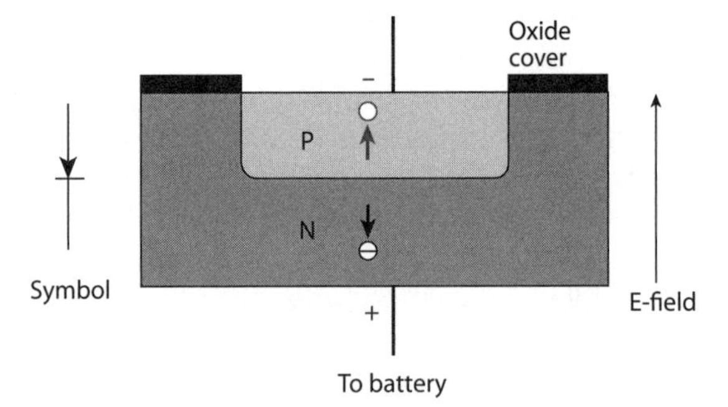

Figure Ap.7.3 A reverse-biased PN junction. Current does not flow. The junction acts as a rectifier.

A memory aid is given here. Consider the holes, O, in the P-silicon and the electrons, e, in the N-silicon. If OP and eN are moving toward each other, the circuit is closed. If OP and eN are moving away from each other, the circuit is open.

Consider the following as a side note: When the holes and electrons flow toward each other in a forward-biased PN junction, electron-hole recombination generates photons. This explains light emitting diodes (LEDs), such as those in TV controllers. When light shines on a reverse-biased PN junction depletion region, the generation of pairs of electrons and holes result in conduction. In this way, the device serves as a photodiode.

A7.4 Spacecraft Anomaly

Microelectronics, such as PN junctions, are sensitive to electric fields, electrons, ions, and photons. Differential charging to thousands of volts between neighboring surfaces on a satellite as a result of keV space plasma interactions creates high electric fields. If energetic (MeV) electrons and ions, originating from cosmic rays or the radiation environment in space,

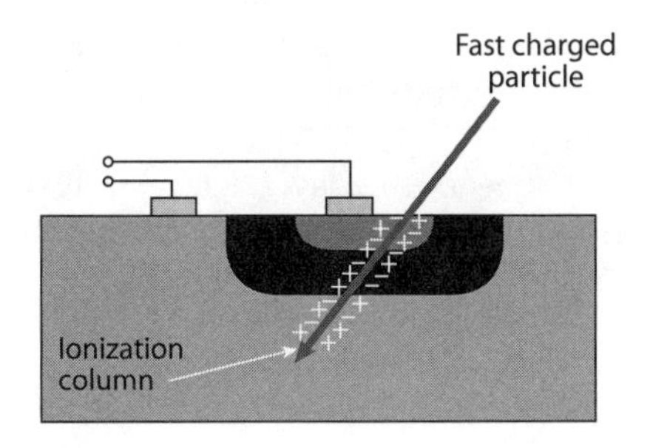

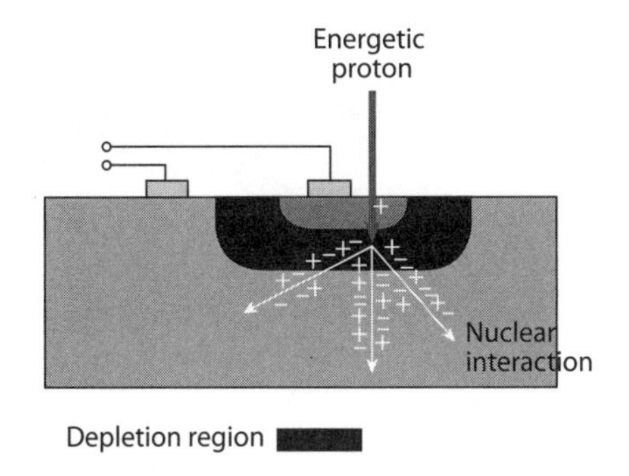

Figure Ap.7.4 Single event upsets (SEUs) . Energetic electrons and ions can penetrate into matter and disrupt the function of microelectronics. (Adapted from reference 5.)

penetrate into the silicon chips of the microelectronics, the charges deposited directly inside, or generated indirectly by ionization, may cause malfunction of the PN junction. The malfunction often manifests as false commands or spurious signals. However, such a malfunction is likely temporary only. With a reset, the microelectronics probably returns to normal function shortly afterward. This kind of anomaly[5] is called a single event upset (SEU) (figure Ap.7.4). It is a soft anomaly.

Very rarely, an incoming energetic heavy particle scores a head-on collision with a nucleus in the lattice structure of a solid, causing a recoil of the nucleus. The recoiling nucleus, in turn, interacts with more electrons and ions in the solid, resulting in more damage. If an ion lattice structure is damaged, forming lattice defects,[6,7] the spacecraft anomalies resulting are difficult to recover or simply nonrecoverable. This kind of anomaly is a hard anomaly. If the incoming energetic particle is relativistic, it may cause nuclear reactions, generating mesons, which can produce further damage.

A7.5 References

1. Kittel, C., *Introduction to Solid State Physics*, 8th ed., John Wiley and Sons New York (1986).
2. Solymar, L., and D. Walsh, *Lectures on the Electrical Properties of Materials*, 2nd ed., Oxford University Press, Oxford, UK (1979).
3. Holmes, R. A., *Physical Principles of Solid State Devices*, Holt-Reinholt Winston, New York (1970).
4. Bube, R. H., *Electrons in Solids*, 2nd ed., Academic Press, San Diego, CA (1981).
5. Lauriente, M., and A. L. Vampola, "Spacecraft anomalies due to radiation environment in space," presented at NASDA/JAERI 2nd International Workshop on Radiation Effects of Semiconductor Devices for Space Applications, Tokyo, Japan (1996).

6. Lai, S. T., B. D. Nener, L. Faraone, A. G. Nassibian, and M.A.C. Hotchkis, "Characterization of deep-level defects in GaAs irradiated by 1 MeV electrons," *J. Appl. Phys.* 73, no. 2: 640–647 (1993).
7. Lai, S. T., D. Alexiev, and B. D. Nener, "Comparison between deep level defects in GaAs induced by gamma, 1 MeV electron, and neutron irradiation," *J. Appl. Phys.* 78, no. 6: 3686–3690 (1995).

APPENDIX 8

Probability Function

Let λdt be the probability of one hit in the interval $(t, t + dt)$. Suppose that the probability is small and that each hit is independent of the others. Let $P_N(t)$ be the probability of N hits in the interval $(0, t)$.

We have

$$P_N(t + dt) = P_N(t)(1 - \lambda dt) + P_{N-1}(t)\lambda dt \qquad \text{(Ap.8.1)}$$

and

$$P_0(t + dt) = P_0(t)(1 - \lambda dt) \qquad \text{(Ap.8.2)}$$

From the preceding equations, we obtain

$$\frac{dP_N(t)}{dt} = -\lambda P_N(t) + \lambda P_{N-1}(t) \qquad \text{(Ap.8.3)}$$

and

$$\frac{dP_0(t)}{dt} = -\lambda P_0(t) \qquad \text{(Ap.8.4)}$$

Thus, $P_0(t) = \exp(-\lambda t)$, and

$$P_N(t) = \frac{(\lambda t)^N \exp(-\lambda t)}{N!} \qquad \text{(Ap.8.5)}$$

which is the Poisson distribution.

ADDENDUM 1

Computer Software for Spacecraft Charging Calculations

AD1.1 The Need for Numerical Calculations

Analytical formulations are the fundamental building blocks of physical models. Analytical solutions often offer insights for better understanding of the underlying physics. However, analytical formulations are not easy for large or complex systems. Often, numerical calculations and simulations can be provided only by using computers.

In spacecraft charging, the surface potentials of complex surface geometry can be tackled only by means of computer modeling. For example, Poisson's equation, which governs the electrostatic potentials, with given boundary conditions can be solved analytically in simple geometry only. One needs to consider numerical methods using, for example, discrete spatial grids, discrete time steps, finite difference schemes, or other numerical schemes. In practice, therefore, one needs computer software for spacecraft charging calculations. This addendum discusses very briefly some examples of computer software available for spacecraft charging.

AD1.2 Surface Charging

There are several computer programs available for spacecraft surface charging calculations. Computer modeling offers enormous advantages and is very much needed in practice. There is one common drawback, viz, software can be analogous to a black box—the user may not know what is inside. The algorithms built in may be very elegant and the physics approximations may be very efficient, but the user may have no inside knowledge. As an analogy, one uses Microsoft Windows but does not know what is inside, although Microsoft reveals existence of loopholes, or issues new versions, from time to time. Thus, it is important to have documentation, verification, and validation for every software produced.

In the 1960s and 1970s, there was early computer software developed for spacecraft charging. Two landmark examples were Laframboise's thesis[1] on plasma probes and the Parker and Whipple probe theory based on tracing reverse particle trajectories.[2] The first major computer software for spacecraft surface charging calculations was NASCAP (NASA Charging Analyzer Program)[3,4] developed by S-Cubed (Systems, Science, and Software) for NASA and the U.S. Air Force around 1980. Its primary purpose was for modeling the surface charging of the SCATHA satellite. The software could also be adapted for modeling other satellites. Fourteen material properties, such as secondary emission coefficients, backscattered electron coefficients, photoemission yields, and conductivity, were built in. Beam emission models, although crude, were also included.

NASCAP has been used as the major spacecraft charging software for many years. The suite of modified NASCAP programs includes NASCAP/GEO for the geosynchronous environment, NASCAP/LEO for the low Earth environment, POLAR for large spacecraft in the polar environment, and DynaPAC for dynamic plasma analysis. Whereas NASCAP/GEO uses grids and linear interpolants of potentials and electrostatic fields, NASCAP/GEO and NASCAP/POLAR use trilinear finite-element approximation. Furthermore, NASCAP/GEO, NASCAP/POLAR, and NASCAP/DynaPAC focused on self-consistent solutions to Poisson's equations and so had changed the charged particle algorithms.

The SENSIT (Space Environment System Interactions Toolbox) code was developed by N. John Stevens of NASA as a simpler version of NASCAP for personal computers. It is relatively unknown and has probably no update beyond version 2.0 of 1995.

NASCAP-2K (N2K)[5-7] replaced the NASCAP suite at the turn of this century, drawing heavily on NASCAP/DynaPAC. NASCAP-2K was developed by SAIC (Science Applications International Corporation) for NASA and the U.S. Air Force. It has many improvements over NASCAP/GEO, including improved algorithms for solving Poisson's equation with boundary elements. N2K is able to use arbitrary and recursively nested Cartesian grids to resolve smooth geometry object while encompassing the large region of space needed to contain sheath and wake. N2K features continuous electric field interpolants. It is capable of calculating potentials in sheaths, in wakes, and in the neighborhood of surfaces. It can solve charging problems in GEO, LEO, and polar environments. N2K is gridless when solving Laplace's equation. For Poisson's equation, N2K still needs grids. It also includes effects of space charge and will include buried charge in the future. The computed results are shown with high-resolution graphics capabilities. The same 14 material properties of NASCAP are built in and can be updated. It can be used to model charging on geosynchronous, low Earth orbit, polar, and interplanetary satellites.[6,8] Currently, N2K is the most popular, and most capable, spacecraft charging software in the United States. However, a detailed manual of what is inside is not published. Papers and a preliminary documentation cover most of this information. NASCAP-2K is export controlled, meaning that it is unavailable to non-U.S. citizens.

The SEE (Space Environment and Effects) spacecraft charging handbook[5] was released at about the same time as NASCAP-2K. It is a much simplified version of NASCAP-2K and is intended for nonexperts. It is easier to use than NASCAP-2K and suitable for demonstrations.

The SPIS (Spacecraft Plasma Interaction System) software[9,10] is being developed by the SPINE (Spacecraft Plasma Interactions Network in Europe) community. It evolved from PicUp3D,[11] a three-dimensional particle-in-cell simulation software of electron and ion dynamics. The SPIS software includes spacecraft charging computation features, including secondary emission, backscattering electrons, photoemission, etc. It can calculate spacecraft surface potentials, sheath potentials, and so on for complex surface geometries. SPIS is poised to become a strong competition to NASCAP-2K. Detailed documentation (user and developer) is available. There are updates from time to time. The software is free. To access the software, the user needs to register for an account.

MUSCAT (Multi-Utility Spacecraft Charging Analysis Tool)[12] is being developed in Japan. It is a particle-in-cell code with particle tracking capabilities. Fast Fourier transforms are used for solving the Poisson equation in a finite difference scheme. The software uses parallel computation techniques for the advantage of fast computation. It includes most of the spacecraft charging computation capabilities of NASCAP and SPIS. It has good potential to rival NASCAP-2K, SPIS, and more. Updates are available from time to time. The software is available commercially.

A hybrid grid-immersed finite element particle-in-cell (HG-IFE-PIC) algorithm[13] is being developed at Virginia Polytech. It attempts to improve the efficiency of modeling spacecraft interactions with reduced memory requirements and computational speed. It focuses on ion engines.

SPARCS[14] is another spacecraft charging computation software. The early version was for charging at geosynchronous altitudes. It is being developed by Thales Alenia Space of France.

AD1.3 Deep Dielectric Charging

Software is available for calculating penetration of high-energy electrons, protons, and ions into materials. References 15 to 18 list some software websites that use the analytical equation of Bethe-Bloch. The equation gives results that are generally accurate above 0.3 MeV (see figure 1 of reference 18 for a comparison with experimental data).

The SRIM software[19] offers calculations of the penetration range of ions into matter. Under development are some capabilities to calculate soft errors in electronics by cosmic rays. Some of the buttons on the website are not yet active at this time.

Oak Ridge National Laboratory (ORNL) has developed a series of sophisticated computer codes called the Integrated Tiger Series (ITS).[20] Monte Carlo methods are used to calculate penetration of electrons, ions, and photons into materials with or without applied electric or magnetic fields. These codes are sophisticated, with capabilities for tackling many interactions such as radiation shielding and not specifically for spacecraft deep dielectric charging. A brief description of a Tiger Monte Carlo code package is given at the website listed in reference 20. For availability, one has to contact ORNL.

Frederickson had mentioned that he used his own software NUMIT[21] for calculating electric fields in one dimension inside dielectrics. It included conductivity, charge density, electron deposition rate, and conduction current, etc. Its availability is unknown.

CERN (The European Organization for Nuclear Research) has developed the GEANT4 software,[22] which simulates by means of Monte Carlo methods high-energy particle penetration into materials and all the interactions involved. It is an open source object-oriented simulation software. Its capability is enormous and includes many reactions at much higher energies than necessary for deep dielectric charging on spacecraft. GEANT4[23,24] is already useful in nuclear physics, high-energy particle physics, and nuclear medicine. Space users are beginning to develop GEANT4-based simulation tools[25,26] for space radiation shielding and space environmental effects analysis. It is expected to be useful for deep dielectric charging on spacecraft.

ESA (European Space Agency) has developed a deep dielectric charging software called DICTAT (Dielectric Internal Charging Threat Assessment Tool)[27] for deep dielectric charging on spacecraft. It calculates, using analytical equations, the currents passing through any shielding to be deposited in the dielectric. The calculated electric field (deep dielectric charging) is compared with the critical electric field for dielectric breakdown, offering assessment for the likelihood of deep dielectric charging breakdown and spacecraft anomaly threats. Descriptions are given in references 27 to 29. The software has also been included in the ESA SPENVIS (Space Environment Information Systems) software.[28,29]

ESA has a software, ESADDC (European Space Agency Deep Dielectric Charging). It uses the ITS (Integrated Tiger Series) Monte Carlo software to calculate the deposition of charges into the materials and then calculates the leakage of the charges depending on the material conductivity.[30]

AD1.4 References

1. Laframboise, J. G., "Theory of spherical and cylindrical langmuir probes in a collisionless, maxwellian plasma at rest," Report No. 160, University of Toronto, Inst. for Aerospace Studies (1966).
2. Parker, L. W., and E. C. Whipple Jr., "Theory of a satellite electrostatic probe," *Ann. Phys.* 44: 126–161 (1967).
3. Katz, I., D. E. Parks, M. J. Mandell, J. M. Harvey, D. H. Bronwell, S. S. Wang, and M. Rotenberg, "A three dimensional dynamic study of electrostatic charging in materials," NASA Report No. NASA CR-135256, NASA Lewis Research Center, Cleveland, OH (1977).
4. Stannard, P. R., I. Katz, M. J. Mandell, J. J. Cassidy, D. E. Parks, M. Rotenberg, and P. G. Steen, "Analysis of the charging of the SCATHA (P78-2) satellite," NASA Report No. NASA CR-165348, NASA Lewis Research Center, Cleveland, OH (1981).

5. Davis, V. A., L. F. Neergaard, M. J. Mandell, I. Katz, B. M. Gardner, J. M. Hilton, and J. Minor, "Spacecraft charging calculations: NASCAP-2K and SEE spacecraft charging handbook," AIAA 2002-0626, presented at 40th AIAA Aerospace Science Meeting, Reno, NV (2002).
6. Mandell, M. J., V. A. Davis, D. L. Cooke, A. T. Wheelock, and C. J. Roth, "NASCAP-2K spacecraft charging code overview," *IEEE Trans. Plasma Sci.* 34, no. 5: 2084–2093 (2006).
7. Mandell, M. J., V. A. Davis, and I. G. Mikellides, "NASCAP-2K preliminary documentation," Scientific Report No. 2, Report No. A555024, AFRL-VS-TR-2002-1676, DTIC ADA420555, Science Applications International Corp. (SAIC), San Diego, CA (2002).
8. Mandell, M. J., D. L. Cooke, V. A. Davis, G. A. Jongeward, B. M. Gardner, R. A. Hilmer, K. P. Ray, S. T. Lai, and L. H. Krause, "Modeling the charging of geosynchronous and interplanetary spacecraft using NASCAP-2K," *Adv. Space Res.* 36: 2511–2515 (2005).
9. Roussel, J. F., F. Rogier, D. Volpert, J. Forest, G. Rousseau, and A. Hilgers, "Spacecraft plasma interaction software (SPIS): numerical solvers—methods and architecture," presented at 9th Spacecraft Charging Technology Conference, Tsukuba, Japan (2005).
10. Roussel, J.-F., F. Rogier, G. Dufour, J.-C. Mateo-Velez, J. Forest, A. Hilgers, D. Rodgers, L. Girard, and D. Payan, "SPIS open-source code: methods, capabilities, achievements, and prospects," *IEEE Trans. Plasma Sci.* 36, no. 5-2: 2360–2368 (2008).
11. Forest, J., A. Hilgers, B. Thiebault, L. Eliasson, J.-J. Bertelier, and H. de Feraudy, "An open-source spacecraft interactions simulation code PicUp3D: tests and validations," *IEEE Trans. Plasma Sci.* 34, no. 5-2: 2103–2113 (2006).
12. Muranaka, T., S. Hosoda, J.-H. Kim, S. Hatta, K. Ikeda, T. Hamanaga, M. Cho, H. Usui, H. O. Ueda, K. Koga, and T. Goka, "Development of Multi-Utility Spacecraft Charging Analysis Tool (MUSCAT)," *IEEE Trans. Plasma Sci.*, 36, no. 5-2, 2336–2349 (2008).
13. Kafafy, R., and J. Wang, "A hybrid grid immersed finite element particle-in-cell algorithm for modeling spacecraft-plasma interactions," *IEEE Trans. Plasma Sci.* 34, no. 5-2, 2114–2124 (2006).
14. Clerc, S., S. Brosse, and M. Chane-Yook, "SPARCS: an advanced software for spacecraft charging analyses," presented at the 8th Spacecraft Charging Technology Conference, Huntsville, AL (2003).
15. Bethe-Bloch website at http://www.hpcalc.org/details.php?id=2207.
16. Brookhaven National Laboratory website at http://tvdg10.phy.bnl.gov/LETCalc.html/.
17. National Institute of Standards and Technology website at http://physics.nist.gov/PhysRefData/Star/Text/contents.html.
18. Bethe, H., and J. Ashkin, *Experimental Nuclear Physics*, ed. E. Segré, p. 253, J. Wiley, New York (1953).
19. Ziegler, J. F., website at http://www.srim.org/.
20. Integrated Tiger code series website at http://rsicc.ornl.gov/codes/ccc/ccc4/ccc-467.html.
21. Frederickson, A. R., "Progress in high-energy electron and x-irradiation of insulating dielectrics," *Braz. J. Phys.* 29, no. 2:241–253 (1999).
22. CERN high-energy particle penetration website at http://geant4.web.cern.ch/geant4/.
23. Agostinelli, S., et al., "Geant4: a simulation toolkit," *Nucl. Instrum. Meth.* A506: 250–303 (2003).
24. Santinin, G., et al., "New Geant4 based simulation tools for space radiation shielding and effects analysis," *Nucl. Phys. B (Proc. Suppl.)* 125: 69–74 (2003).
25. Santinin, G., V. Ivanchenko, H. Evans, P. Niemann, and E. Daly, "GRAS: a general-purpose 3-D modular simulation tool for space environment effects analysis," *IEEE Trans. Nucl. Sci.* 52, no. 6: 2294–2299 (2006).

26. Lei, F., P. R. Truscott, C. S. Dyer, B. Quaghebeur, D. Heynderickx, P. Nieminen, H. Evans, and E. Daly, "MULASSIS: a Geant4 based multilayered shielding simulation tool," *IEEE Trans. Nucl. Sci.* 49, no. 6: 2788–2793 (2002).
27. Rodgers, D. J., L. Levy, P. M. Latham, K. A. Ryden, J. Sorensen, and G. L. Wrenn, "Prediction of internal dielectric charging using the DICTAT code," presented at ESA Workshop on Space Weather, Noordwijk, The Netherlands (1998); available at http://esa.spaceweather.net/spweather/workshops/proceedings _w1/posters/rogers16.pdf.
28. DICTAT website at http://www.space.qinetiq.com/idc/dictat2_sum_v0.0.pdf.
29. SPENVIS website at http://www.spenvis.oma.be/spenvis/help/background/charging/dictat/dictatman.html.
30. Sorensen, J., "An engineering specification of internal charging," *Environ. Model. Space-Based Appl.*, Symposium Proceedings (ESA SP-392), eds. W. Burke and T.-D. Guvenne, p. 129, ESTEC Noordwijk, The Netherlands (1996); available at http://adsabs.harvard.edu/full/1996ESASP.392..129S.

ADDENDUM 2

Spacecraft Charging at Jupiter and Saturn

At present, we have been concerned with spacecraft charging in the space environment of our own Earth more than other planetary space environments. In the future, as explorations of the planets and beyond are inevitable, spacecraft charging research in the space environments of other planets in our Solar system, and perhaps even in other solar systems, will likely be of great interest. Since the moon lacks a magnetosphere and has no keV plasma that would associate with it, natural charging to high voltages does not occur there. The most explored planetary environments to-date are those of Jupiter and Saturn. This appendix discusses briefly the space environments, spacecraft surface charging, and deep dielectric charging at these two planets.

AD2.1 Space Environments at Jupiter and Saturn

Jupiter and Saturn are approximately the same size. Each is about 10 times the size of the Earth. Their magnetic moments are respectively 10^5 and 10^3 times stronger than the Earth's. They rotate much faster than the Earth. One Jovian (Jupiter's) day is about 10 hours, and one Saturn day is nearly the same (10.23 hours). They are much farther from the Sun. While the Earth is at a distance of 1 AU, Jupiter is at about 5 AU and Saturn at 10 AU. Because of the long distances, the photoelectron flux is 25 ($= 5^2$) times weaker at the Earth's magnetosphere and that of Saturn is about 100 ($= 10^2$) times weaker than at Earth.

Like the Earth, they have bow shocks and magnetospheres.Their magnetospheres include magnetosheath, magnetopause, cusp region, ionosphere, neutral sheet, plasma region, and trapped radiation region. Their plasmaspheres are commonly called *plasma disks*, because the rapid rotations (with 10-hour days) of Jupiter and Saturn generate large centrifugal force shaping the disks. The plasmas inside the plasma disks are of low energy (below 1 keV) and high density (10^3 cm^{-3}), and therefore no significant spacecraft charging is expected. Outside the disks, the plasma in the energy range of about 1 keV to about 60 keV can cause surface charging. In the trapped radiation region, the high-energy electrons and ions (MeV and above) can cause deep dielectric charging and radiation damage on spacecraft.

Each of the two planets has a significant moon. Jupiter has Io at about 5 R_j (Jupiter radii). Saturn has Titan at about 20 R_S (Saturn radii). Each of the two moons generates abundant neutral gas accompanied by low-energy plasma in their vicinity. The low-energy plasma is generated by charge exchange between the higher energy plasma and the low-energy gas. The low-energy plasma is accelerated by the high electric and magnetic fields and contributes to the abundance of high-energy electrons and ions in the trapped radiation region.

The corotation speeds of their ionospheres are much faster than orbital spacecraft velocities there. As a result, wakes can form, but they do not necessarily trail a spacecraft. As a final remark, Saturn has a flat disk of many beautiful rings, which almost every child knows. In the future, more knowledge will be gained as more space missions will explore and study the planets.

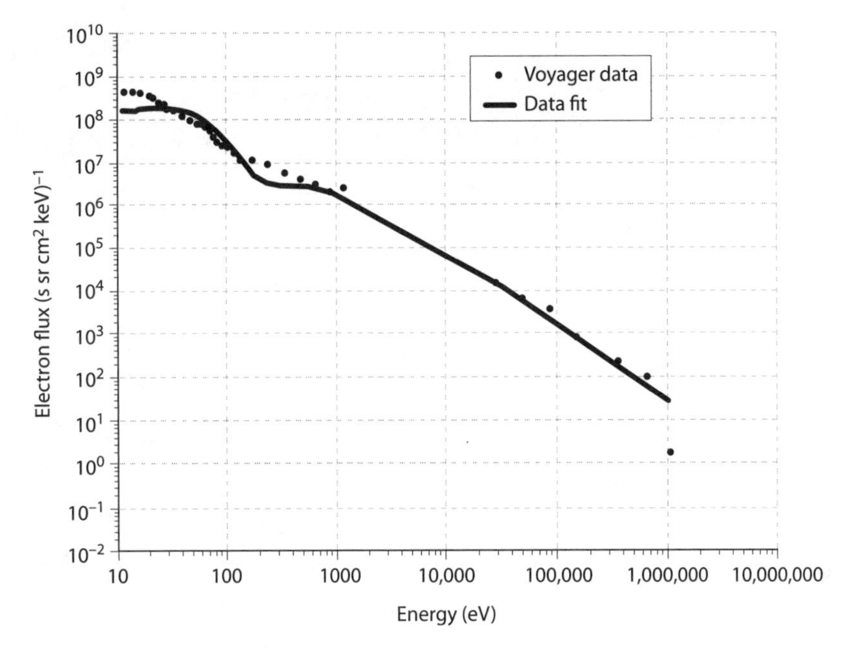

Figure Ad.2.1 Voyager Saturn (L = 11.59) data and data fit (Maxwellian below about 1 keV and kappa above about 20 keV). A kappa fit is also shown between, despite absence of data there. (Adapted from reference 1.)

AD2.2 Spacecraft Surface Charging

Using the Voyager spacecraft data obtained on Voyager 2 at magnetic latitude $L = 11.59$ at Saturn, Garrett and Hoffmann[1] fitted the electron distribution by means of Maxwellian and Kappa distributions. They found that the Maxwellian distribution fitted well in the low-energy region (below 1 keV), there were no data from about 1 keV to about 22 keV, and the kappa distribution fitted well above about 22 keV (figure Ad.2.1).

Using a current balance model assuming no sunlight, their calculation showed that the electrons of figure Ad.2.1 would charge the spacecraft to about −480 V. With photoemission in sunlight, the magnitude of charging would be lower.

According to the estimates of Garrett and Hoffmann,[1] typical spacecraft surface potentials at Jupiter and at Saturn are about one-third lower than at Earth. They concluded that "the Earth represents the worst threat to spacecraft" for surface charging.

AD2.3 Deep Dielectric Charging

High-energy electrons and ions are abundant at Jupiter and Saturn. Figure Ad.2.2 shows x-ray aurorae observed at Jupiter by NASA's Chandra x-ray observatory spacecraft. We cite the NASA Chandra press release[2-3] as follows: "Electric voltages of about 10 million volts, and currents of 10 million amps—a hundred times greater than the most powerful lightning bolts—are required to explain the x-ray observations. These voltages would also explain the radio emission from energetic electrons observed near Jupiter by the Ulysses spacecraft."

The high-energy (about MeV or higher) electrons and ions penetrate into dielectrics and accumulate inside. After days and weeks of bombardment by high fluencies of high-energy electrons and ions, the internal electric field built up may cause discharges. The discharges, in

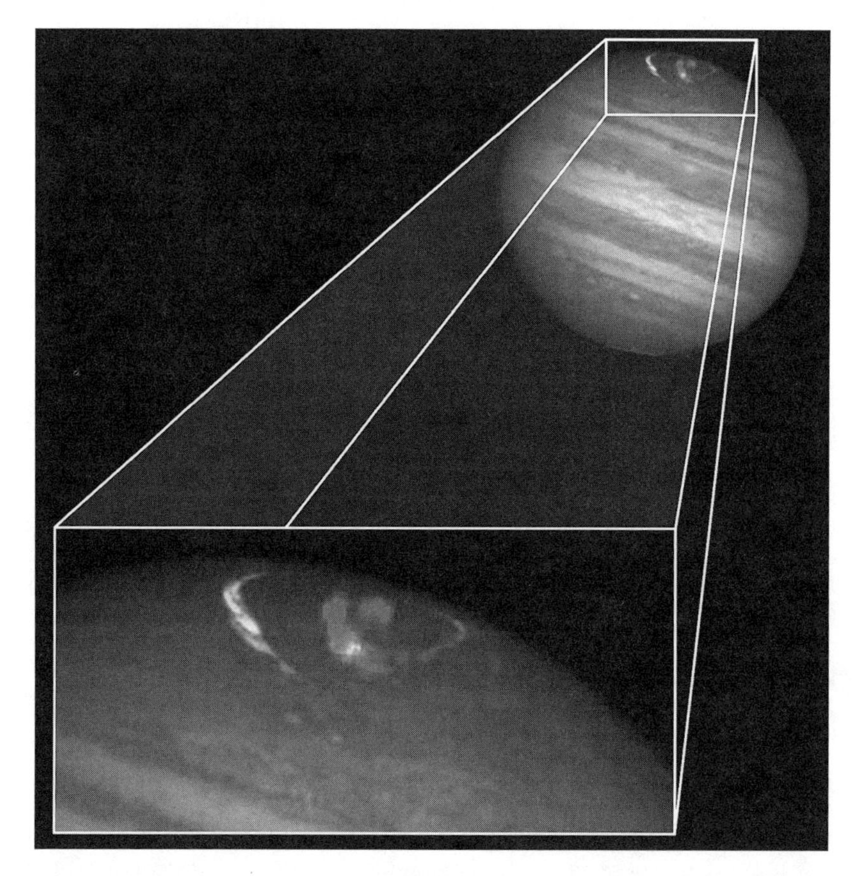

Figure Ad.2.2 X-ray aurora at Jupiter, a telltale sign of the presence of high-energy electrons and ions. (Adapted from reference 3. X-ray: NASA/SWRI/R. Gladstone et al.; UV: NASA/HST/J. Clarke et al.; Optical: NASA/HST/R. Beebe et al.)

turn, may cause anomalies in the electronics on spacecraft. It has been documented[4] that the Pioneer and Voyager spacecraft experienced anomalies during their encounters with Jupiter. These anomalies occurred during low spacecraft surface potentials in periods of high fluencies of high-energy electrons. These symptoms rule out surface charging as the cause of the anomalies but suggest that deep dielectric charging was likely. Figure Ad.2.3 shows modeled[5] contours of MeV electron fluence at Jupiter for a 10-hour period.

AD2.4 Conclusion

Modeling results[1] indicate that the surface charging level on spacecraft at Jupiter and Saturn is expected to be about one-third lower than that at Earth. That is, spacecraft surface charging is not as severe as in Earth's space environment. However, Jupiter and Saturn are much bigger, have higher magnetic moments, and rotate more than twice as fast as Earth. There are abundant high-energy (MeV or higher) electrons and ions in the planets' magnetospheres. The brief encounters of spacecraft with Jupiter already experienced several anomalies attributable to deep dielectric charging. Future spacecraft encountering such hazardous space radiation environment should prepare for possible deep dielectric charging, internal discharges, and anomalies.

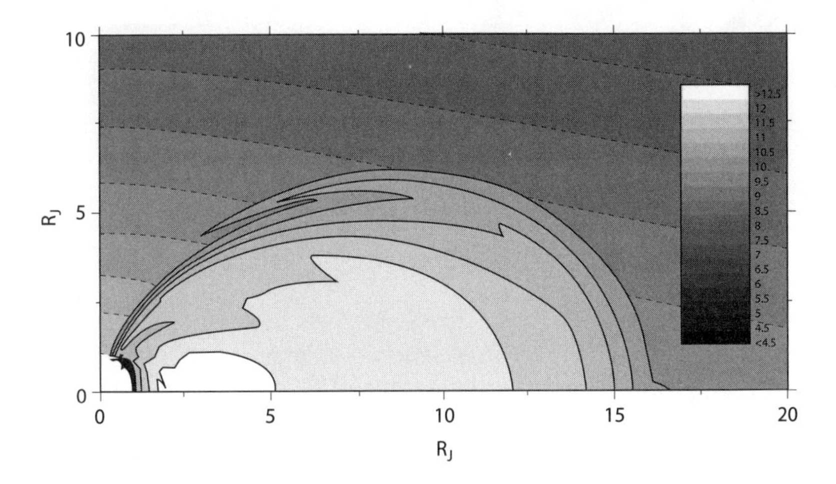

Figure Ad.2.3 Modeled contour plot of the high-energy (> 1 MeV) electron fluence for 10 hours at Jupiter. (Adapted from reference 5.)

AD2.5 References

1. Garrett, H. B., and A. Hoffman, "Comparison of spacecraft charging environments at the Earth, Jupiter, and Saturn," *IEEE Trans. Plasma Sci.* 28, no. 6, 2017–2028 (2000).
2. "Chandra probes high-voltage auroras on Jupiter," March 2, 2005; available at http://chandra.harvard.edu/press/05_releases/press_030205.html.
3. Available at http://chandra.harvard.edu/photo/2002/0001/0001_xray_opt_uv_zoom.jpg.
4. Leung, P., A. C. Whittlesey, H. B. Garrett, and P. A. Robinson Jr., "Environmental-induced electrostatic discharges as the cause of Voyager I power-on resets," *J. Spacecraft* 23, no. 3: 323–331 (1996).
5. Evans, R. W., and H. B. Garrett, "Modeling Jupiter's internal electrostatic discharge environment," *J. Spacecraft and Rockets* 39, no. 6: 926–932 (2002).

ADDENDUM 3

Physical Constants and Conversions

AD3.1 Constants

TABLE AD.3.1
Fundamental Constants

Constant	*Symbol*	*Value*
Speed of light	c	3.00×10^{8} m/s
Elementary charge	e	1.60×10^{-19} C
Planck's constant	h	6.63×10^{-34} Js
Gravitational constant	G	6.67×10^{-11} m^{3}s^{-2} kg^{-1}
Boltzmann's constant	k	1.38×10^{-23} J/K
Electron rest mass	m	9.11×10^{-31} kg
Proton mass	M	1.68×10^{-27} kg
Avogadro's number	N	6.02×10^{23} mol^{-1}
Permittivity constant	ε_0	8.85×10^{-12} F/m
Permeability constant	μ_0	$4\pi \times 10^{-7}$ Hm^{-1}
Electron volt	eV	1.6×10^{-19} J = 11,600 K
Bohr radius	a	5.2917×10^{-11} m
Lyman Alpha wavelength	Lyα	1200 Å = 120 nm
Lyman Alpha in energy	$E_{\mathrm{Ly}\alpha}$	10 eV

TABLE AD.3.2
Geophysical Constants

Parameter	*Symbol*	*Value*
Earth radius	R_E	6370 km
Earth orbital speed	V_O	29.783 km/s
Escape velocity	V_E	11.186 km/s
Sun–Earth distance (astronomical unit)	AU	1.496×10^{11} m
Geosynchronous orbit radius	R_G	6.5 R_E = 42,200 km
Geosynchronous orbit altitude	h_G	5.5 R_E = 35,800 km
Sun radius	R_S	6.96×10^{5} km

AD3.2 Conversions

1 eV = 1.6×10^{-12} erg = 1.6×10^{-19} Coul Volt
1 Coul.Volt = 10^{7} erg = 10^{7} gm cm^2/sec^2 = 1 J
1 keV = 10^{3} eV
1 MeV = 10^{6} eV
1 nm = 10^{-9} m

$1\ \text{km} = 10^3\ \text{m}$
$1\ \mu\text{m} = 10^{-6}\ \text{m}$
$1\ \text{ms} = 10^{-3}\ \text{s}$
$1\ \mu\text{s} = 10^{-6}\ \text{s}$
$1\ \text{ns} = 10^{-9}\ \text{s}$
$1\ \text{kg} = 10^3\ \text{g}$
$1\ \text{newton} = 1\ \frac{\text{Coul.Volt}}{\text{meter}}$
$1\ \text{T} = 10^4\ \text{G}$
$1\ \text{Gauss} = 10^{-4}\ \text{Weber/m}^2 = 10^{-4}\ \text{Volt sec m}^{-2}$
$1\ \text{Weber} = 1\ \text{Volt sec}$
$c = (\varepsilon_o \mu_o)^{-1/2} = 3 \times 10^8\ \text{m/sec}$

ACKNOWLEDGMENTS

The author pays respect to his former mentor, Charlie Pike, who not only organized the first Spacecraft Charging Technology Conference in 1978 but also foresaw the importance of this field. In the early days of spacecraft charging research, the author was inspired by the immense wisdom of Arthur Besse, who was working two offices down the hall from the author. The author has benefited from, and/or has been inspired by, the publications of, and discussions with, many colleagues, including Allen Rubin, Henry Garrett, Elden C. Whipple, Richard C. Olsen, James Laframboise, Paul Kellogg, Susan Gussenhoven, Herbert Cohen, William J. Burke, David Cooke, Maurice Tautz, Edmond Murad, Robb Frederickson, Daniel Hastings, Ira Katz, Myron Mendell, Victoria Davis, Michelle Thomsen, Mengu Cho, Dale Ferguson, J. R. Dennison, Allan Tribble, Boris Vagner, Jean-Francois Roussel, Alain Hilgers, David Rodgers, John Sorensen, Frank Zimmermann, and many more, in no particular order. He thanks them deeply. He also thanks Edmond Murad, William J. McNeil, Ingvar Axnäs, Nils Brenning, Dennis Papadopoulos, Gerhard Haerendel, Roy Torbert, Hans Stenbaek-Nielsen, Eugene Westcott, and the late Chris Goertz, for collaboration or discussions on meteoric impacts and critical ionization velocity studies.

The author has had the honor and privilege of presenting invited lectures on spacecraft charging at the Swedish Institute of Space Physics, Kiruna, Sweden, from 1999 to the present. He is grateful to Johnny Ejemalm, Asta Pellinin-Wannberg, and Lars Eliasson for their invitation and hospitality. He also thanks the many students who asked so many good questions over the years and encouraged him to publish this book.

The author is very grateful to his colleagues Donald Hunton, David Cooke, Edmond Murad, William J. McNeil, and Allan Tribble for reading and scrutinizing many chapters of the book manuscript. He also thanks Juan Sanmartin for reading the section on tethers carefully with helpful comments. They deserve all their credit, but any errors are solely the author's.

During the last six months of this work, the author was at the Space Propulsion Laboratory, Department of Aeronautics and Astronautics, Massachusetts Institute of Technology, Cambridge Massachusetts. He thanks Professor Manuel Martinez-Sanchez for hospitality.

Index